STUDENT SOLUTIONS MANUAL

Richard N. Aufmann
Palomar College

Vernon C. Barker
Palomar College

Joanne S. Lockwood
Plymouth State University

BASIC COLLEGE MATHEMATICS: AN APPLIED APPROACH

EIGHTH EDITION

Aufmann/Barker/Lockwood

BROOKS/COLE
CENGAGE Learning™

Australia • Brazil • Japan • Korea • Mexico • Singapore • Spain • United Kingdom • United States

BROOKS/COLE
CENGAGE Learning™

Student Solutions Manual: Basic College Mathematics: An Applied Approach, Eighth Edition
Richard N. Aufmann, Vernon C. Barker, Joanne S. Lockwood

Senior Sponsoring Editor: Lynn Cox

Associate Editor: Melissa Parkin

Editorial Assistant: Noel Kamm

Senior Project Editor: Carol Merrigan

Editorial Assistant: Eric Moore

Assistant Manufacturing Coordinator: Karmen Chong

Senior Marketing Manager: Ben Rivera

For product information and technology assistance, contact us at
Cengage Learning Customer & Sales Support, 1-800-354-9706

For permission to use material from this text or product, submit all requests online at **www.cengage.com/permissions**
Further permissions questions can be emailed to
permissionrequest@cengage.com

ISBN-13: 978-0-618-52010-7

ISBN-10: 0-618-52010-4

Brooks/Cole
10 Davis Drive
Belmont, CA 94002
USA

Cengage Learning is a leading provider of customized learning solutions with office locations around the globe, including Singapore, the United Kingdom, Australia, Mexico, Brazil, and Japan. Locate your local office at **international.cengage.com/region**

Cengage Learning products are represented in Canada by Nelson Education, Ltd.

To learn more about Brooks/Cole, visit **www.cengage.com/brookscole**

Purchase any of our products at your local college store or at our preferred online store **www.ichapters.com**

Printed in the United States of America
8 11 10 09 08

Contents

Student Solutions Manual

Chapter 1: Whole Numbers

Prep Test

1. 8

2. 1 2 3 4 5 6 7 8 9 10

3. a and D; b and E; c and A; d and B; e and F; f and C

Go Figure

On the first trip, the two children row over. The second trip, one child returns with the boat. The third trip, one adult rows to the other side. The fourth trip, one child returns with the boat. At this point, one adult has crossed the river. Repeat the first four trips an additional four times, one time for each adult. After completing the fifth time, there have been twenty trips taken, and the only people waiting to cross the river are the two children. On the twenty-first trip, the two children cross the river. There is a minimum of 21 trips.

Section 1.1

Objective A Exercises

1.

3.

5. $37 < 49$

7. $101 > 87$

9. $245 > 158$

11. $0 < 45$

13. $815 < 928$

Objective B Exercises

15. Millions

17. Hundred-thousands

19. Three thousand seven hundred ninety

21. Fifty-eight thousand four hundred seventy-three

23. Four hundred ninety-eight thousand five hundred twelve

25. Six million eight hundred forty-two thousand seven hundred fifteen

27. 357

29. 63,780

31. 7,024,709

Objective C Exercises

33. $6000 + 200 + 90 + 5$

35. $400,000 + 50,000 + 3000 + 900 + 20 + 1$

37. $300,000 + 1000 + 800 + 9$

39. $3,000,000 + 600 + 40 + 2$

Objective D Exercises

41. 850

43. 4000

45. 53,000

47. 630,000

49. 250,000

51. 72,000,000

Applying the Concepts

53. 999; 10,000

Section 1.2

Objective A Exercises

1. 28

3. 125

5. 102

7. 154

9. 1489

11. 828

13.
$$\begin{array}{r} \overset{1}{8}59 \\ +\ 725 \\ \hline 1584 \end{array}$$

15.
$$\begin{array}{r} \overset{1}{4}70 \\ +\ 749 \\ \hline 1219 \end{array}$$

17.
$$\begin{array}{r} \overset{1\ \ 1}{3}6,925 \\ +\ 65,392 \\ \hline 102,317 \end{array}$$

19.
$$\begin{array}{r} \overset{1\ \ 1}{5}0,873 \\ +\ 28,453 \\ \hline 79,326 \end{array}$$

21.
```
  2 2
  878
  737
+ 189
 1804
```

23.
```
    1
  319
  348
+ 912
 1579
```

25.
```
      1 2
  9409
  3253
+ 7078
 19,740
```

27.
```
    1 2
  2038
  2243
+ 3139
  7420
```

29.
```
  1 1  1 1
  67,428
  32,171
+ 20,971
 120,570
```

31.
```
  1 1  1
  76,290
  43,761
+ 87,402
 207,453
```

33.
```
   1  1 1
  20,958
   3,218
+     42
  24,218
```

35.
```
     1 1 1
     392
      37
  10,924
+    621
  11,974
```

37.
```
  1 2 2
    294
   1029
   7935
+    65
   9323
```

39.
```
  1 1  2 1
      97
   7,234
  69,532
+    276
  77,139
```

41. 14,383

43. 9473

45. 33,247

47. 5058

49. 1992

51. 68,263

53.
1234	≈	1200
9780	≈	9800
+ 6740	≈	+ 6700
Cal.: 17,754		Est.: 17,700

55.
241	≈	200
569	≈	600
390	≈	400
+ 1672	≈	+ 1700
Cal.: 2872		Est.: 2900

57.
32,461	≈	32,000
9,844	≈	10,000
+ 59,407	≈	+ 59,000
Cal.: 101,712		Est.: 101,000

59.
25,432	≈	25,000
62,941	≈	63,000
+ 70,390	≈	+ 70,000
Cal.: 158,763		Est.: 158,000

61.
67,421	≈	70,000
82,984	≈	80,000
66,361	≈	70,000
10,792	≈	10,000
+ 34,037	≈	+ 30,000
Cal.: 261,595		Est.: 260,000

63.
281,421	≈	280,000
9,874	≈	10,000
34,394	≈	30,000
526,398	≈	530,000
+ 94,631	≈	+ 90,000
Cal.: 946,718		Est.: 940,000

65.
28,627,052	≈	29,000,000
983,073	≈	1,000,000
+ 3,081,496	≈	+ 3,000,000
Cal.: 32,691,621		Est.: 33,000,000

67.
12,377,491	≈	12,000,000
3,409,723	≈	3,000,000
7,928,026	≈	8,000,000
+ 10,705,682	≈	+ 11,000,000
Cal.: 34,420,922		Est.: 34,000,000

Objective B Exercises

69. **Strategy** To find the total number of multiple births, add the four amounts (110,670, 6919, 627, and 79).

Solution
$$
\begin{array}{r}
110,670 \\
6,919 \\
627 \\
+ \quad 79 \\
\hline
118,295
\end{array}
$$
The were 118,295 multiple births during the year.

71. **Strategy** To estimate the total income from the first four *Star Wars* movies, estimate each movie to the nearest hundred million and then add the estimates.

Solution
$$
\begin{array}{rcr}
461,000,000 & \approx & 500,000,000 \\
290,200,000 & \approx & 300,000,000 \\
309,100,000 & \approx & 300,000,000 \\
431,100,000 & \approx & + \ 400,000,000 \\
\hline
& & \$1,500,000,000
\end{array}
$$
The estimated income from the first four *Star Wars* movies was $1,500,000,000.

73a. **Strategy** To find the total income from the two movies with the lowest box-office incomes, add the incomes from Episode V ($290,200,000) and Episode VI ($309,100,000).

Solution
$$
\begin{array}{r}
\$290,200,000 \\
+ \ 309,100,000 \\
\hline
\$599,300,000
\end{array}
$$
The income from the two movies with the lowest box-office returns is $599,300,000.

b. The income from the 1977 *Star Wars* production is $461,000,000. The income from the two movies with the lowest box-office returns is $599,300,000 (from Exercise 73a). Yes, this income exceeds the income from the 1977 *Star Wars* production.

75a. **Strategy** To find the total number of miles driven during the three days, add the three amounts (515, 492, and 278 miles).

Solution
$$
\begin{array}{r}
515 \\
492 \\
+ \ 278 \\
\hline
1285
\end{array}
$$
1285 miles will be driven during the three days.

b. **Strategy** To find what the odometer reading will be by the end of the trip, add the total number of miles driven during the three days (1285) to the original odometer reading (68,692).

Solution
$$
\begin{array}{r}
1,285 \\
+ \ 68,692 \\
\hline
69,977
\end{array}
$$
At the end of the trip, the odometer will read 69,977 miles.

77. **Strategy** To find the total average amount invested for Americans ages 16 to 34, add the amounts in checking accounts ($375), savings accounts ($1155), and U.S. Savings Bonds ($266).

Solution
$$
\begin{array}{r}
\$375 \\
1155 \\
+ \ 266 \\
\hline
\$1796
\end{array}
$$
Americans ages 16 to 34 have invested $1796 in these three investments.

79. **Strategy** To find whether the sum of the average amount invested in home equity and retirement for all Americans is greater than or less than the sum of all categories for Americans between the ages of 16 and 34:

• Add the amount invested in home equity ($43,070) and retirement ($9016) for all Americans.
• Add all the values for all categories for Americans 16 to 34.
• Compare these sums.

Solution
$$
\begin{array}{rr}
\$43,070 & \$375 \\
+ \ 9,016 & 1,155 \\
\hline
\$52,086 & 266 \\
& 4,427 \\
& 1,615 \\
& 17,184 \\
& + \ 4,298 \\
\cline{2-2}
& \$29,320
\end{array}
$$
$52,086 > $29,320

The sum of the average amounts invested in home equity and retirement for all Americans is greater than the sum of all categories for Americans ages 16 to 34.

Applying the Concepts

81. There are 6 possible outcomes for each die (1, 2, 3, 4, 5, and 6). The smallest sum on two dice is $1 + 1 = 2$. The largest sum on two dice is $6 + 6 = 12$. There are 11 different sums from 2 to 12 (2, 3, 4, 5, 6, 7, 8, 9, 10, 11, and 12).

83. No; $0 + 0 = 0$

85. Ten numbers that are less than 100 end in a 7. They are 7, 17, 27, 37, 47, 57, 67, 77, 87, and 97.

Section 1.3

Objective A Exercises

1. 4

3. 4

5. 10

7. 4

9. 9

11. 22

13. 60

15. 66

17. 31

19. 901

21. 791

23. 1125

25. 3131

27. 47

29. 925

31. 4561

33. 3205

35. 1222

37. 3021

39. 3022

41. 3040

43. 212

45. 60,245

Objective B Exercises

47.
$$\begin{array}{r} {}^{8\,13}\!\!\not{9}\not{3} \\ -\ 28 \\ \hline 65 \end{array}$$

49.
$$\begin{array}{r} {}^{3\,14}\!\not{4}\not{4} \\ -\ 27 \\ \hline 17 \end{array}$$

51.
$$\begin{array}{r} {}^{4\,10}\!\not{5}\not{0} \\ -\ 27 \\ \hline 23 \end{array}$$

53.
$$\begin{array}{r} {}^{8\,13}\!\,9\not{9}\not{3} \\ -\ 537 \\ \hline 456 \end{array}$$

55.
$$\begin{array}{r} {}^{13}\!\,{}^{7\,\not{8}\,10}\!\not{8}\not{4}\not{0} \\ -\ 783 \\ \hline 57 \end{array}$$

57.
$$\begin{array}{r} {}^{6\,16\,10}\!\not{7}\not{7}\not{0} \\ -\ 395 \\ \hline 375 \end{array}$$

59.
$$\begin{array}{r} {}^{11}\!\,{}^{4\,\not{1}\,16}\!3\not{5}\not{2}\not{6} \\ -\ 387 \\ \hline 3139 \end{array}$$

61.
$$\begin{array}{r} {}^{3\,13\,4\,10}\!\not{4}\not{3}\not{5}\not{0} \\ -\ 729 \\ \hline 3621 \end{array}$$

63.
$$\begin{array}{r} {}^{159}\!\,{}^{0\,\not{5}\,10\,17}\!\not{1}\not{6}\not{0}\not{7} \\ -\ 869 \\ \hline 738 \end{array}$$

65.
$$\begin{array}{r} {}^{6\,12\,8\,13}\!\not{7}\not{2}\not{9}\not{3} \\ -\ 3748 \\ \hline 3545 \end{array}$$

67.
$$\begin{array}{r} {}^{169}\!\,{}^{2\,\not{6}\,10\,16}\!\not{3}\not{7}\not{0}\not{6} \\ -\ 2957 \\ \hline 749 \end{array}$$

69.
$$\begin{array}{r} {}^{7\,10\,4\,12}\!\not{8}\not{0}\not{5}\not{2} \\ -\ 2709 \\ \hline 5343 \end{array}$$

71.
$$\begin{array}{r} {}^{9}\!\,{}^{6\,10\,6\,10\,12}\!\not{7}\not{0}\not{7}\not{0}\not{2} \\ -\ 4,239 \\ \hline 66,463 \end{array}$$

73.
$$\begin{array}{r} {}^{9}\!\,{}^{9}\!\,{}^{7\,10\,10\,10}\!\not{8}\not{0}\not{0}\not{0}9 \\ -63,419 \\ \hline 16,590 \end{array}$$

75.
$$\begin{array}{r} {}^{9}\!\,{}^{7\,10\,10\,4\,13}\!\not{8}\not{0}\not{0}\not{5}\not{3} \\ -27,649 \\ \hline 52,404 \end{array}$$

77.
$$\begin{array}{r} {}^{9}\!\,{}^{7\,10\,7\,10\,10}\!\not{8}\not{0}\not{8}\not{0}\not{0} \\ -42,023 \\ \hline 38,777 \end{array}$$

79.
$$\begin{array}{r} {}^{139}\!\,{}^{7\,\not{3}\,10\,10}\!\not{8}\not{4}\not{0}\not{0} \\ -3762 \\ \hline 4638 \end{array}$$

81.
$$\begin{array}{r} {}^{9}\!\,{}^{5\,10\,10}\!\not{6}\not{0}\not{0}4 \\ -2392 \\ \hline 3612 \end{array}$$

83.
$$\begin{array}{r} {}^{6\,10\,4\,10}\!\not{7}\not{0}\not{5}\not{0} \\ -4137 \\ \hline 2913 \end{array}$$

85.
$$\begin{array}{r} {}^{11}\!\,{}^{3\,\not{1}\,10}\!4\not{2}\not{0}\not{7} \\ -1624 \\ \hline 2583 \end{array}$$

87.
$$\begin{array}{r} {}^{9}\!\,{}^{9}\!\,{}^{7\,10\,10\,13}\!\not{8}\not{0}\not{0}\not{3} \\ -2735 \\ \hline 5268 \end{array}$$

89.
$$\begin{array}{r} {}^{9}\!\,{}^{9}\!\,{}^{9}\!\,{}^{7\,10\,10\,10\,14}\!\not{8}\not{0}\not{0}\not{0}\not{4} \\ -\ 8,237 \\ \hline 71,767 \end{array}$$

91.
$$\begin{array}{r} {}^{6\,9\,12\,11}\!\not{1}\not{7}\not{0}\not{3}\not{1} \\ -\ 5,792 \\ \hline 11,239 \end{array}$$

93.
$$\begin{array}{r} {}^{7\,17}\!29,\not{8}\not{7}4 \\ -21,392 \\ \hline 8,482 \end{array}$$

95.
$$\begin{array}{r} {}^{6\,9\,9\,9\,14}\\ 70,\cancel{0}\cancel{0}\cancel{0}4\\ -69,379 \\ \hline 625 \end{array}$$

97.
$$\begin{array}{r} {}^{7\,15\,16\,9\,11}\\ 86,7\cancel{0}\cancel{1}\\ -\ 9,976 \\ \hline 76,725 \end{array}$$

99. Strategy   To find the amount that completes the statement, subtract the addend (67) from the sum (90).

    Solution
$$\begin{array}{r} 90\\ -67 \\ \hline 23 \end{array}$$
Therefore 23 completes the statement, $67 + 23 = 90$.

101. Strategy   To find the amount that completes the statement, subtract the addend (253) from the sum (4901).

    Solution
$$\begin{array}{r} 4901\\ -\ 253 \\ \hline 4648 \end{array}$$
Therefore 4648 completes the statement, $253 + 4648 = 4901$.

103.
$$\begin{array}{rcr} 90,765 & \approx & 90,000\\ -60,928 & \approx & -60,000 \\ \hline \end{array}$$
Cal.: 29,837   Est.: 30,000

105.
$$\begin{array}{rcr} 96,430 & \approx & 100,000\\ -59,762 & \approx & -\ 60,000 \\ \hline \end{array}$$
Cal.: 36,668   Est.: 40,000

107.
$$\begin{array}{rcr} 300,712 & \approx & 300,000\\ -198,714 & \approx & -200,000 \\ \hline \end{array}$$
Cal.: 101,998   Est.: 100,000

Objective C Exercises

109. Strategy   To find the increase in the number of identity theft complaints from 2001 to 2002, subtract the number of complaints for 2001 (86,198) from the number of complaints for 2002 (161,819).

    Solution
$$\begin{array}{r} 161,819\\ -\ 86,198 \\ \hline 75,621 \end{array}$$
The increase was 75,621 complaints.

111. Strategy   To find the amount that remains to be paid, subtract the down payment ($950) from the cost ($11,225).

    Solution
$$\begin{array}{r} \$11,225\\ -\ \ \ 950 \\ \hline \$10,275 \end{array}$$
The amount that remains to be paid is $10,275.

113. Strategy   To find how much higher the Giant erupts than Old Faithful, subtract the height of Old Faithful (175) from the height of the Giant (200).

    Solution
$$\begin{array}{r} 200\\ -175 \\ \hline 25 \end{array}$$
The Giant erupts 25 feet higher than Old Faithful.

115. Strategy   To find the expected increase over 10 years, subtract the population expected in 2010 (129,000) from the population expected in 2020 (235,000).

    Solution
$$\begin{array}{r} 235,000\\ -129,000 \\ \hline 106,000 \end{array}$$
The expected increase over 10 years is 106,000.

117. Strategy   To find the amount remaining in the sales executive's monthly expense account, subtract the sum of amounts already spent from the beginning account balance ($1500).

    Solution

$479	transportation
268	food
317	lodging
$1064	

$$\begin{array}{r} \$1500\\ -1064 \\ \hline \$436 \end{array}$$
The amount remaining in the expense account is $436.

Applying the Concepts

119a. True

  b. False

  c. False

Section 1.4

Objective A Exercises

1. 6×2 or $6 \cdot 2$

3. 4×7 or $4 \cdot 7$

5. 12

7. 35

9. 25

11. 0

13. 72

15.
$$\begin{array}{r} \overset{1}{66} \\ \times\ 3 \\ \hline 198 \end{array}$$

17.
$$\begin{array}{r} \overset{3}{67} \\ \times\ 5 \\ \hline 335 \end{array}$$

19.
$$\begin{array}{r} \overset{1}{623} \\ \times\ 4 \\ \hline 2492 \end{array}$$

21.
$$\begin{array}{r} \overset{6}{607} \\ \times\ 9 \\ \hline 5463 \end{array}$$

23.
$$\begin{array}{r} 600 \\ \times\ 7 \\ \hline 4200 \end{array}$$

25.
$$\begin{array}{r} \overset{2}{703} \\ \times\ 9 \\ \hline 6327 \end{array}$$

27.
$$\begin{array}{r} 632 \\ \times\ 3 \\ \hline 1896 \end{array}$$

29.
$$\begin{array}{r} \overset{21}{632} \\ \times\ 8 \\ \hline 5056 \end{array}$$

31.
$$\begin{array}{r} \overset{13}{337} \\ \times\ 5 \\ \hline 1685 \end{array}$$

33.
$$\begin{array}{r} \overset{4\ 6}{6709} \\ \times\ 7 \\ \hline 46{,}963 \end{array}$$

35.
$$\begin{array}{r} \overset{3\,4\,5}{8568} \\ \times\ 7 \\ \hline 59{,}976 \end{array}$$

37.
$$\begin{array}{r} \overset{3\,3}{4780} \\ \times\ 4 \\ \hline 19{,}120 \end{array}$$

39.
$$\begin{array}{r} \overset{1\,1\,1}{9895} \\ \times\ 2 \\ \hline 19{,}790 \end{array}$$

41. $6 \times 2 \times 9 = 108$

43.
$$\begin{array}{r} 458 \\ \times\ 8 \\ \hline 3664 \end{array}$$

45.
$$\begin{array}{r} 5009 \\ \times\ 4 \\ \hline 20{,}036 \end{array}$$

47.
$$\begin{array}{r} 8957 \\ \times\ 8 \\ \hline 71{,}656 \end{array}$$

Objective B Exercises

49.
$$\begin{array}{r} 18 \\ \times\ 24 \\ \hline 72 \\ 36\ \\ \hline 432 \end{array}$$

51.
$$\begin{array}{r} 27 \\ \times\ 72 \\ \hline 54 \\ 189\ \\ \hline 1944 \end{array}$$

53.
$$\begin{array}{r} 581 \\ \times\ 72 \\ \hline 1162 \\ 4067\ \\ \hline 41{,}832 \end{array}$$

55.
$$\begin{array}{r} 727 \\ \times\ 60 \\ \hline 43{,}620 \end{array}$$

57.
$$\begin{array}{r} 9577 \\ \times\ 35 \\ \hline 47885 \\ 28731\ \\ \hline 335{,}195 \end{array}$$

59.
$$\begin{array}{r} 8875 \\ \times\ 67 \\ \hline 62125 \\ 53250\ \\ \hline 594{,}625 \end{array}$$

61.
$$\begin{array}{r} 6702 \\ \times\ 48 \\ \hline 53616 \\ 26808\ \\ \hline 321{,}696 \end{array}$$

63.
$$\begin{array}{r} 6003 \\ \times\ 57 \\ \hline 42021 \\ 30015\ \\ \hline 342{,}171 \end{array}$$

65.
$$\begin{array}{r} 607 \\ \times\ 460 \\ \hline 36420 \\ 2428\ \\ \hline 279{,}220 \end{array}$$

67.
$$\begin{array}{r} 700 \\ \times\ 274 \\ \hline 2800 \\ 4900\ \\ 1400\ \ \\ \hline 191{,}800 \end{array}$$

69.
$$\begin{array}{r} 688 \\ \times\ 674 \\ \hline 2752 \\ 4816\ \\ 4128\ \ \\ \hline 463{,}712 \end{array}$$

71.
$$\begin{array}{r} 423 \\ \times\ 427 \\ \hline 2961 \\ 846\ \\ 1692\ \ \\ \hline 180{,}621 \end{array}$$

73.
$$\begin{array}{r} 684 \\ \times\ 700 \\ \hline 478{,}800 \end{array}$$

75.
$$\begin{array}{r} 758 \\ \times\ 209 \\ \hline 6822 \\ 15160\ \ \\ \hline 158{,}422 \end{array}$$

77.
$$\begin{array}{r} 5207 \\ \times\ 902 \\ \hline 10414 \\ 468630\ \ \\ \hline 4{,}696{,}714 \end{array}$$

79.
$$\begin{array}{r} 6327 \\ \times\ 876 \\ \hline 37962 \\ 44289\ \\ 50616\ \ \\ \hline 5{,}542{,}452 \end{array}$$

81.
$$\begin{array}{r} 7349 \\ \times\ 27 \\ \hline 51443 \\ 14698\ \\ \hline 198{,}423 \end{array}$$

83. $6 \times 73 = 438$

$$
\begin{array}{r}
438 \\
\times\ 43 \\
\hline
1314 \\
1752\ \\
\hline
18,834
\end{array}
$$

85.
$$
\begin{array}{r}
842 \\
\times\ 309 \\
\hline
7578 \\
2526\ \ \\
\hline
260,178
\end{array}
$$

87.
$$
\begin{array}{r}
34,985 \\
\times\ 9007 \\
\hline
244895 \\
314865\ \ \ \\
\hline
315,109,895
\end{array}
$$

89.
$$
\begin{array}{rcr}
4732 & \approx & 5000 \\
\times\ 93 & \approx & \times\ 90 \\
\end{array}
$$
Cal. : 440,076   Est. : 450,000

91.
$$
\begin{array}{rcr}
8941 & \approx & 9000 \\
\times\ 726 & \approx & \times\ 700 \\
\end{array}
$$
Cal. : 6,491,166   Est. : 6,300,000

93.
$$
\begin{array}{rcr}
6379 & \approx & 6000 \\
\times\ 2936 & \approx & \times\ 3000 \\
\end{array}
$$
Cal. : 18,728,744   Est. : 18,000,000

95.
$$
\begin{array}{rcr}
62,504 & \approx & 60,000 \\
\times\ 923 & \approx & \times\ 900 \\
\end{array}
$$
Cal. : 57,691,192   Est. : 54,000,000

Objective C Exercises

97. Strategy To find the number of gallons of fuel used on a 6-hour flight, multiply the number of gallons used in 1 hour (865) by 6.

Solution
$$
\begin{array}{r}
865 \\
\times\ 6 \\
\hline
5190
\end{array}
$$
The plane used 5190 gallons of fuel in a 6-hour flight.

99. Strategy To find the area, multiply the length of one side (16 miles) by itself (16 miles).

Solution
$$
\begin{array}{r}
16 \\
\times 16 \\
\hline
96 \\
16\ \\
\hline
256
\end{array}
$$
The area is 256 square miles.

101. Strategy To determine which company offers the lights for the lower total price:
● Find the total price each company would charge.
● Compare the two prices.

Solution

Company A	Company B
$43 \times 2 = 86$	$43 \times 3 = 129$
$15 \times 6 = 90$	$15 \times 4 = 60$
$20 \times 12 = 240$	$20 \times 11 = 220$
$1 \times 998 = 998$	$1 \times 1089 = 1089$
Total = 1414	Total = 1498

$1414 < 1498$
Company A offers the lower total price.

103. Strategy To estimate the cost for the electricians' labor, multiply the number of electricians (3) by the number of hours each works (50) by the wage per hour (34).

Solution
Total = no. of × no. of × wages
cost   electricians hours each   per
                    works   hour
$= 3 \times 50 \times 34$
$= \$5100$
The estimated cost of the electricians' labor is $5100.

105. Strategy To find the total cost for the four components:
● Determine the costs for the electrician, the plumber, the clerical work, and the bookkeeper.
● Add to find the sum of the four costs.

Solution
$$
\begin{aligned}
\text{Electrician} &= 1 \times 30 \times \$34 = \$1020 \\
\text{Plumber} &= 1 \times 33 \times \$30 = \ \ \$990 \\
\text{Clerk} &= 1 \times 3 \times \$16 = \ \ \ \ \ \$48 \\
\text{Bookkeeper} &= 1 \times 4 \times \$20 = \ \ \ \ \ \$80 \\
\hline
&\qquad\qquad\qquad \text{Total} = \$2138
\end{aligned}
$$
The total cost is $2138.

Applying the Concepts

107. There is one accidental death every 5 minutes.
There are 60 minutes in an hour.
$5 \times 12 = 60$
There are 12 accidental deaths in an hour.
There are 24 hours per day.
$12 \times 24 = 288$
There are 288 accidental deaths in a day.
There are 365 days in a year.
$288 \times 365 = 105,120$
There are 105,120 accidental deaths in a year.

109. $3 \times 37,037 = 111,111$
By the Multiplication
Property of One, a number
times 1 equals the number.
Because the product of 3 and
37,037 is 111,111, the
product of 111,111 and a
number between 1 and 9 will
be a six-digit number in
which all digits equal the
number between 1 and 9.

Section 1.5

Objective A Exercises

1. 2

3. 6

5. 7

7.
```
      16
  6)96
    −6
    36
   −36
     0
```

9.
```
     210
  4)840
   −8
    04
   − 4
    00
   − 0
     0
```

11.
```
      44
  7)308
   −28
    28
   −28
     0
```

13.
```
     703
  9)6327
   −63
    02
   − 0
    27
   −27
     0
```

15.
```
     910
  8)7280
   −72
    08
   − 8
    00
   − 0
     0
```

17.
```
     21,560
  3)64,680
   −6
    04
   − 3
    16
   −15
    18
   − 18
    00
   − 0
     0
```

19.
```
      3,580
  6)21,480
   −18
    34
   −30
    48
   −48
    00
   − 0
     0
```

21.
```
     482
  3)1446
   −12
    24
   −24
    06
   − 6
     0
```

23.
```
     1075
  7)7525
   −7
    05
   − 0
    52
   −49
    35
   −35
     0
```

25.
```
      52
  7)364
   −35
    14
   −14
     0
```

27.
```
      34
  5)170
   −15
    20
   −20
     0
```

Objective B Exercises

29.
```
     2   r1
  4)9
   −8
    1
```

31.
```
     5   r2
  5)27
   −25
    2
```

33.
```
     13   r1
  3)40
   −3
    10
   − 9
    1
```

35.
```
     10   r3
  8)83
   −8
    03
   − 0
    3
```

37.
```
      90   r2
  7)632
   −63
    02
   − 0
    2
```

39.
```
     230   r1
  4)921
   −8
    12
   −12
    01
   − 0
    1
```

41.
```
      204   r3
  8)1635
   −16
    03
   − 0
    35
   −32
     3
```

43.
```
    1347  r3
7)9432
   -7
    24
   -21
    33
   -28
    52
   -49
     3
```

45.
```
    1720  r2
3)5162
   -3
    21
   -21
    06
   - 6
    02
   - 0
     2
```

47.
```
    409  r2
8)3274
  -32
   07
  - 0
   74
  -72
    2
```

49.
```
   6,214  r2
7)43,500
  -42
   15
  -14
   10
  - 7
   30
  -28
    2
```

51.
```
   8,708  r2
5)43,542
  -40
   35
  -35
   04
  - 0
   42
  -40
    2
```

53.
```
    1080  r2
8)8642
  -8
   06
  - 0
   64
  -64
   02
  - 0
    2
```

55.
```
   4,210  r6
9)37,896
  -36
   18
  -18
   09
  - 9
   06
  - 0
    6
```
Round to 4200.

57.
```
   19,586  r1
4)78,345
  -4
   38
  -36
   23
  -20
   34
  -32
   25
  -24
    1
```
Round to 19,600.

Objective C Exercises

59.
```
    1  r38
44)82
  -44
   38
```

61.
```
    1  r26
67)93
  -67
   26
```

63.
```
    21  r21
32)693
  -64
   53
  -32
   21
```

65.
```
    30  r22
25)772
  -75
   22
  - 0
   22
```

67.
```
    5  r40
92)500
  -460
   40
```

69.
```
    9  r17
50)467
  -450
   17
```

71.
```
    200  r21
44)8821
  -88
   02
  - 0
   21
  - 0
   21
```

73.
```
    303  r1
32)9697
  -96
   09
  - 0
   97
  -96
    1
```

75.
```
    67  r13
92)6177
  -552
   657
  -644
   13
```

77.
```
    176  r13
27)4765
  -27
   206
  -189
   175
  -162
   13
```

79.
```
   1,086  r7
77)83,629
  -77
   66
  - 0
   662
  -616
   469
  -462
     7
```

81.
$$
\begin{array}{r}
403 \\
78\overline{)31{,}434} \\
-312 \\
\hline
23 \\
-0 \\
\hline
234 \\
-234 \\
\hline
0
\end{array}
$$

83.
$$
\begin{array}{r}
12 \quad r456 \\
504\overline{)6504} \\
-504 \\
\hline
1464 \\
-1008 \\
\hline
456
\end{array}
$$

85.
$$
\begin{array}{r}
4 \quad r160 \\
546\overline{)2344} \\
-2184 \\
\hline
160
\end{array}
$$

87.
$$
\begin{array}{r}
160 \quad r27 \\
53\overline{)8507} \\
-53 \\
\hline
320 \\
-318 \\
\hline
27
\end{array}
$$

89.
$$
\begin{array}{r}
1{,}669 \quad r14 \\
46\overline{)76{,}788} \\
-46 \\
\hline
307 \\
-276 \\
\hline
318 \\
-276 \\
\hline
428 \\
-414 \\
\hline
14
\end{array}
$$

91.
$$
\begin{array}{r}
7{,}948 \quad r17 \\
43\overline{)341{,}781} \\
-301 \\
\hline
407 \\
-387 \\
\hline
208 \\
-172 \\
\hline
361 \\
-344 \\
\hline
17
\end{array}
$$
Round to 7950.

93. Cal.: $53\overline{)117{,}925}$ → $2{,}225$ Est.: $50\overline{)100{,}000}$ → $2{,}000$

95. Cal.: $67\overline{)738{,}072}$ → $11{,}016$ Est.: $70\overline{)700{,}000}$ → $10{,}000$

97. Cal.: $34\overline{)906{,}304}$ → $26{,}656$ Est.: $30\overline{)900{,}000}$ → $30{,}000$

99. Cal.: $642\overline{)323{,}568}$ → 504 Est.: $600\overline{)300{,}000}$ → 500

101. Cal.: $614\overline{)332{,}174}$ → 541 Est.: $600\overline{)300{,}000}$ → 500

103. Cal.: $374\overline{)7{,}712{,}254}$ → $20{,}621$ Est.: $400\overline{)8{,}000{,}000}$ → $20{,}000$

Objective D Exercises

105. **Strategy**   To find the monthly expense for housing, divide annual housing expense ($11,713) by the number of months (12).

Solution
$$
\begin{array}{r}
976 \\
12\overline{)11{,}713} \\
-108 \\
\hline
91 \\
-84 \\
\hline
73 \\
-72 \\
\hline
1
\end{array}
$$
The average monthly expense for housing is $976.

107. **Strategy**   To find the average monthly claim for theft, divide the annual claim for theft ($300,000) by the number of months (12).

Solution
$$
\begin{array}{r}
25{,}000 \\
12\overline{)300{,}000} \\
-24 \\
\hline
60 \\
-60 \\
\hline
00 \\
-0 \\
\hline
00 \\
-0 \\
\hline
00 \\
-0 \\
\hline
0
\end{array}
$$
The average monthly claim for theft is $25,000.

109. **Strategy**   To find the average hours worked by employees in Britain, divide the annual hours worked (1731) by the number of weeks (50).

Solution
$$
\begin{array}{r}
34 \\
50\overline{)1731} \\
-150 \\
\hline
231 \\
-200 \\
\hline
31
\end{array}
$$
Since 31 is greater than half of 50, the average number of hours worked by employees in Britain is 35 hours.

111. **Strategy** To find the approximate number of pennies per person, divide the number of pennies in circulation (114,000,000,000) by the number of people (300,000,000).

Solution 380 pennies are in circulation for each person.

Applying the Concepts

113. **Strategy** To find the number of cases of eggs produced during the year, add all the values read from the graph.

Solution

111,100,000	Retail Stores
61,600,000	Non-shell Products
24,100,000	Food Service Use
1,600,000	Exported
198,400,000	

198,400,000 cases of eggs were produced during the year.

115. **Strategy** To find the estimated amount the average U.S. household will spend on gasoline in 2020, divide the annual amount ($1562) by 12.

Solution

$$\begin{array}{r} 130 \\ 12\overline{)1562} \\ \underline{12} \\ 36 \\ \underline{36} \\ 2 \end{array}$$

Since 2 is less than half of 12, the average U.S. household will spend $130 (to the nearest dollar) on gasoline each month in 2020.

117. **Strategy** To find how much greater the average starting salary of an accounting major is than a psychology major, subtract the psychology major's starting salary ($27,454) from the accounting major's starting salary ($40,546).

Solution

$$\begin{array}{r} \$40{,}546 \\ -\$27{,}454 \\ \hline \$13{,}092 \end{array}$$

The average starting salary of an accounting major is $13,092 more than a psychology major.

119. **Strategy** To find the difference between a colonel's annual pay and a lieutenant colonel's annual pay:
• Subtract the lieutenant colonel's monthly pay ($6329) from the colonel's monthly pay ($7233) to find the monthly difference.
• Multiply the monthly difference by 12 to find the annual difference.

Solution

$$\begin{array}{r} \$7233 \\ -\$6329 \\ \hline \$904 \end{array} \qquad \begin{array}{r} \$904 \\ \times 12 \\ \hline 1808 \\ 904 \\ \hline \$10848 \end{array}$$

The difference in annual pay for a colonel and a lieutenant colonel is $10,848.

121. **Strategy** To find the total pay:
• Multiply the overtime rate ($13) by the number of extra hours worked (9).
• Add the extra wages to the amount for working a 40 hour week ($374).

Solution

$$\begin{array}{r} \$13 \\ \times 9 \\ \hline \$117 \end{array} \text{ additional pay} \qquad \begin{array}{r} \$374 \\ +\$117 \\ \hline \$491 \end{array}$$

The sales associate's total pay for the week was $491.

123. The smallest three-digit palindrome number that is divisible by 4 is 212. Consider the list of the palindromic numbers 101, 111, 121, 131, ... 191, 202, 212, ...
The first ten numbers have an odd final digit and are thus not divisible by 4.

Section 1.6

Objective A Exercises

1. 2^3

3. $6^3 \cdot 7^4$

5. $2^3 \cdot 3^3$

7. $5 \cdot 7^5$

9. $3^3 \cdot 6^4$

11. $3^3 \cdot 5 \cdot 9^3$

13. $2 \cdot 2 \cdot 2 = 8$

15. $2 \cdot 2 \cdot 2 \cdot 2 \cdot 5 \cdot 5 = 16 \cdot 25 = 400$

17. $3 \cdot 3 \cdot 10 \cdot 10 = 9 \cdot 100 = 900$

19. $6 \cdot 6 \cdot 3 \cdot 3 \cdot 3 = 36 \cdot 27 = 972$

21. $5 \cdot 2 \cdot 2 \cdot 2 \cdot 3 = 5 \cdot 8 \cdot 3 = 120$

23. $2 \cdot 2 \cdot 3 \cdot 3 \cdot 10 = 4 \cdot 9 \cdot 10 = 360$

25. $0 \cdot 0 \cdot 4 \cdot 4 \cdot 4 = 0 \cdot 64 = 0$

27. $3 \cdot 3 \cdot 10 \cdot 10 \cdot 10 \cdot 10 = 9 \cdot 10,000 = 90,000$

29. $2 \cdot 2 \cdot 3 \cdot 3 \cdot 3 \cdot 5 = 4 \cdot 27 \cdot 5 = 540$

31. $2 \cdot 3 \cdot 3 \cdot 3 \cdot 3 \cdot 5 \cdot 5 = 2 \cdot 81 \cdot 25 = 4050$

33. $5 \cdot 5 \cdot 3 \cdot 3 \cdot 7 \cdot 7 = 25 \cdot 9 \cdot 49 = 11,025$

35. $3 \cdot 3 \cdot 3 \cdot 3 \cdot 2 \cdot 2 \cdot 2 \cdot 2 \cdot 2 \cdot 2 \cdot 5 = 81 \cdot 64 \cdot 5$
$\qquad = 25,920$

37. $4 \cdot 4 \cdot 3 \cdot 3 \cdot 3 \cdot 10 \cdot 10 \cdot 10 \cdot 10 = 16 \cdot 27 \cdot 10,000$
$\qquad = 4,320,000$

Objective B Exercises

39. $6 - 3 + 2 = 3 + 2 = 5$

41. $8 \div 4 + 8 = 2 + 8 = 10$

43. $5 \cdot 9 + 2 = 45 + 2 = 47$

45. $5^2 - 17 = 25 - 17 = 8$

47. $3 + (4 + 2) \div 3 \ = \ 3 + 6 \div 3$
$\qquad\qquad\qquad = \ 3 + 2 = 5$

49. $8 - 2^2 + 4 \ = \ 8 - 4 + 4$
$\qquad\qquad\quad = \ 4 + 4 = 8$

51. $12 \cdot (1 + 5) \div 12 \ = \ 12 \cdot 6 \div 12$
$\qquad\qquad\qquad\quad = \ 72 \div 12 = 6$

53. $5 \cdot 3^2 + 8 \ = \ 5 \cdot 9 + 8$
$\qquad\qquad\quad = \ 45 + 8 = 53$

55. $12 + 4 \cdot 2^3 \ = \ 12 + 4 \cdot 8$
$\qquad\qquad\quad = \ 12 + 32 = 44$

57. $7 + (9 - 5) \cdot 3 \ = \ 7 + 4 \cdot 3$
$\qquad\qquad\qquad = \ 7 + 12 = 19$

59. $3^3 + 5 \cdot (8 - 6)^3 \ = \ 3^3 + 5 \cdot 2^3$
$\qquad\qquad\qquad\quad = \ 27 + 5 \cdot 8$
$\qquad\qquad\qquad\quad = \ 27 + 40 = 67$

61. $4 \cdot 6 + 3^2 \cdot 4^2 \ = \ 4 \cdot 6 + 9 \cdot 16$
$\qquad\qquad\qquad = \ 24 + 9 \cdot 16$
$\qquad\qquad\qquad = \ 24 + 144 = 168$

63. $12 + 3 \cdot 5 = 12 + 15 = 27$

65. $5 \cdot (8 - 4) - 6 = 5 \cdot 4 - 6 = 20 - 6 = 14$

67. $12 - (12 - 4) \div 4 = 12 - 8 \div 4 = 12 - 2 = 10$

69. $10 + 1 - 5 \cdot 2 \div 5 \ = \ 10 + 1 - 10 \div 5$
$\qquad\qquad\qquad\quad = \ 10 + 1 - 2$
$\qquad\qquad\qquad\quad = \ 11 - 2 = 9$

71. $(7 - 3)^2 \div 2 - 4 + 8 \ = \ 4^2 \div 2 - 4 + 8$
$\qquad\qquad\qquad\qquad = \ 16 \div 2 - 4 + 8$
$\qquad\qquad\qquad\qquad = \ 8 - 4 + 8$
$\qquad\qquad\qquad\qquad = \ 4 + 8 = 12$

73. $12 \div 3 \cdot 2^2 + (7 - 3)^2 \ = \ 12 \div 3 \cdot 2^2 + 4^2$
$\qquad\qquad\qquad\qquad\quad = \ 12 \div 3 \cdot 4 + 16$
$\qquad\qquad\qquad\qquad\quad = \ 4 \cdot 4 + 16$
$\qquad\qquad\qquad\qquad\quad = \ 16 + 16 = 32$

75. $18 - 2 \cdot 3 + (4 - 1)^3 \ = \ 18 - 2 \cdot 3 + 3^3$
$\qquad\qquad\qquad\qquad\quad = \ 18 - 2 \cdot 3 + 27$
$\qquad\qquad\qquad\qquad\quad = \ 18 - 6 + 27$
$\qquad\qquad\qquad\qquad\quad = \ 12 + 27 = 39$

Applying the Concepts

77. $2^{10} = 1024$

Section 1.7

Objective A Exercises

1. $4 \div 1 = 4$
$4 \div 2 = 2$
Factors are 1, 2, and 4.

3. $10 \div 1 = 10$
$10 \div 2 = 5$
$10 \div 5 = 2$
Factors are 1, 2, 5, and 10.

5. $7 \div 1 = 7$
$7 \div 7 = 1$
Factors are 1 and 7.

7. $9 \div 1 = 9$
$9 \div 3 = 3$
Factors are 1, 3, and 9.

9. $13 \div 1 = 13$
$13 \div 13 = 1$
Factors are 1 and 13.

11. $18 \div 1 = 18$
$18 \div 2 = 9$
$18 \div 3 = 6$
$18 \div 6 = 3$
Factors are 1, 2, 3, 6, 9, and 18.

13. $56 \div 1 = 56$
$56 \div 2 = 28$
$56 \div 4 = 14$
$56 \div 7 = 8$
$56 \div 8 = 7$
Factors are 1, 2, 4, 7, 8, 14, 28, and 56.

15. $45 \div 1 = 45$
$45 \div 3 = 15$
$45 \div 5 = 9$
Factors are 1, 3, 5, 9, 15, and 45.

17. $29 \div 1 = 29$
$29 \div 29 = 1$
Factors are 1 and 29.

19. $22 \div 1 = 22$
$22 \div 2 = 11$
$22 \div 11 = 2$
Factors are 1, 2, 11, and 22.

21. $52 \div 1 = 52$
$52 \div 2 = 26$
$52 \div 4 = 13$
$52 \div 13 = 4$
Factors are 1, 2, 4, 13, 26, and 52.

23. $82 \div 1 = 82$
$82 \div 2 = 41$
$82 \div 41 = 2$
Factors are 1, 2, 41, and 82.

25. $57 \div 1 = 57$
$57 \div 3 = 19$
$57 \div 19 = 3$
Factors are 1, 3, 19, and 57.

27. $48 \div 1 = 48$
$48 \div 2 = 24$
$48 \div 3 = 16$
$48 \div 4 = 12$
$48 \div 6 = 8$
$48 \div 8 = 6$
Factors are 1, 2, 3, 4, 6, 8, 12, 16, 24, and 48.

29. $95 \div 1 = 95$
$95 \div 5 = 19$
$95 \div 19 = 5$
Factors are 1, 5, 19, and 95.

31. $54 \div 1 = 54$
$54 \div 2 = 27$
$54 \div 3 = 18$
$54 \div 6 = 9$
$54 \div 9 = 6$
Factors are 1, 2, 3, 6, 9, 18, 27, and 54.

33. $66 \div 1 = 66$
$66 \div 2 = 33$
$66 \div 3 = 22$
$66 \div 6 = 11$
$66 \div 11 = 6$
Factors are 1, 2, 3, 6, 11, 22, 33, and 66.

35. $80 \div 1 = 80$
$80 \div 2 = 40$
$80 \div 4 = 20$
$80 \div 5 = 16$
$80 \div 8 = 10$
$80 \div 10 = 8$
Factors are 1, 2, 4, 5, 8, 10, 16, 20, 40, and 80.

37. $96 \div 1 = 96$
$96 \div 2 = 48$
$96 \div 3 = 32$
$96 \div 4 = 24$
$96 \div 6 = 16$
$96 \div 8 = 12$
$96 \div 12 = 8$
Factors are 1, 2, 3, 4, 6, 8, 12, 16, 24, 32, 48, and 96.

39. $90 \div 1 = 90$
$90 \div 2 = 45$
$90 \div 3 = 30$
$90 \div 5 = 18$
$90 \div 6 = 15$
$90 \div 9 = 10$
$90 \div 10 = 9$
Factors are 1, 2, 3, 5, 6, 9, 10, 15, 18, 30, 45, and 90.

Objective B Exercises

41. $6 = 2 \cdot 3$

43. 17 is prime.

45. $24 = 2 \cdot 2 \cdot 2 \cdot 3$

47. $27 = 3 \cdot 3 \cdot 3$

49. $36 = 2 \cdot 2 \cdot 3 \cdot 3$

51. 19 is prime.

53. $90 = 2 \cdot 3 \cdot 3 \cdot 5$

55. $115 = 5 \cdot 23$

57. $18 = 2 \cdot 3 \cdot 3$

59. $28 = 2 \cdot 2 \cdot 7$

61. 31 is prime.

63. $62 = 2 \cdot 31$

65. $22 = 2 \cdot 11$

67. 101 is prime.

69. $66 = 2 \cdot 3 \cdot 11$

71. $74 = 2 \cdot 37$

73. 67 is prime.

75. $55 = 5 \cdot 11$

77.
$$\begin{array}{r|r} & 120 \\ 2 & 60 \\ 2 & 30 \\ 2 & 15 \\ 3 & 5 \\ 5 & 1 \end{array}$$
$120 = 2 \cdot 2 \cdot 2 \cdot 3 \cdot 5$

79.
$$\begin{array}{r|r} & 160 \\ 2 & 80 \\ 2 & 40 \\ 2 & 20 \\ 2 & 10 \\ 2 & 5 \\ 5 & 1 \end{array}$$
$160 = 2 \cdot 2 \cdot 2 \cdot 2 \cdot 2 \cdot 5$

81.
$$\begin{array}{r|r} & 216 \\ 2 & 108 \\ 2 & 54 \\ 2 & 27 \\ 3 & 9 \\ 3 & 3 \\ 3 & 1 \end{array}$$
$216 = 2 \cdot 2 \cdot 2 \cdot 3 \cdot 3 \cdot 3$

83.
$$\begin{array}{r|r} & 625 \\ 5 & 125 \\ 5 & 25 \\ 5 & 5 \\ 5 & 1 \end{array}$$
$625 = 5 \cdot 5 \cdot 5 \cdot 5$

Applying the Concepts

85. 3, 5; 5, 7; 11, 13; 41, 43; 71, 73; and 17, 19 are all the twin primes less than 100.

87. 2 is the *only* even prime number because all other even numbers have 2 as a factor and thus cannot be prime.

Chapter 1 Review Exercises

1. $3 \cdot 2^3 \cdot 5^2 \quad = \quad 3 \cdot 8 \cdot 25$
$\qquad\qquad = \quad 24 \cdot 25 = 600$

2. $10,000 + 300 + 20 + 7$

3. $18 \div 1 = 18$
$18 \div 2 = 9$
$18 \div 3 = 6$
$18 \div 6 = 3$
Factors are 1, 2, 3, 6, 9, and 18.

4.
$$\begin{array}{r} {\scriptstyle 1\,1\,1} \\ 5894 \\ 6301 \\ +\quad 298 \\ \hline 12,493 \end{array}$$

5.
$$\begin{array}{r} {\scriptstyle 8\,11\,16} \\ 4926 \\ -3177 \\ \hline 1749 \end{array}$$

6.
$$\begin{array}{r} 2,135 \\ 7)\overline{14,945} \\ -14 \\ \hline 09 \\ -\,7 \\ \hline 24 \\ -21 \\ \hline 35 \\ -35 \\ \hline 0 \end{array}$$

7. $101 > 87$

8. $5 \cdot 5 \cdot 7 \cdot 7 \cdot 7 \cdot 7 \cdot 7 = 5^2 \cdot 7^5$

9.
$$\begin{array}{r} {\scriptstyle 6} \\ 2019 \\ \times\quad 307 \\ \hline 14133 \\ 60570 \\ \hline 619,833 \end{array}$$

10.
$$\begin{array}{r} {\scriptstyle 9\,0\,10\,11\,12\,14} \\ 10,134 \\ -\quad 4,725 \\ \hline 5,409 \end{array}$$

11.
$$\begin{array}{r} {\scriptstyle 1\,1} \\ 298 \\ 461 \\ +\quad 322 \\ \hline 1081 \end{array}$$

12. $2^3 - 3 \cdot 2 = 8 - 3 \cdot 2 = 8 - 6 = 2$

13. 45,700

14. Two hundred seventy-six thousand fifty-seven

15.
$$\begin{array}{r} 1306 \quad \text{r}59 \\ 84)\overline{109,763} \\ -84 \\ \hline 257 \\ -252 \\ \hline 56 \\ -\,0 \\ \hline 563 \\ -504 \\ \hline 59 \end{array}$$

16. 2,011,044

17.
```
    488  r2
8)3906
 −32
   70
  −64
   66
  −64
    2
```

18. $3^2 + 2^2 \cdot (5-3) = 3^2 + 2^2 \cdot (2)$
$= 9 + 4 \cdot 2$
$= 9 + 8 = 17$

19. $8 \cdot (6-2)^2 \div 4 = 8 \cdot 4^2 \div 4$
$= 8 \cdot 16 \div 4$
$= 128 \div 4 = 32$

20. $72 = 2 \cdot 2 \cdot 2 \cdot 3 \cdot 3$
```
    72
2 | 36
2 | 18
2 | 9
3 | 3
3 | 1
```

21. 2133

22.
```
  32
 843
× 27
5901
1686
22,761
```

23. Strategy To find the total pay for last week's work:
• Multiply the overtime rate ($24) by the number of hours worked (12).
• Add the total earned as overtime to the assistant's salary ($480).

Solution
```
 $24      $480
×12      + 288
 48       $768
 24
$288
```
The total pay for last week's work is $768.

24. Strategy To find the number of miles driven per gallon of gasoline, divide the total number of miles driven (351) by the number of gallons used (13).

Solution
```
    27
13)351
 −26
   91
  −91
    0
```
He drove 27 miles per gallon of gasoline.

25. Strategy To find the monthly car payment:
• Subtract the down payment ($3000) from the cost of the car ($17,880) to find the balance.
• Divide the balance by the number of equal payments (48).

Solution
```
 $17,880        310
 − 3,000    48)14,880
 $14,880      −144
                48
               −48
                00
               − 0
                 0
```
Each monthly car payment is $310.

26. Strategy To find the total income from commissions, add the amounts received for each of the 4 weeks ($723, $544, $812, and $488).

Solution
```
 $723
  544
  812
 +488
$2567
```
The total income from commissions is $2567.

27. Strategy To find the total amount deposited, add the two deposits ($88 and $213). To find the new checking account balance, add the total amount deposited ($301) to the original balance ($516).

Solution
```
 $88
+213
$301
```
The total amount deposited is $301.
```
 $301
+ 516
 $817
```
The new checking balance is $817.

28. Strategy To find the total of the car payments over a 12-month period, multiply the amount of each payment ($246) by the number of payments (12).

Solution
```
 $246
 × 12
  492
  246
$2952
```
The total of the car payments is $2952.

29. **Strategy** To find the year that there were more males involved in sports, read the values from the table and determine which number is larger.

   Solution $208{,}866 > 170{,}384$
   Since 208,866 is associated with the year 2001, there were more males involved in college sports in 2001 than in 1972.

30. **Strategy** To find the difference between the number of males involved in sports and the number of females involved in sports at U.S. colleges in 1972, subtract the values given in the table.

   Solution
$$\begin{array}{r} 170{,}384 \quad \text{males} \\ -\ 29{,}977 \quad \text{females} \\ \hline 140{,}407 \end{array}$$
   The difference between the numbers of male and female athletes in 1972 was 140,407 students.

31. **Strategy** To find the increase in the number of females involved in sports in U.S. colleges from 1972 to 2001, subtract the number in 1972 (29,977) from the number in 2001 (150,916).

   Solution
$$\begin{array}{r} 150{,}916 \\ -\ 29{,}977 \\ \hline 120{,}939 \end{array}$$
   The number of female athletes increased by 120,939 students from 1972 to 2001.

32. **Strategy** To find how many more U.S. college students were involved in athletics in 2001 than in 1972:
   • Add the number of male and female athletes in 1972.
   • Add the number of male and female athletes in 2001.
   • Subtract these two sums to find the increase.

   Solution
$$\begin{array}{r} \underline{1972} \\ 170{,}384 \quad \text{males} \\ \underline{29{,}977} \quad \text{females} \\ 200{,}361 \end{array}$$

$$\begin{array}{r} \underline{2001} \\ 208{,}866 \quad \text{males} \\ \underline{150{,}916} \quad \text{females} \\ 359{,}782 \end{array}$$

$$\begin{array}{r} 359{,}782 \quad 2001 \\ -200{,}361 \quad 1972 \\ \hline 159{,}421 \end{array}$$
   159,421 more students were involved in athletics in 2001 than in 1972.

Chapter 1 Test

1. $3^3 \cdot 4^2 = 27 \cdot 16 = 432$

2. Two hundred seven thousand sixty-eight

3.
$$\begin{array}{r} {}^{0\,17} \\ \cancel{1}7{,}495 \\ -8{,}162 \\ \hline 9{,}333 \end{array}$$

4. $20 \div 1 = 20$
   $20 \div 2 = 10$
   $20 \div 4 = 5$
   $20 \div 5 = 4$
   Factors are 1, 2, 4, 5, 10, and 20.

5.
$$\begin{array}{r} 9736 \\ \times\ \ 704 \\ \hline 38{,}944 \\ 681{,}520 \ \ \\ \hline 6{,}854{,}144 \end{array}$$

6. $\begin{aligned} 4^2 \cdot (4-2) \div 8 + 5 &= 4^2 \cdot (2) \div 8 + 5 \\ &= 16 \cdot (2) \div 8 + 5 \\ &= 32 \div 8 + 5 \\ &= 4 + 5 = 9 \end{aligned}$

7. $900{,}000 + 6000 + 300 + 70 + 8$

8. $75{,}000$

9.
$$\begin{array}{r} 1121 \quad \text{r}27 \\ 97{\overline{)108{,}764}} \\ -\ 97\phantom{{,}000} \\ \hline 117\phantom{{,}00} \\ -\ 97\phantom{{,}00} \\ \hline 206\phantom{{,}0} \\ -194\phantom{{,}0} \\ \hline 124 \\ -\ 97 \\ \hline 27 \end{array}$$

10. $3 \cdot 3 \cdot 3 \cdot 7 \cdot 7 = 3^3 \cdot 7^2$

11.
$$\begin{array}{r} {}^{2\ \ 21} \\ 8{,}756 \\ 9{,}094 \\ +37{,}065 \\ \hline 54{,}915 \end{array}$$

12. $84 = 2 \cdot 2 \cdot 3 \cdot 7$
$$\begin{array}{r|l} & 84 \\ \hline 2 & 42 \\ 2 & 21 \\ 3 & 7 \\ 7 & 1 \end{array}$$

13. $\begin{aligned} 16 \div 4 \cdot 2 - (7-5)^2 &= 16 \div 4 \cdot 2 - 2^2 \\ &= 16 \div 4 \cdot 2 - 4 \\ &= 4 \cdot 2 - 4 \\ &= 8 - 4 = 4 \end{aligned}$

14.
$$\begin{array}{r} \overset{\scriptstyle 6\;52}{90{,}763} \\ \times\qquad 8 \\ \hline 726{,}104 \end{array}$$

15. 1,204,006

16.
$$\begin{array}{r} 8710 \quad r2 \\ 7\overline{)60972} \\ \underline{-56} \\ 49 \\ \underline{-49} \\ 07 \\ \underline{-7} \\ 02 \\ \underline{-0} \\ 2 \end{array}$$

17. 21 > 19

18.
$$\begin{array}{r} 703 \\ 8\overline{)5624} \\ \underline{-56} \\ 02 \\ \underline{-0} \\ 24 \\ \underline{-24} \\ 0 \end{array}$$

19.
$$\begin{array}{r} 25{,}492 \\ +71{,}306 \\ \hline 96{,}798 \end{array}$$

20.
$$\begin{array}{r} \overset{\scriptstyle 1\;1817}{29{,}736} \\ -\quad 9{,}814 \\ \hline 19{,}922 \end{array}$$

21. **Strategy**   To find the difference between the total enrollment in 2012 and 2009:
● Add the numbers in the two columns for 2009 to find the total enrollment for 2009.
● Add the numbers in the two columns for 2012 to find the total enrollment for 2012.
● Subtract the two values to find the difference.

Solution
$$\begin{array}{ll} \underline{2009} & \\ 37{,}726{,}000 & \text{K--8} \\ +15{,}812{,}000 & \text{9--12} \\ \hline 53{,}538{,}000 & \end{array}$$

$$\begin{array}{ll} \underline{2012} & \\ 38{,}258{,}000 & \text{K--8} \\ +15{,}434{,}000 & \text{9--12} \\ \hline 53{,}692{,}000 & \end{array}$$

$$\begin{array}{ll} 53{,}692{,}000 & 2012 \\ -53{,}538{,}000 & 2009 \\ \hline 154{,}000 & \end{array}$$

The difference in projected total enrollment between 2012 and 2009 is 154,000 students.

22. **Strategy**   To find the average enrollment in each of grades 9 through 12 in 2012, divide the total enrollment (15,434,000) in the four grades by 4.

Solution
$$\begin{array}{r} 3{,}858{,}500 \\ 4\overline{)15{,}434{,}000} \\ \underline{12} \\ 34 \\ \underline{32} \\ 23 \\ \underline{20} \\ 34 \\ \underline{32} \\ 20 \\ \underline{20} \\ 0 \end{array}$$

The average enrollment in each of grades 9 through 12 in 2012 is projected to be 3,858,500 students.

23. **Strategy** To find how many boxes were needed
      to pack the lemons:
      ● Find the total number of lemons
      harvested by adding the amounts
      harvested from the two groves (48,290
      and 23,710 pounds).
      ● Divide the total number of pounds
      harvested by the number of pounds of
      lemons that can be packed in each box
      (24).

Solution
$$
\begin{array}{r}
48,290 \\
+23,710 \\
\hline
72,000
\end{array}
$$

$$
\begin{array}{r}
3000 \\
24)\overline{72,000} \\
-72 \\
\hline
00 \\
-\,0 \\
\hline
00 \\
-\,0 \\
\hline
00 \\
-\,0 \\
\hline
0
\end{array}
$$

3000 boxes were needed to pack the
lemons.

24. **Strategy** To find the amount that the investor
      receives over the period, multiply the
      amount she receives each month
      ($237) by the number of months (12).

Solution
$$
\begin{array}{r}
\$237 \\
\times\ 12 \\
\hline
474 \\
237 \\
\hline
\$2844
\end{array}
$$

The investor receives $2844 over the
12-month period.

25a. **Strategy** To find the total number of miles
      driven during the 3 days, add the
      amounts driven each day (425, 187,
      and 243 miles).

Solution
$$
\begin{array}{r}
425 \\
187 \\
+243 \\
\hline
855
\end{array}
$$

855 miles were driven during the 3
days.

b.   **Strategy** To find the odometer reading at the
      end of the 3 days, add the number of
      miles driven during the 3 days (855) to
      the odometer reading at the start of
      the vacation (47,626).

Solution
$$
\begin{array}{r}
47,626 \\
+\ \ 855 \\
\hline
48,481
\end{array}
$$

The odometer reading at the end of
the 3 days is 48,481 miles.

Chapter 2: Fractions

Prep Test

1. 20

2. 120

3. 9

4. 10

5. 7

6.
$$\begin{array}{r} 2\ r3 \\ 30\overline{)63} \\ -60 \\ \hline 3 \end{array}$$

7. 1, 2, 3, 4, 6, 12

8. $8 \cdot 7 + 3 = 56 + 3 = 59$

9. 7

10. <

Go Figure

One lap is down to the end of the pool and back. If you swim one lap every 4 minutes and your friend swims one lap every 5 minutes, then a visual time reference might look like this

4	4	4	4	4

5	5	5	5

where each section represents one lap. As you can see, you meet up with your friend after you swim 5 laps and your friend swims 4 laps. Five 4-minute laps are 20 minutes ($5 \times 4 = 20$). From the visual reference, you can see that you pass each other 4 times for each lap during the 20 minutes. Since one lap is down and back, you pass each other twice per lap. So you pass each other 8 times ($4 \times 2 = 8$).

Section 2.1

Objective A Exercises

1.
$$5 = \qquad 5$$
$$8 = 2 \cdot 2 \cdot 2$$
$$\text{LCM} = 2 \cdot 2 \cdot 2 \cdot 5 = 40$$

3.
$$3 = \qquad 3$$
$$8 = 2 \cdot 2 \cdot 2$$
$$\text{LCM} = 2 \cdot 2 \cdot 2 \cdot 3 = 24$$

5.
$$5 = \qquad\qquad 5$$
$$6 = 2 \cdot 3$$
$$\text{LCM} = 2 \cdot 3 \cdot 5 = 30$$

7.
$$4 = 2 \cdot 2$$
$$6 = 2 \qquad 3$$
$$\text{LCM} = 2 \cdot 2 \cdot 3 = 12$$

9.
$$8 = 2 \cdot 2 \cdot 2$$
$$12 = 2 \cdot 2 \qquad 3$$
$$\text{LCM} = 2 \cdot 2 \cdot 2 \cdot 3 = 24$$

11.
$$5 = \qquad\qquad 5$$
$$12 = 2 \cdot 2 \quad 3$$
$$\text{LCM} = 2 \cdot 2 \cdot 3 \cdot 5 = 60$$

13.
$$8 = 2 \cdot 2 \cdot 2$$
$$14 = 2 \qquad 7$$
$$\text{LCM} = 2 \cdot 2 \cdot 2 \cdot 7 = 56$$

15.
$$3 = \qquad 3$$
$$9 = 3 \cdot 3$$
$$\text{LCM} = 3 \cdot 3 = 9$$

17.
$$8 = 2 \cdot 2 \cdot 2$$
$$32 = 2 \cdot 2 \cdot 2 \cdot 2 \cdot 2$$
$$\text{LCM} = 2 \cdot 2 \cdot 2 \cdot 2 \cdot 2 = 32$$

19.
$$9 = \qquad 3 \cdot 3$$
$$36 = 2 \cdot 2 \quad 3 \cdot 3$$
$$\text{LCM} = 2 \cdot 2 \cdot 3 \cdot 3 = 36$$

21.
$$44 = 2 \cdot 2 \qquad\qquad 11$$
$$60 = 2 \cdot 2 \quad 3 \quad 5$$
$$\text{LCM} = 2 \cdot 3 \cdot 5 \cdot 11 = 660$$

23.
$$102 = 2 \qquad 3 \quad 17$$
$$184 = 2 \cdot 2 \cdot 2 \qquad\qquad 23$$
$$\text{LCM} = 2 \cdot 2 \cdot 2 \cdot 3 \cdot 17 \cdot 23 = 9384$$

25.
$$4 = 2 \cdot 2$$
$$8 = 2 \cdot 2 \cdot 2$$
$$12 = 2 \cdot 2 \qquad 3$$
$$\text{LCM} = 2 \cdot 2 \cdot 2 \cdot 3 = 24$$

27.
$$3 = \qquad 3$$
$$5 = \qquad\qquad 5$$
$$10 = 2 \qquad\qquad 5$$
$$\text{LCM} = 2 \cdot 3 \cdot 5 = 30$$

29.

	2	3
3 =		3
8 =	(2 · 2 · 2)	
12 =	2 · 2	③

LCM = 2 · 2 · 2 · 3 = 24

31.

	2	3
9 =		3 · 3
36 =	2 · 2	(3 · 3)
64 =	(2 · 2 · 2 · 2 · 2 · 2)	

LCM = 2 · 2 · 2 · 2 · 2 · 2 · 3 · 3 = 576

33.

	2	3	5	7
16 =	(2 · 2 · 2 · 2)			
30 =	2	3	⑤	
84 =	2 · 2	③		⑦

LCM = 2 · 2 · 2 · 2 · 3 · 5 · 7 = 1680

Objective B Exercises

35.

	3	5
3 =	3	
5 =		5

GCF = 1

37.

	2	3
6 =	2	③
9 =		3 · 3

GCF = 3

39.

	3	5
15 =	3	⑤
25 =		5 · 5

GCF = 5

41.

	2	5
25 =		5 · 5
100 =	2 · 2	(5 · 5)

GCF = 5 · 5 = 25

43.

	2	3	17
32 =	2 · 2 · 2 · 2 · 2		
51 =		3	17

GCF = 1

45.

	2	3	5
12 =	(2 · 2)	3	
80 =	2 · 2 · 2 · 2		5

GCF = 2 · 2 = 4

47.

	2	5	7
16 =	2 · 2 · 2 · 2		
140 =	(2 · 2)	5	7

GCF = 2 · 2 = 4

49.

	2	3	5
24 =	2 · 2 · 2	3	
30 =	②	③	5

GCF = 2 · 3 = 6

51.

	2	3	11
44 =	(2 · 2)		11
96 =	2 · 2 · 2 · 2 · 2	3	

GCF = 2 · 2 = 4

53.

	3	5	11
3 =	3		
5 =		5	
11 =			11

GCF = 1

55.

	2	7
7 =		7
14 =	2	⑦
49 =		7 · 7

GCF = 7

57.

	2	3	5
10 =	2		5
15 =		3	5
20 =	2 · 2		⑤

GCF = 5

59.

	2	3	5
24 =	2 · 2 · 2	3	
40 =	2 · 2 · 2		5
72 =	(2 · 2 · 2)	3 · 3	

GCF = 2 · 2 · 2 = 8

61.

	3	17	31
17 =		17	
31 =			31
81 =	3 · 3 · 3 · 3		

GCF = 1

63.

	5
25 =	(5 · 5)
125 =	5 · 5 · 5
625 =	5 · 5 · 5 · 5

GCF = 25

65.

	2	5	7
28 =	2 · 2		7
35 =		5	7
70 =	2	5	⑦

GCF = 7

67.

	2	3	7
32 =	2 · 2 · 2 · 2 · 2		
56 =	2 · 2 · 2		7
72 =	(2 · 2 · 2)	3 · 3	

GCF = 2 · 2 · 2 = 8

Applying the Concepts

69. Relatively prime numbers are numbers with no common factors except 1. Examples: 4 and 5, 8 and 9, and 16 and 21

71. The LCM of 2 and 3 is 6. The LCM of 5 and 7 is 35. The LCM of 11 and 19 is 209. The LCM of two prime numbers is the product of the two numbers. The LCM of three prime numbers is the product of the three numbers.

73. Yes, the LCM of the two numbers is always divisible by the GCF of the two numbers. Two numbers are factors of their LCM. The GCF of the two numbers is a factor of the LCM of the same numbers. That is, the LCM of two numbers always is divisible by the GCF of the two numbers. For example, the GCF of 4 and 6 is 2, and the LCM of 4 and 6 is 12. 12 is divisible by 2.

Section 2.2

Objective A Exercises

1. Improper fraction

3. Proper fraction

5. $\dfrac{3}{4}$

7. $\dfrac{7}{8}$

9. $1\dfrac{1}{2}$

11. $2\dfrac{5}{8}$

13. $3\dfrac{3}{5}$

15. $\dfrac{5}{4}$

17. $\dfrac{8}{3}$

19. $\dfrac{28}{8}$

21.

23.

25.

27. $4\overline{)11}$ $\dfrac{-8}{3}$ $\dfrac{11}{4}=2\dfrac{3}{4}$

29. $4\overline{)20}$ $\dfrac{-20}{0}$ $\dfrac{20}{4}=5$

31. $8\overline{)9}$ $\dfrac{-8}{1}$ $\dfrac{9}{8}=1\dfrac{1}{8}$

33. $10\overline{)23}$ $\dfrac{-20}{3}$ $\dfrac{23}{10}=2\dfrac{3}{10}$

35. $16\overline{)48}$ $\dfrac{-48}{0}$ $\dfrac{48}{16}=3$

37. $7\overline{)8}$ $\dfrac{-7}{1}$ $\dfrac{8}{7}=1\dfrac{1}{7}$

39. $3\overline{)7}$ $\dfrac{-6}{1}$ $\dfrac{7}{3}=2\dfrac{1}{3}$

41. $1\overline{)16}$ $\dfrac{-1}{06}$ $\dfrac{-6}{0}$ $\dfrac{16}{1}=16$

43. $8\overline{)17}$ $\dfrac{-16}{1}$ $\dfrac{17}{8}=2\dfrac{1}{8}$

45. $5\overline{)12}$ $\dfrac{-10}{2}$ $\dfrac{12}{5}=2\dfrac{2}{5}$

47. $9\overline{)9}$ $\dfrac{-9}{0}$ $\dfrac{9}{9}=1$

49. $8\overline{)72}$ $\dfrac{-72}{0}$ $\dfrac{72}{8}=9$

51. $2\dfrac{1}{3}=\dfrac{6+1}{3}=\dfrac{7}{3}$

53. $6\dfrac{1}{2}=\dfrac{12+1}{2}=\dfrac{13}{2}$

55. $6\dfrac{5}{6}=\dfrac{36+5}{6}=\dfrac{41}{6}$

57. $9\dfrac{1}{4}=\dfrac{36+1}{4}=\dfrac{37}{4}$

59. $10\dfrac{1}{2}=\dfrac{20+1}{2}=\dfrac{21}{2}$

61. $8\dfrac{1}{9}=\dfrac{72+1}{9}=\dfrac{73}{9}$

63. $5\dfrac{3}{11}=\dfrac{55+3}{11}=\dfrac{58}{11}$

65. $2\dfrac{5}{8}=\dfrac{16+5}{8}=\dfrac{21}{8}$

67. $1\dfrac{5}{8}=\dfrac{8+5}{8}=\dfrac{13}{8}$

69. $11\dfrac{1}{9}=\dfrac{99+1}{9}=\dfrac{100}{9}$

71. $3\dfrac{3}{8}=\dfrac{24+3}{8}=\dfrac{27}{8}$

73. $6\dfrac{7}{13}=\dfrac{78+7}{13}=\dfrac{85}{13}$

Applying the Concepts

75. Students might mention any of the following: fractional parts of an hour, as in three-quarters of an hour; lengths of nails, as in $\frac{3}{4}$-inch nail; lengths of fabric, as in $1\frac{5}{8}$ yards of material; lengths of lumber, as in $2\frac{1}{2}$ feet of pine; ingredients in a recipe, as in $1\frac{1}{2}$ cups sugar; or innings pitched, as in four and two-thirds innings.

Section 2.3

Objective A Exercises

1. $\dfrac{1\cdot5}{2\cdot5}=\dfrac{5}{10}$

3. $\dfrac{3\cdot3}{16\cdot3}=\dfrac{9}{48}$

5. $\dfrac{3\cdot4}{8\cdot4}=\dfrac{12}{32}$

7. $\dfrac{3\cdot3}{17\cdot3}=\dfrac{9}{51}$

22   Chapter 2   Fractions

9. $\dfrac{3 \cdot 4}{4 \cdot 4} = \dfrac{12}{16}$

11. $\dfrac{3 \cdot 9}{1 \cdot 9} = \dfrac{27}{9}$

13. $\dfrac{1 \cdot 20}{3 \cdot 20} = \dfrac{20}{60}$

15. $\dfrac{11 \cdot 4}{15 \cdot 4} = \dfrac{44}{60}$

17. $\dfrac{2 \cdot 6}{3 \cdot 6} = \dfrac{12}{18}$

19. $\dfrac{5 \cdot 7}{7 \cdot 7} = \dfrac{35}{49}$

21. $\dfrac{5 \cdot 2}{9 \cdot 2} = \dfrac{10}{18}$

23. $\dfrac{7 \cdot 3}{1 \cdot 3} = \dfrac{21}{3}$

25. $\dfrac{7 \cdot 5}{9 \cdot 5} = \dfrac{35}{45}$

27. $\dfrac{15 \cdot 4}{16 \cdot 4} = \dfrac{60}{64}$

29. $\dfrac{3 \cdot 7}{14 \cdot 7} = \dfrac{21}{98}$

31. $\dfrac{5 \cdot 6}{8 \cdot 6} = \dfrac{30}{48}$

33. $\dfrac{5 \cdot 3}{14 \cdot 3} = \dfrac{15}{42}$

35. $\dfrac{17 \cdot 6}{24 \cdot 6} = \dfrac{102}{144}$

37. $\dfrac{3 \cdot 51}{8 \cdot 51} = \dfrac{153}{408}$

39. $\dfrac{17 \cdot 20}{40 \cdot 20} = \dfrac{340}{800}$

Objective B Exercises

41. $\dfrac{4}{12} = \dfrac{\cancel{2} \cdot \cancel{2}}{\cancel{2} \cdot \cancel{2} \cdot 3} = \dfrac{1}{3}$

43. $\dfrac{22}{44} = \dfrac{\cancel{2} \cdot \cancel{11}}{2 \cdot 2 \cdot \cancel{11}} = \dfrac{1}{2}$

45. $\dfrac{2}{12} = \dfrac{\cancel{2}}{\cancel{2} \cdot 2 \cdot 3} = \dfrac{1}{6}$

47. $\dfrac{40}{36} = \dfrac{\cancel{2} \cdot \cancel{2} \cdot 2 \cdot 5}{\cancel{2} \cdot \cancel{2} \cdot 3 \cdot 3} = \dfrac{10}{9} = 1\dfrac{1}{9}$

49. $\dfrac{0}{30} = 0$

51. $\dfrac{9}{22} = \dfrac{3 \cdot 3}{2 \cdot 11} = \dfrac{9}{22}$

53. $\dfrac{75}{25} = \dfrac{3 \cdot \cancel{5} \cdot \cancel{5}}{\cancel{5} \cdot \cancel{5}} = 3$

55. $\dfrac{16}{84} = \dfrac{\cancel{2} \cdot \cancel{2} \cdot 2 \cdot 2}{\cancel{2} \cdot \cancel{2} \cdot 3 \cdot 7} = \dfrac{4}{21}$

57. $\dfrac{12}{35} = \dfrac{2 \cdot 2 \cdot 3}{5 \cdot 7} = \dfrac{12}{35}$

59. $\dfrac{28}{44} = \dfrac{\cancel{2} \cdot \cancel{2} \cdot 7}{\cancel{2} \cdot \cancel{2} \cdot 11} = \dfrac{7}{11}$

61. $\dfrac{16}{12} = \dfrac{\cancel{2} \cdot \cancel{2} \cdot 2 \cdot 2}{\cancel{2} \cdot \cancel{2} \cdot 3} = \dfrac{4}{3} = 1\dfrac{1}{3}$

63. $\dfrac{24}{40} = \dfrac{\cancel{2} \cdot \cancel{2} \cdot \cancel{2} \cdot 3}{\cancel{2} \cdot \cancel{2} \cdot \cancel{2} \cdot 5} = \dfrac{3}{5}$

65. $\dfrac{8}{88} = \dfrac{\cancel{2} \cdot \cancel{2} \cdot \cancel{2}}{\cancel{2} \cdot \cancel{2} \cdot \cancel{2} \cdot 11} = \dfrac{1}{11}$

67. $\dfrac{144}{36} = \dfrac{\cancel{2} \cdot \cancel{2} \cdot 2 \cdot 2 \cdot \cancel{3} \cdot \cancel{3}}{\cancel{2} \cdot \cancel{2} \cdot \cancel{3} \cdot \cancel{3}} = 4$

69. $\dfrac{48}{144} = \dfrac{\cancel{2} \cdot \cancel{2} \cdot \cancel{2} \cdot \cancel{2} \cdot \cancel{3}}{\cancel{2} \cdot \cancel{2} \cdot \cancel{2} \cdot \cancel{2} \cdot \cancel{3} \cdot 3} = \dfrac{1}{3}$

71. $\dfrac{60}{100} = \dfrac{\cancel{2} \cdot \cancel{2} \cdot 3 \cdot \cancel{5}}{\cancel{2} \cdot \cancel{2} \cdot 5 \cdot \cancel{5}} = \dfrac{3}{5}$

73. $\dfrac{36}{16} = \dfrac{\cancel{2} \cdot \cancel{2} \cdot 3 \cdot 3}{\cancel{2} \cdot \cancel{2} \cdot 2 \cdot 2} = \dfrac{9}{4} = 2\dfrac{1}{4}$

75. $\dfrac{32}{160} = \dfrac{\cancel{2} \cdot \cancel{2} \cdot \cancel{2} \cdot \cancel{2} \cdot 2}{\cancel{2} \cdot \cancel{2} \cdot \cancel{2} \cdot \cancel{2} \cdot 2 \cdot 5} = \dfrac{1}{5}$

Applying the Concepts

77. $\dfrac{3}{1}, \dfrac{6}{2}, \dfrac{9}{3}, \dfrac{12}{4}, \dfrac{15}{5}$ are fractions that are equal to 3.

79a. $\dfrac{8}{50} = \dfrac{4}{25}$
Maine, Maryland, Massachusetts, Michigan, Minnesota, Mississippi, Missouri, Montana

b. $\dfrac{8}{50} = \dfrac{4}{25}$
Alabama, Alaska, Arizona, Idaho, Indiana, Iowa, Ohio, Oklahoma

Section 2.4

Objective A Exercises

1. $\begin{array}{r} \dfrac{2}{7} \\ + \dfrac{1}{7} \\ \hline \dfrac{3}{7} \end{array}$

3. $\begin{array}{r} \dfrac{1}{2} \\ + \dfrac{1}{2} \\ \hline \dfrac{2}{2} = 1 \end{array}$

5. $\begin{array}{r} \dfrac{8}{11} \\ + \dfrac{7}{11} \\ \hline \dfrac{15}{11} = 1\dfrac{4}{11} \end{array}$

7. $\begin{array}{r} \dfrac{8}{5} \\ + \dfrac{9}{5} \\ \hline \dfrac{17}{5} = 3\dfrac{2}{5} \end{array}$

9. $\begin{array}{r} \dfrac{3}{5} \\ \dfrac{8}{5} \\ + \dfrac{3}{5} \\ \hline \dfrac{14}{5} = 2\dfrac{4}{5} \end{array}$

11.
$$\frac{3}{4}$$
$$\frac{1}{4}$$
$$+\frac{5}{4}$$
$$\frac{9}{4}=2\frac{1}{4}$$

13.
$$\frac{3}{8}$$
$$\frac{7}{8}$$
$$+\frac{1}{8}$$
$$\frac{11}{8}=1\frac{3}{8}$$

15.
$$\frac{4}{15}$$
$$\frac{7}{15}$$
$$+\frac{11}{15}$$
$$\frac{22}{15}=1\frac{7}{15}$$

17.
$$\frac{3}{16}$$
$$\frac{5}{16}$$
$$+\frac{7}{16}$$
$$\frac{15}{16}$$

19.
$$\frac{3}{11}$$
$$\frac{5}{11}$$
$$+\frac{7}{11}$$
$$\frac{15}{11}=1\frac{4}{11}$$

21. $\frac{4}{9}+\frac{5}{9}=\frac{9}{9}=1$

23. $\frac{5}{8}+\frac{3}{8}+\frac{7}{8}=\frac{15}{8}=1\frac{7}{8}$

Objective B Exercises

25.
$$\frac{1}{2}=\frac{3}{6}$$
$$+\frac{2}{3}=\frac{4}{6}$$
$$\frac{7}{6}=1\frac{1}{6}$$

27.
$$\frac{3}{14}=\frac{3}{14}$$
$$+\frac{5}{7}=\frac{10}{14}$$
$$\frac{13}{14}$$

29.
$$\frac{8}{15}=\frac{32}{60}$$
$$+\frac{7}{20}=\frac{21}{60}$$
$$\frac{53}{60}$$

31.
$$\frac{3}{8}=\frac{21}{56}$$
$$+\frac{9}{14}=\frac{36}{56}$$
$$\frac{57}{56}=1\frac{1}{56}$$

33.
$$\frac{3}{20}=\frac{9}{60}$$
$$+\frac{7}{30}=\frac{14}{60}$$
$$\frac{23}{60}$$

35.
$$\frac{2}{3}=\frac{38}{57}$$
$$+\frac{6}{19}=\frac{18}{57}$$
$$\frac{56}{57}$$

37.
$$\frac{1}{3}=\frac{6}{18}$$
$$\frac{5}{6}=\frac{15}{18}$$
$$+\frac{7}{9}=\frac{14}{18}$$
$$\frac{35}{18}=1\frac{17}{18}$$

39.
$$\frac{5}{6}=\frac{40}{48}$$
$$\frac{1}{12}=\frac{4}{48}$$
$$+\frac{5}{16}=\frac{15}{48}$$
$$\frac{59}{48}=1\frac{11}{48}$$

41.
$$\frac{2}{3}=\frac{40}{60}$$
$$\frac{1}{5}=\frac{12}{60}$$
$$+\frac{7}{12}=\frac{35}{60}$$
$$\frac{87}{60}=1\frac{27}{60}=1\frac{9}{20}$$

43.
$$\frac{1}{4}=\frac{45}{180}$$
$$\frac{4}{5}=\frac{144}{180}$$
$$+\frac{5}{9}=\frac{100}{180}$$
$$\frac{289}{180}=1\frac{109}{180}$$

45.
$$\frac{5}{16}=\frac{45}{144}$$
$$\frac{11}{18}=\frac{88}{144}$$
$$+\frac{17}{24}=\frac{102}{144}$$
$$\frac{235}{144}=1\frac{91}{144}$$

47.
$$\frac{2}{3}=\frac{48}{72}$$
$$\frac{5}{8}=\frac{45}{72}$$
$$+\frac{7}{9}=\frac{56}{72}$$
$$\frac{149}{72}=2\frac{5}{72}$$

49.
$$\frac{3}{8}=\frac{15}{40}$$
$$+\frac{3}{5}=\frac{24}{40}$$
$$\frac{39}{40}$$

51.
$$\frac{3}{8}=\frac{9}{24}$$
$$\frac{5}{6}=\frac{20}{24}$$
$$+\frac{7}{12}=\frac{14}{24}$$
$$\frac{43}{24}=1\frac{19}{24}$$

53.
$$\frac{1}{2}=\frac{36}{72}$$
$$\frac{5}{8}=\frac{45}{72}$$
$$+\frac{7}{9}=\frac{56}{72}$$
$$\frac{137}{72}=1\frac{65}{72}$$

Objective C Exercises

55.
$$1\frac{1}{2}=1\frac{3}{6}$$
$$+2\frac{1}{6}=2\frac{1}{6}$$
$$3\frac{4}{6}=3\frac{2}{3}$$

57.
$$4\frac{1}{2} = 4\frac{6}{12}$$
$$+5\frac{7}{12} = 5\frac{7}{12}$$
$$9\frac{13}{12} = 10\frac{1}{12}$$

59.
$$4$$
$$+5\frac{2}{7}$$
$$9\frac{2}{7}$$

61.
$$3\frac{5}{8} = 3\frac{25}{40}$$
$$+2\frac{11}{20} = 2\frac{22}{40}$$
$$5\frac{47}{40} = 6\frac{7}{40}$$

63.
$$7\frac{5}{12} = 7\frac{20}{48}$$
$$+2\frac{9}{16} = 2\frac{27}{48}$$
$$9\frac{47}{48}$$

65.
$$6$$
$$+2\frac{3}{13}$$
$$8\frac{3}{13}$$

67.
$$8\frac{29}{30} = 8\frac{116}{120}$$
$$+7\frac{11}{40} = 7\frac{33}{120}$$
$$15\frac{149}{120} = 16\frac{29}{120}$$

69.
$$17\frac{3}{8} = 17\frac{15}{40}$$
$$+\ 7\frac{7}{20} = 7\frac{14}{40}$$
$$24\frac{29}{40}$$

71.
$$5\frac{7}{8} = \ 5\frac{21}{24}$$
$$+27\frac{5}{12} = 27\frac{10}{24}$$
$$32\frac{31}{24} = 33\frac{7}{24}$$

73.
$$7\frac{5}{9} = 7\frac{20}{36}$$
$$+2\frac{7}{12} = 2\frac{21}{36}$$
$$9\frac{41}{36} = 10\frac{5}{36}$$

75.
$$2\frac{1}{2} = 2\frac{6}{12}$$
$$3\frac{2}{3} = 3\frac{8}{12}$$
$$+4\frac{1}{4} = 4\frac{3}{12}$$
$$9\frac{17}{12} = 10\frac{5}{12}$$

77.
$$3\frac{1}{2} = 3\frac{45}{90}$$
$$3\frac{1}{5} = 3\frac{18}{90}$$
$$+8\frac{1}{9} = 8\frac{10}{90}$$
$$14\frac{73}{90}$$

79.
$$2\frac{3}{8} = 2\frac{18}{48}$$
$$4\frac{7}{12} = 4\frac{28}{48}$$
$$+3\frac{5}{16} = 3\frac{15}{48}$$
$$9\frac{61}{48} = 10\frac{13}{48}$$

81.
$$6\frac{5}{6} = 6\frac{45}{54}$$
$$17\frac{2}{9} = 17\frac{12}{54}$$
$$+18\frac{5}{27} = 18\frac{10}{54}$$
$$41\frac{67}{54} = 42\frac{13}{54}$$

83.
$$2\frac{4}{9} = 2\frac{16}{36}$$
$$+5\frac{7}{12} = 5\frac{21}{36}$$
$$7\frac{37}{36} = 8\frac{1}{36}$$

85.
$$4\frac{3}{4} = 4\frac{9}{12}$$
$$+9\frac{1}{3} = 9\frac{4}{12}$$
$$13\frac{13}{12} = 14\frac{1}{12}$$

87.
$$2 \ = 2$$
$$4\frac{5}{8} = 4\frac{45}{72}$$
$$+2\frac{2}{9} = 2\frac{16}{72}$$
$$8\frac{61}{72}$$

Objective D Exercises

89. **Strategy** To find the length of the shaft, add the three distances $\left(\frac{3}{8}, \frac{11}{16}, \text{and } \frac{1}{4} \text{ inch}\right)$.

Solution
$$\frac{3}{8} = \frac{6}{16}$$
$$\frac{11}{16} = \frac{11}{16}$$
$$+\frac{1}{4} = \frac{4}{16}$$
$$\frac{21}{16} = 1\frac{5}{16}$$

The length of the shaft is $1\frac{5}{16}$ inches.

91. **Strategy** To find the total thickness of the table after the veneer has been applied, add the table-top thickness $\left(1\frac{1}{8} \text{ inch}\right)$ to the veneer thickness $\left(\frac{3}{16} \text{ inch}\right)$.

Solution
$$1\frac{1}{8} = 1\frac{2}{16}$$
$$+\frac{3}{16} = \frac{3}{16}$$
$$1\frac{5}{16}$$

The total thickness is $1\frac{5}{16}$ inches.

93a. **Strategy** To find the total number of hours worked, add the five amounts $\left(5, 3\frac{3}{4}, 2\frac{1}{3}, 1\frac{1}{4}, \text{and } 7\frac{2}{3} \text{ hours}\right)$.

Solution
$$5 = 5$$
$$3\frac{3}{4} = 3\frac{9}{12}$$
$$2\frac{1}{3} = 2\frac{4}{12}$$
$$1\frac{1}{4} = 1\frac{3}{12}$$
$$+7\frac{2}{3} = 7\frac{8}{12}$$
$$18\frac{24}{12} = 20$$

A total of 20 hours was worked.

b. **Strategy** To find the total salary for the week, multiply the number of hours worked (20) by the pay for 1 hour ($11).

Solution
$$\begin{array}{r} 11 \\ \times\ 20 \\ \hline \$220 \end{array}$$

Your total salary for the week is $220.

95. **Strategy** To find the total length of the wood beams, add the three lengths $\left(25\frac{3}{4}, 12\frac{1}{2}, \text{and } 17\frac{1}{2} \text{ feet}\right)$.

Solution
$$25\frac{3}{4} = 25\frac{3}{4}$$
$$12\frac{1}{2} = 12\frac{2}{4}$$
$$+17\frac{1}{2} = 17\frac{2}{4}$$
$$54\frac{7}{4} = 55\frac{3}{4}$$

The total length of wood needed is $55\frac{3}{4}$ feet.

97. **Strategy** To find what fractional part of those who changed homes moved outside the county, add the fractional part of those who moved to a different state $\left(\frac{1}{7}\right)$ to the fractional part of those who moved to a different county in the same state $\left(\frac{4}{21}\right)$.

Solution
$$\frac{1}{7} = \frac{3}{21}$$
$$+\frac{4}{21} = \frac{4}{21}$$
$$\frac{7}{21} = \frac{1}{3}$$

Those who changed homes outside the county were $\frac{1}{3}$ of the people who changed homes.

Applying the Concepts

99. We can use 3 dimes and 1 quarter to model adding $\frac{3}{10}$ and $\frac{1}{4}$. Because the dime and the quarter are not the same unit, we change them to nickels and add.

$$\begin{array}{r} 3 \text{ dimes} = 6 \text{ nickels} \\ +1 \text{ quarter} = 5 \text{ nickels} \\ \hline 11 \text{ nickels} \end{array}$$

Now, consider the above problem in terms of dollars.

$$\frac{3}{10} = \frac{6}{20}$$
$$+\frac{1}{4} = \frac{5}{20}$$
$$\frac{11}{20}$$

Note that 1 dime = $\frac{1}{10}$ of a dollar,

1 quarter = $\frac{1}{4}$ of a dollar, and

1 nickel = $\frac{1}{20}$ of a dollar.

Section 2.5

Objective A Exercises

1.
$$\frac{9}{17}$$
$$-\frac{7}{17}$$
$$\frac{2}{17}$$

3.
$$\frac{11}{12}$$
$$-\frac{7}{12}$$
$$\frac{4}{12}=\frac{1}{3}$$

5.
$$\frac{9}{20}$$
$$-\frac{7}{20}$$
$$\frac{2}{20}=\frac{1}{10}$$

7.
$$\frac{42}{65}$$
$$-\frac{17}{65}$$
$$\frac{25}{65}=\frac{5}{13}$$

9.
$$\frac{23}{30}$$
$$-\frac{13}{30}$$
$$\frac{10}{30}=\frac{1}{3}$$

11.
$$\frac{13}{14}$$
$$-\frac{5}{14}$$
$$\frac{8}{14}=\frac{4}{7}$$

13.
$$\frac{7}{8}$$
$$-\frac{5}{8}$$
$$\frac{2}{8}=\frac{1}{4}$$

15.
$$\frac{18}{23}$$
$$-\frac{9}{23}$$
$$\frac{9}{23}$$

17.
$$\frac{17}{24}$$
$$-\frac{11}{24}$$
$$\frac{6}{24}=\frac{1}{4}$$

Objective B Exercises

19.
$$\frac{2}{3}=\frac{4}{6}$$
$$-\frac{1}{6}=\frac{1}{6}$$
$$\frac{3}{6}=\frac{1}{2}$$

21.
$$\frac{5}{8}=\frac{35}{56}$$
$$-\frac{2}{7}=\frac{16}{56}$$
$$\frac{19}{56}$$

23.
$$\frac{5}{7}=\frac{10}{14}$$
$$-\frac{3}{14}=\frac{3}{14}$$
$$\frac{7}{14}=\frac{1}{2}$$

25.
$$\frac{8}{15}=\frac{32}{60}$$
$$-\frac{7}{20}=\frac{21}{60}$$
$$\frac{11}{60}$$

27.
$$\frac{9}{14}=\frac{36}{56}$$
$$-\frac{3}{8}=\frac{21}{56}$$
$$\frac{15}{56}$$

29.
$$\frac{46}{51}=\frac{46}{51}$$
$$-\frac{3}{17}=\frac{9}{51}$$
$$\frac{37}{51}$$

31.
$$\frac{21}{35}=\frac{42}{70}$$
$$-\frac{5}{14}=\frac{25}{70}$$
$$\frac{17}{70}$$

33.
$$\frac{29}{60}=\frac{58}{120}$$
$$-\frac{3}{40}=\frac{9}{120}$$
$$\frac{49}{120}$$

35.
$$\frac{11}{15}=\frac{33}{45}$$
$$-\frac{5}{9}=\frac{25}{45}$$
$$\frac{8}{45}$$

37.
$$\frac{9}{14}=\frac{27}{42}$$
$$-\frac{5}{42}=\frac{5}{42}$$
$$\frac{22}{42}=\frac{11}{21}$$

39.
$$\frac{17}{20}=\frac{51}{60}$$
$$-\frac{7}{15}=\frac{28}{60}$$
$$\frac{23}{60}$$

41.
$$\frac{5}{6}=\frac{15}{18}$$
$$-\frac{7}{9}=\frac{14}{18}$$
$$\frac{1}{18}$$

Objective C Exercises

43.
$$16\frac{11}{15}$$
$$-11\frac{8}{15}$$
$$5\frac{3}{15}=5\frac{1}{5}$$

45.
$$19\frac{16}{17}$$
$$-9\frac{7}{17}$$
$$10\frac{9}{17}$$

47.
$$5\frac{7}{8}$$
$$-1$$
$$4\frac{7}{8}$$

49.
$$3\quad=2\frac{21}{21}$$
$$-2\frac{5}{21}=2\frac{5}{21}$$
$$\frac{16}{21}$$

51.
$$16\frac{3}{8} = 15\frac{11}{8}$$
$$-10\frac{7}{8} = 10\frac{7}{8}$$
$$\overline{\phantom{-10\frac{7}{8}=10}5\frac{4}{8} = 5\frac{1}{2}}$$

53.
$$8\frac{3}{7} = 7\frac{10}{7}$$
$$-2\frac{6}{7} = 2\frac{6}{7}$$
$$\overline{\phantom{-2\frac{6}{7}=2}5\frac{4}{7}}$$

55.
$$23\frac{7}{8} = 23\frac{21}{24}$$
$$-16\frac{2}{3} = 16\frac{16}{24}$$
$$\overline{\phantom{-16\frac{2}{3}=16}7\frac{5}{24}}$$

57.
$$65\frac{8}{35} = 65\frac{16}{70} = 64\frac{86}{70}$$
$$-16\frac{11}{14} = 16\frac{55}{70} = 16\frac{55}{70}$$
$$\overline{\phantom{-16\frac{11}{14}=16\frac{55}{70}=}48\frac{31}{70}}$$

59.
$$101\frac{2}{9}$$
$$-\ \ 16$$
$$\overline{85\frac{2}{9}}$$

61.
$$17\ \ \ = 16\frac{13}{13}$$
$$-7\frac{8}{13} = 7\frac{8}{13}$$
$$\overline{\phantom{-7\frac{8}{13}=}9\frac{5}{13}}$$

63.
$$23\frac{3}{20} = 23\frac{3}{20} = 22\frac{23}{20}$$
$$-7\frac{3}{5} = 7\frac{12}{20} = 7\frac{12}{20}$$
$$\overline{\phantom{-7\frac{3}{5}=7\frac{12}{20}=}15\frac{11}{20}}$$

65.
$$12\frac{3}{8} = 12\frac{9}{24} = 11\frac{33}{24}$$
$$-7\frac{5}{12} = 7\frac{10}{24} = 7\frac{10}{24}$$
$$\overline{\phantom{-7\frac{5}{12}=7\frac{10}{24}=}4\frac{23}{24}}$$

67.
$$6\frac{1}{3} = 6\frac{5}{15} = 5\frac{20}{15}$$
$$-3\frac{3}{5} = 3\frac{9}{15} = 3\frac{9}{15}$$
$$\overline{\phantom{-3\frac{3}{5}=3\frac{9}{15}=}2\frac{11}{15}}$$

Objective D Exercises

69.
$$12\frac{3}{8} = 11\frac{11}{8}$$
$$-2\frac{7}{8} = 2\frac{7}{8}$$
$$\overline{\phantom{-2\frac{7}{8}=}9\frac{4}{8} = 9\frac{1}{2}}$$

The missing dimension is $9\frac{1}{2}$ inches.

71. Strategy   To find the difference between Meyfarth's distance and Coachman's distance, subtract Coachman's distance $\left(66\frac{1}{8} \text{ inches}\right)$ from Meyfarth's distance $\left(75\frac{1}{2} \text{ inches}\right)$.

Solution
$$75\frac{1}{2} = 75\frac{4}{8}$$
$$-66\frac{1}{8} = 66\frac{1}{8}$$
$$\overline{\phantom{-66\frac{1}{8}=}9\frac{3}{8}}$$

The difference between Meyfarth's distance and Coachman's distance was $9\frac{3}{8}$ inches.

Strategy   To find the difference between Kostadinova's distance and Meyfarth's distance, subtract Meyfarth's distance $\left(75\frac{1}{2} \text{ inches}\right)$ from Kostadinova's distance $\left(80\frac{3}{4} \text{ inches}\right)$.

Solution
$$80\frac{3}{4} = 80\frac{3}{4}$$
$$-75\frac{1}{2} = 75\frac{2}{4}$$
$$\overline{\phantom{-75\frac{1}{2}=}5\frac{1}{4}}$$

The difference between Kostadinova's distance and Meyfarth's distance was $5\frac{1}{4}$ inches.

73a.   Strategy   To find the distance, add the distance from the starting point to the first checkpoint $\left(3\frac{3}{8}\text{ miles}\right)$ to the distance from the first checkpoint to the second checkpoint $\left(4\frac{1}{3}\text{ miles}\right)$.

   Solution
$$3\frac{3}{8}=3\frac{9}{24}$$
$$+4\frac{1}{3}=4\frac{8}{24}$$
$$\overline{\phantom{+4\frac{1}{3}=}\,7\frac{17}{24}}$$

The distance from the starting point to the second checkpoint is $7\frac{17}{24}$ miles.

b.   Strategy   To find the distance, subtract the distance from the starting point to the second checkpoint $\left(7\frac{17}{24}\text{ miles}\right)$ from the total distance (12 miles).

   Solution
$$12\phantom{\frac{17}{24}}=11\frac{24}{24}$$
$$-7\frac{17}{24}=7\frac{17}{24}$$
$$\overline{\phantom{-7\frac{17}{24}=}\,4\frac{7}{24}}$$

The distance from the second checkpoint to the finish line is $4\frac{7}{24}$ miles.

75a.   The wrestler has lost $5\frac{1}{4}$ pounds the first week and $4\frac{1}{4}$ pounds the second week. Thus the wrestler has lost more than 9 pounds the first two weeks. Since less than 13 pounds needs to be lost, the wrestler can attain the weight class by losing less than 4 pounds. Yes, this is less than the $4\frac{1}{4}$ pounds lost in the second week.

b.   Strategy   To find how much weight must be lost to reach the desired weight:
● Add the amounts of weight lost during the first 2 weeks $\left(5\frac{1}{4}\text{ and }4\frac{1}{4}\text{ pounds}\right)$.
● Subtract the total weight lost so far from the amount that is required $\left(12\frac{3}{4}\text{ pounds}\right)$.

Solution
$$5\frac{1}{4}\qquad\qquad 12\frac{3}{4}=12\frac{3}{4}$$
$$+4\frac{1}{4}\qquad\quad -9\frac{1}{2}=\,9\frac{2}{4}$$
$$\overline{9\frac{2}{4}=9\frac{1}{2}}\qquad\overline{\phantom{-9\frac{1}{2}=}\,3\frac{1}{4}}$$

The wrestler needs to lose $3\frac{1}{4}$ pounds to reach the desired weight.

Applying the Concepts

77.   To find the missing number, subtract $2\frac{1}{2}$ from $5\frac{1}{3}$.
$$5\frac{1}{3}=5\frac{2}{6}=4\frac{8}{6}$$
$$-2\frac{1}{2}=2\frac{3}{6}=2\frac{3}{6}$$
$$\overline{\phantom{-2\frac{1}{2}=2\frac{3}{6}=}\,2\frac{5}{6}}$$

79.   Right diagonal: $\dfrac{3}{4}+\dfrac{5}{8}+\dfrac{1}{2}=\dfrac{6}{8}+\dfrac{5}{8}+\dfrac{4}{8}=\dfrac{15}{8}$

Left diagonal: $\dfrac{5}{8}+\dfrac{7}{8}=\dfrac{12}{8};\dfrac{15}{8}-\dfrac{12}{8}=\dfrac{3}{8}$

Top across: $\dfrac{3}{8}+\dfrac{3}{4}=\dfrac{3}{8}+\dfrac{6}{8}=\dfrac{9}{8};\dfrac{15}{8}-\dfrac{9}{8}=\dfrac{6}{8}=\dfrac{3}{4}$

Left down: $\dfrac{3}{8}+\dfrac{1}{2}=\dfrac{3}{8}+\dfrac{4}{8}=\dfrac{7}{8};\dfrac{15}{8}-\dfrac{7}{8}=\dfrac{8}{8}=1$

Middle across: $1+\dfrac{5}{8}=\dfrac{8}{8}+\dfrac{5}{8}=\dfrac{13}{8};$
$$\dfrac{15}{8}-\dfrac{13}{8}=\dfrac{2}{8}=\dfrac{1}{4}$$

Bottom across: $\dfrac{1}{2}+\dfrac{7}{8}=\dfrac{4}{8}+\dfrac{7}{8}=\dfrac{11}{8};$
$$\dfrac{15}{8}-\dfrac{11}{8}=\dfrac{4}{8}=\dfrac{1}{2}$$

$\frac{3}{8}$	$\frac{3}{4}$	$\frac{3}{4}$
1	$\frac{5}{8}$	$\frac{1}{4}$
$\frac{1}{2}$	$\frac{1}{2}$	$\frac{7}{8}$

Section 2.6

Objective A Exercises

1. $\dfrac{2\cdot 7}{3\cdot 8}=\dfrac{\overset{1}{\cancel{2}}\cdot 7}{3\cdot\underset{1}{2}\cdot 2\cdot 2}=\dfrac{7}{12}$

3. $\dfrac{5\cdot 7}{16\cdot 15}=\dfrac{\overset{1}{\cancel{5}}\cdot 7}{2\cdot 2\cdot 2\cdot 2\cdot 3\cdot\underset{1}{\cancel{5}}}=\dfrac{7}{48}$

5. $\dfrac{1\cdot 1}{6\cdot 8}=\dfrac{1\cdot 1}{2\cdot 3\cdot 2\cdot 2\cdot 2}=\dfrac{1}{48}$

7. $\dfrac{11 \cdot 6}{12 \cdot 7} = \dfrac{11 \cdot \overset{1}{\cancel{2}} \cdot \overset{1}{\cancel{3}}}{\underset{1}{\cancel{2}} \cdot 2 \cdot \underset{1}{\cancel{3}} \cdot 7} = \dfrac{11}{14}$

9. $\dfrac{1 \cdot 6}{6 \cdot 7} = \dfrac{1 \cdot \overset{1}{\cancel{2}} \cdot \overset{1}{\cancel{3}}}{\underset{1}{\cancel{2}} \cdot \underset{1}{\cancel{3}} \cdot 7} = \dfrac{1}{7}$

11. $\dfrac{1 \cdot 5}{5 \cdot 8} = \dfrac{1 \cdot \overset{1}{\cancel{5}}}{\underset{1}{\cancel{5}} \cdot 2 \cdot 2 \cdot 2} = \dfrac{1}{8}$

13. $\dfrac{8 \cdot 27}{9 \cdot 4} = \dfrac{\overset{1}{\cancel{2}} \cdot \overset{1}{\cancel{2}} \cdot 2 \cdot \overset{1}{\cancel{3}} \cdot \overset{1}{\cancel{3}} \cdot 3}{\underset{1}{\cancel{3}} \cdot \underset{1}{\cancel{3}} \cdot \underset{1}{\cancel{2}} \cdot \underset{1}{\cancel{2}}} = 6$

15. $\dfrac{5 \cdot 1}{6 \cdot 2} = \dfrac{5 \cdot 1}{2 \cdot 3 \cdot 2} = \dfrac{5}{12}$

17. $\dfrac{16 \cdot 27}{9 \cdot 8} = \dfrac{\overset{1}{\cancel{2}} \cdot \overset{1}{\cancel{2}} \cdot \overset{1}{\cancel{2}} \cdot 2 \cdot \overset{1}{\cancel{3}} \cdot \overset{1}{\cancel{3}} \cdot 3}{\underset{1}{\cancel{3}} \cdot \underset{1}{\cancel{3}} \cdot \underset{1}{\cancel{2}} \cdot \underset{1}{\cancel{2}} \cdot \underset{1}{\cancel{2}}} = 6$

19. $\dfrac{3 \cdot 4}{2 \cdot 9} = \dfrac{\overset{1}{\cancel{3}} \cdot \overset{1}{\cancel{2}} \cdot 2}{\underset{1}{\cancel{2}} \cdot \underset{1}{\cancel{3}} \cdot 3} = \dfrac{2}{3}$

21. $\dfrac{7 \cdot 3}{8 \cdot 14} = \dfrac{\overset{1}{\cancel{7}} \cdot 3}{2 \cdot 2 \cdot 2 \cdot 2 \cdot \underset{1}{\cancel{7}}} = \dfrac{3}{16}$

23. $\dfrac{1 \cdot 3}{10 \cdot 8} = \dfrac{3}{2 \cdot 5 \cdot 2 \cdot 2 \cdot 2} = \dfrac{3}{80}$

25. $\dfrac{15 \cdot 16}{8 \cdot 3} = \dfrac{\overset{1}{\cancel{3}} \cdot 5 \cdot \overset{1}{\cancel{2}} \cdot \overset{1}{\cancel{2}} \cdot \overset{1}{\cancel{2}} \cdot 2}{\underset{1}{\cancel{2}} \cdot \underset{1}{\cancel{2}} \cdot \underset{1}{\cancel{2}} \cdot \underset{1}{\cancel{3}}} = 10$

27. $\dfrac{1 \cdot 2}{2 \cdot 15} = \dfrac{1 \cdot \overset{1}{\cancel{2}}}{\underset{1}{\cancel{2}} \cdot 3 \cdot 5} = \dfrac{1}{15}$

29. $\dfrac{5 \cdot 14}{7 \cdot 15} = \dfrac{\overset{1}{\cancel{5}} \cdot 2 \cdot \overset{1}{\cancel{7}}}{\underset{1}{\cancel{7}} \cdot 3 \cdot \underset{1}{\cancel{5}}} = \dfrac{2}{3}$

31. $\dfrac{5 \cdot 42}{12 \cdot 65} = \dfrac{\overset{1}{\cancel{5}} \cdot \overset{1}{\cancel{2}} \cdot \overset{1}{\cancel{3}} \cdot 7}{2 \cdot \underset{1}{\cancel{2}} \cdot \underset{1}{\cancel{3}} \cdot \underset{1}{\cancel{5}} \cdot 13} = \dfrac{7}{26}$

33. $\dfrac{12 \cdot 5}{5 \cdot 3} = \dfrac{2 \cdot 2 \cdot \overset{1}{\cancel{3}} \cdot \overset{1}{\cancel{5}}}{\underset{1}{\cancel{5}} \cdot \underset{1}{\cancel{3}}} = 4$

35. $\dfrac{16 \cdot 125}{85 \cdot 84} = \dfrac{\overset{1}{\cancel{2}} \cdot \overset{1}{\cancel{2}} \cdot 2 \cdot 2 \cdot \overset{1}{\cancel{5}} \cdot 5 \cdot 5}{\underset{1}{\cancel{5}} \cdot 17 \cdot \underset{1}{\cancel{2}} \cdot \underset{1}{\cancel{2}} \cdot 3 \cdot 7} = \dfrac{100}{357}$

37. $\dfrac{7 \cdot 15}{12 \cdot 42} = \dfrac{\overset{1}{\cancel{7}} \cdot \overset{1}{\cancel{3}} \cdot 5}{2 \cdot 2 \cdot 3 \cdot 2 \cdot \underset{1}{\cancel{3}} \cdot \underset{1}{\cancel{7}}} = \dfrac{5}{24}$

39. $\dfrac{5 \cdot 3}{9 \cdot 20} = \dfrac{\overset{1}{\cancel{5}} \cdot \overset{1}{\cancel{3}}}{\underset{-1}{\cancel{3}} \cdot 3 \cdot 2 \cdot 2 \cdot \underset{1}{\cancel{5}}} = \dfrac{1}{12}$

41. $\dfrac{1 \cdot 8}{2 \cdot 15} = \dfrac{1 \cdot \overset{1}{\cancel{2}} \cdot 2 \cdot 2}{\underset{1}{\cancel{2}} \cdot 3 \cdot 5} = \dfrac{4}{15}$

Objective B Exercises

43. $\dfrac{4 \cdot 3}{1 \cdot 8} = \dfrac{\overset{1}{\cancel{2}} \cdot \overset{1}{\cancel{2}} \cdot 3}{1 \cdot \underset{1}{\cancel{2}} \cdot \underset{1}{\cancel{2}} \cdot 2} = \dfrac{3}{2} = 1\dfrac{1}{2}$

45. $\dfrac{2 \cdot 6}{3 \cdot 1} = \dfrac{2 \cdot 2 \cdot \overset{1}{\cancel{3}}}{\underset{1}{\cancel{3}} \cdot 1} = 4$

47. $\dfrac{1}{3} \times 1\dfrac{1}{3} = \dfrac{1}{3} \times \dfrac{4}{3} = \dfrac{1 \cdot 2 \cdot 2}{3 \cdot 3} = \dfrac{4}{9}$

49. $1\dfrac{7}{8} \times \dfrac{4}{15} = \dfrac{15}{8} \times \dfrac{4}{15} = \dfrac{\overset{1}{\cancel{3}} \cdot \overset{1}{\cancel{5}} \cdot \overset{1}{\cancel{2}} \cdot \overset{1}{\cancel{2}}}{\underset{1}{\cancel{2}} \cdot \underset{1}{\cancel{2}} \cdot 2 \cdot \underset{1}{\cancel{3}} \cdot \underset{1}{\cancel{5}}} = \dfrac{1}{2}$

51. $55 \times \dfrac{3}{10} = \dfrac{55 \cdot 3}{1 \cdot 10} = \dfrac{\overset{1}{\cancel{5}} \cdot 11 \cdot 3}{1 \cdot 2 \cdot \underset{1}{\cancel{5}}} = \dfrac{33}{2} = 16\dfrac{1}{2}$

53. $4 \times 2\dfrac{1}{2} = \dfrac{4}{1} \times \dfrac{5}{2} = \dfrac{2 \cdot \overset{1}{\cancel{2}} \cdot 5}{1 \cdot \underset{1}{\cancel{2}}} = 10$

55. $2\dfrac{1}{7} \times 3 = \dfrac{15}{7} \times \dfrac{3}{1} = \dfrac{3 \cdot 5 \cdot 3}{7 \cdot 1} = \dfrac{45}{7} = 6\dfrac{3}{7}$

57. $3\dfrac{2}{3} \times 5 = \dfrac{11}{3} \times \dfrac{5}{1} = \dfrac{11 \cdot 5}{3 \cdot 1} = \dfrac{55}{3} = 18\dfrac{1}{3}$

59. $\dfrac{1}{2} \times 3\dfrac{3}{7} = \dfrac{1}{2} \times \dfrac{24}{7} = \dfrac{1 \cdot \overset{1}{\cancel{2}} \cdot 2 \cdot 2 \cdot 3}{\underset{1}{\cancel{2}} \cdot 7} = \dfrac{12}{7} = 1\dfrac{5}{7}$

61. $6\dfrac{1}{8} \times \dfrac{4}{7} = \dfrac{49}{8} \times \dfrac{4}{7} = \dfrac{\overset{1}{\cancel{7}} \cdot 7 \cdot \overset{1}{\cancel{2}} \cdot \overset{1}{\cancel{2}}}{\underset{1}{\cancel{2}} \cdot \underset{1}{\cancel{2}} \cdot 2 \cdot \underset{1}{\cancel{7}}} = \dfrac{7}{2} = 3\dfrac{1}{2}$

63. $5\dfrac{1}{8} \times 5 = \dfrac{41}{8} \times \dfrac{5}{1} = \dfrac{41 \cdot 5}{2 \cdot 2 \cdot 2 \cdot 1} = \dfrac{205}{8} = 25\dfrac{5}{8}$

65. $\dfrac{3}{8} \times 4\dfrac{1}{2} = \dfrac{3}{8} \times \dfrac{9}{2} = \dfrac{3 \cdot 3 \cdot 3}{2 \cdot 2 \cdot 2 \cdot 2} = \dfrac{27}{16} = 1\dfrac{11}{16}$

67. $6 \times 2\dfrac{2}{3} = \dfrac{6}{1} \times \dfrac{8}{3} = \dfrac{2 \cdot \overset{1}{\cancel{3}} \cdot 2 \cdot 2 \cdot 2}{1 \cdot \underset{1}{\cancel{3}}} = 16$

69. $1\dfrac{1}{3} \times 2\dfrac{1}{4} = \dfrac{4}{3} \times \dfrac{9}{4} = \dfrac{\overset{1}{\cancel{2}} \cdot \overset{1}{\cancel{2}} \cdot \overset{1}{\cancel{3}} \cdot 3}{\underset{1}{\cancel{3}} \cdot \underset{1}{\cancel{2}} \cdot \underset{1}{\cancel{2}}} = 3$

71. $2\dfrac{5}{8} \times 3\dfrac{2}{5} = \dfrac{21}{8} \times \dfrac{17}{5} = \dfrac{3 \cdot 7 \cdot 17}{2 \cdot 2 \cdot 2 \cdot 5} = \dfrac{357}{40} = 8\dfrac{37}{40}$

73. $3\dfrac{1}{7} \times 2\dfrac{1}{8} = \dfrac{22}{7} \times \dfrac{17}{8} = \dfrac{\overset{1}{\cancel{2}} \cdot 11 \cdot 17}{7 \cdot \underset{1}{\cancel{2}} \cdot 2 \cdot 2} = \dfrac{187}{28} = 6\dfrac{19}{28}$

75. $2\dfrac{2}{5} \times 3\dfrac{1}{12} = \dfrac{12}{5} \times \dfrac{37}{12} = \dfrac{\overset{1}{\cancel{2}} \cdot \overset{1}{\cancel{2}} \cdot \overset{1}{\cancel{3}} \cdot 37}{5 \cdot \underset{1}{\cancel{2}} \cdot \underset{1}{\cancel{2}} \cdot \underset{1}{\cancel{3}}} = \dfrac{37}{5} = 7\dfrac{2}{5}$

77. $5\dfrac{1}{5} \times 3\dfrac{1}{13} = \dfrac{26}{5} \times \dfrac{40}{13} = \dfrac{2 \cdot \overset{1}{\cancel{13}} \cdot 2 \cdot 2 \cdot 2 \cdot \overset{1}{\cancel{5}}}{\underset{1}{\cancel{5}} \cdot \underset{1}{\cancel{13}}} = 16$

79. $10\frac{1}{4} \times 3\frac{1}{5} = \frac{41}{4} \times \frac{16}{5} = \frac{41 \cdot \overset{1}{2} \cdot \overset{1}{2} \cdot 2 \cdot 2}{2 \cdot 2 \cdot 5} = \frac{164}{5} = 32\frac{4}{5}$

81. $5\frac{3}{7} \times 5\frac{1}{4} = \frac{38}{7} \times \frac{21}{4} = \frac{\overset{1}{2} \cdot 19 \cdot 3 \cdot \overset{1}{7}}{7 \cdot 2 \cdot 2} = \frac{57}{2} = 28\frac{1}{2}$

83. $2\frac{1}{2} \times 3\frac{3}{5} = \frac{5}{2} \times \frac{18}{5} = \frac{\overset{1}{5} \cdot \overset{1}{2} \cdot 3 \cdot 3}{2 \cdot 5} = 9$

85. $2\frac{1}{8} \times \frac{5}{17} = \frac{17}{8} \times \frac{5}{17} = \frac{\overset{1}{17} \cdot 5}{2 \cdot 2 \cdot 2 \cdot \overset{}{17}} = \frac{5}{8}$

87. $1\frac{3}{8} \times 2\frac{1}{5} = \frac{11}{8} \times \frac{11}{5} = \frac{121}{40} = 3\frac{1}{40}$

Objective C Exercises

89. **Strategy**   To find the cost of the salmon, multiply the amount of salmon $\left(2\frac{3}{4}\text{ pounds}\right)$ by the cost per pound ($4).

   Solution $2\frac{3}{4} \times 4 = \frac{11}{4} \times \frac{4}{1} = \frac{11 \cdot 4}{4 \cdot 1} = 11$
   The salmon costs $11.

91a. No, $\frac{1}{3}$ of 9 feet is approximately 3 feet; therefore, $\frac{1}{3}$ of $9\frac{1}{4}$ feet is approximately 3 feet.

   b. **Strategy**   To find the length cut, multiply the length of the board $\left(9\frac{1}{4}\text{ feet}\right)$ by $\frac{1}{3}$.

   Solution $\frac{1}{3} \times 9\frac{1}{4} = \frac{1}{3} \times \frac{37}{4} = \frac{1 \cdot 37}{3 \cdot 4} = \frac{37}{12} = 3\frac{1}{12}$
   The length of the board cut off is $3\frac{1}{12}$ feet.

93. **Strategy**   To find the area of the square, multiply the length of one side $\left(5\frac{1}{4}\text{ feet}\right)$ by itself $\left(5\frac{1}{4}\text{ feet}\right)$.

   Solution $5\frac{1}{4} \times 5\frac{1}{4} = \frac{21}{4} \times \frac{21}{4} = \frac{21 \cdot 21}{4 \cdot 4}$
   $= \frac{441}{16} = 27\frac{9}{16}$
   The area of the square is $27\frac{9}{16}$ square feet.

95a. **Strategy**   To find the amount budgeted for housing and utilities, multiply the total monthly income ($4200) by $\frac{2}{5}$.

   Solution $\frac{2}{5} \times \$4200 = \frac{2 \cdot 4200}{5} = \$1680$
   The amount budgeted for housing and utilities is $1680.

   b. **Strategy**   To find the amount remaining for other than housing and utilities, subtract the amount for housing and utilities ($1680) from the total monthly income ($4200).

   Solution $\begin{array}{r} \$4200 \\ -\ 1680 \\ \hline \$2520 \end{array}$
   The amount remaining for other than housing and utilities is $2520.

97. **Strategy**   To find the total cost of the capes, multiply the amount of material each cape requires $\left(1\frac{3}{8}\text{ yards}\right)$ by the cost of 1 yard ($12) and by the number of capes needed (22).

   Solution $1\frac{3}{8} \times \$12 \times 22 = \frac{11}{8} \times 12 \times 22$
   $= \frac{11 \times 12 \times 22}{8} = \$363$
   The total cost is $363.

99. $12\frac{7}{12} \times 4\frac{1}{3} = \frac{151}{12} \times \frac{13}{3} = \frac{1963}{36} = 54\frac{19}{36}$
   The weight of the $12\frac{7}{12}$-foot steel rod is $54\frac{19}{36}$ pounds.

101. $\frac{1}{2} \times \frac{3}{8} = \frac{3}{16}$
   $\frac{3}{16}$ of the total portfolio is invested in corporate bonds.

Applying the Concepts

103. Student explanations should include the idea that every 4 years we must add 1 day to the usual 365-day year.

105. *A*. See problem 104.

Section 2.7

Objective A Exercises

1. $\frac{1}{3} \times \frac{5}{2} = \frac{1 \cdot 5}{3 \cdot 2} = \frac{5}{6}$

3. $\frac{3}{7} \times \frac{7}{3} = \frac{\overset{1}{3} \cdot \overset{1}{7}}{7 \cdot 3} = 1$

5. $0 \times \frac{4}{3} = 0$

7. $\frac{5}{24} \times \frac{36}{15} = \frac{\overset{1}{5} \cdot \overset{1}{2} \cdot \overset{1}{2} \cdot \overset{1}{3} \cdot \overset{1}{3}}{2 \cdot 2 \cdot 2 \cdot 3 \cdot 3 \cdot 5} = \frac{1}{2}$

9. $\frac{15}{16} \times \frac{39}{16} = \frac{3 \cdot 5 \cdot 3 \cdot 13}{2 \cdot 2 \cdot 2 \cdot 2 \cdot 2 \cdot 2 \cdot 2 \cdot 2} = \frac{585}{256} = 2\frac{73}{256}$

11. $\dfrac{8}{9} \times \dfrac{5}{4} = \dfrac{2 \cdot \overset{1}{\cancel{2}} \cdot \overset{1}{\cancel{2}} \cdot 5}{3 \cdot 3 \cdot \underset{1}{\cancel{2}} \cdot \underset{1}{\cancel{2}}} = \dfrac{10}{9} = 1\dfrac{1}{9}$

13. $\dfrac{1}{9} \times \dfrac{3}{2} = \dfrac{\overset{1}{\cancel{3}}}{\underset{1}{\cancel{3}} \cdot 3 \cdot 2} = \dfrac{1}{6}$

15. $\dfrac{2}{5} \times \dfrac{7}{4} = \dfrac{\overset{1}{\cancel{2}} \cdot 7}{5 \cdot \underset{1}{\cancel{2}} \cdot 2} = \dfrac{7}{10}$

17. $\dfrac{1}{2} \times \dfrac{4}{1} = \dfrac{2 \cdot \overset{1}{\cancel{2}}}{\underset{1}{\cancel{2}}} = 2$

19. $\dfrac{1}{5} \times \dfrac{10}{1} = \dfrac{2 \cdot \overset{1}{\cancel{5}}}{\underset{1}{\cancel{5}}} = 2$

21. $\dfrac{7}{15} \times \dfrac{5}{14} = \dfrac{\overset{1}{\cancel{7}} \cdot \overset{1}{\cancel{5}}}{3 \cdot \underset{1}{\cancel{5}} \cdot 2 \cdot \underset{1}{\cancel{7}}} = \dfrac{1}{6}$

23. $\dfrac{14}{3} \times \dfrac{9}{7} = \dfrac{2 \cdot \overset{1}{\cancel{7}} \cdot \overset{1}{\cancel{3}} \cdot 3}{\underset{1}{\cancel{3}} \cdot \underset{1}{\cancel{7}}} = 6$

25. $\dfrac{5}{9} \times \dfrac{3}{25} = \dfrac{\overset{1}{\cancel{5}} \cdot \overset{1}{\cancel{3}}}{\underset{1}{\cancel{3}} \cdot 3 \cdot \underset{1}{\cancel{5}} \cdot 5} = \dfrac{1}{15}$

27. $\dfrac{2}{3} \times \dfrac{3}{1} = \dfrac{2 \cdot \overset{1}{\cancel{3}}}{\underset{1}{\cancel{3}}} = 2$

29. $\dfrac{5}{7} \times \dfrac{7}{2} = \dfrac{5 \cdot \overset{1}{\cancel{7}}}{\underset{1}{\cancel{7}} \cdot 2} = \dfrac{5}{2} = 2\dfrac{1}{2}$

31. $\dfrac{2}{3} \times \dfrac{9}{2} = \dfrac{\overset{1}{\cancel{2}} \cdot \overset{1}{\cancel{3}} \cdot 3}{\underset{1}{\cancel{3}} \cdot \underset{1}{\cancel{2}}} = 3$

33. $\dfrac{7}{8} \div \dfrac{3}{4} = \dfrac{7}{8} \times \dfrac{4}{3} = \dfrac{7 \cdot \overset{1}{\cancel{2}} \cdot \overset{1}{\cancel{2}}}{\underset{1}{\cancel{2}} \cdot \underset{1}{\cancel{2}} \cdot 2 \cdot 3} = \dfrac{7}{6} = 1\dfrac{1}{6}$

35. $\dfrac{5}{7} \div \dfrac{3}{14} = \dfrac{5}{7} \times \dfrac{14}{3} = \dfrac{5 \cdot 2 \cdot \overset{1}{\cancel{7}}}{\underset{1}{\cancel{7}} \cdot 3} = \dfrac{10}{3} = 3\dfrac{1}{3}$

Objective B Exercises

37. $\dfrac{4}{1} \times \dfrac{3}{2} = \dfrac{2 \cdot \overset{1}{\cancel{2}} \cdot 3}{\underset{1}{\cancel{2}}} = 6$

39. $\dfrac{3}{2} \times \dfrac{1}{3} = \dfrac{\overset{1}{\cancel{3}}}{2 \cdot \underset{1}{\cancel{3}}} = \dfrac{1}{2}$

41. $\dfrac{5}{6} \times \dfrac{1}{25} = \dfrac{\overset{1}{\cancel{5}}}{2 \cdot 3 \cdot \underset{1}{\cancel{5}} \cdot 5} = \dfrac{1}{30}$

43. $\dfrac{6}{1} \div \dfrac{10}{3} = \dfrac{6}{1} \times \dfrac{3}{10} = \dfrac{\overset{1}{\cancel{2}} \cdot 3 \cdot 3}{\underset{1}{\cancel{2}} \cdot 5} = \dfrac{9}{5} = 1\dfrac{4}{5}$

45. $\dfrac{13}{2} \div \dfrac{1}{2} = \dfrac{13}{2} \times \dfrac{2}{1} = \dfrac{13 \cdot \overset{1}{\cancel{2}}}{\underset{1}{\cancel{2}}} = 13$

47. $\dfrac{5}{12} \div \dfrac{24}{5} = \dfrac{5}{12} \times \dfrac{5}{24} = \dfrac{5 \cdot 5}{2 \cdot 2 \cdot 3 \cdot 2 \cdot 2 \cdot 2 \cdot 3} = \dfrac{25}{288}$

49. $\dfrac{33}{4} \div \dfrac{11}{4} = \dfrac{33}{4} \times \dfrac{4}{11} = \dfrac{3 \cdot \overset{1}{\cancel{11}} \cdot \overset{1}{\cancel{2}} \cdot \overset{1}{\cancel{2}}}{\underset{1}{\cancel{2}} \cdot \underset{1}{\cancel{2}} \cdot \underset{1}{\cancel{11}}} = 3$

51. $\dfrac{21}{5} \div \dfrac{21}{1} = \dfrac{21}{5} \times \dfrac{1}{21} = \dfrac{\overset{1}{\cancel{3}} \cdot \overset{1}{\cancel{7}}}{5 \cdot \underset{1}{\cancel{3}} \cdot \underset{1}{\cancel{7}}} = \dfrac{1}{5}$

53. $\dfrac{11}{12} \div \dfrac{7}{3} = \dfrac{11}{12} \times \dfrac{3}{7} = \dfrac{11 \cdot \overset{1}{\cancel{3}}}{2 \cdot 2 \cdot \underset{1}{\cancel{3}} \cdot 7} = \dfrac{11}{28}$

55. $\dfrac{5}{16} \div \dfrac{43}{8} = \dfrac{5}{16} \times \dfrac{8}{43} = \dfrac{5 \cdot \overset{1}{\cancel{2}} \cdot \overset{1}{\cancel{2}} \cdot \overset{1}{\cancel{2}}}{2 \cdot 2 \cdot \underset{1}{\cancel{2}} \cdot \underset{1}{\cancel{2}} \cdot \underset{1}{\cancel{2}} \cdot 43} = \dfrac{5}{86}$

57. $35 \div \dfrac{7}{24} = \dfrac{35}{1} \times \dfrac{24}{7} = \dfrac{5 \cdot \overset{1}{\cancel{7}} \cdot 2 \cdot 2 \cdot 2 \cdot 3}{\underset{1}{\cancel{7}}} = 120$

59. $\dfrac{11}{18} \div \dfrac{20}{9} = \dfrac{11}{18} \times \dfrac{9}{20} = \dfrac{11 \cdot \overset{1}{\cancel{3}} \cdot \overset{1}{\cancel{3}}}{2 \cdot \underset{1}{\cancel{3}} \cdot \underset{1}{\cancel{3}} \cdot 2 \cdot 5} = \dfrac{11}{40}$

61. $\dfrac{33}{16} \div \dfrac{5}{2} = \dfrac{33}{16} \times \dfrac{2}{5} = \dfrac{3 \cdot 11 \cdot \overset{1}{\cancel{2}}}{2 \cdot 2 \cdot 2 \cdot \underset{1}{\cancel{2}} \cdot 5} = \dfrac{33}{40}$

63. $\dfrac{5}{3} \div \dfrac{3}{8} = \dfrac{5}{3} \times \dfrac{8}{3} = \dfrac{5 \cdot 2 \cdot 2 \cdot 2}{3 \cdot 3} = \dfrac{40}{9} = 4\dfrac{4}{9}$

65. $\dfrac{13}{8} \div \dfrac{4}{1} = \dfrac{13}{8} \times \dfrac{1}{4} = \dfrac{13}{2 \cdot 2 \cdot 2 \cdot 2 \cdot 2} = \dfrac{13}{32}$

67. $16 \div \dfrac{3}{2} = \dfrac{16}{1} \times \dfrac{2}{3} = \dfrac{2 \cdot 2 \cdot 2 \cdot 2 \cdot 2}{3} = \dfrac{32}{3} = 10\dfrac{2}{3}$

69. $\dfrac{133}{8} \div \dfrac{5}{3} = \dfrac{133}{8} \times \dfrac{3}{5} = \dfrac{133 \cdot 3}{2 \cdot 2 \cdot 2 \cdot 5} = \dfrac{399}{40} = 9\dfrac{39}{40}$

71. $\dfrac{4}{3} \div \dfrac{53}{9} = \dfrac{4}{3} \times \dfrac{9}{53} = \dfrac{2 \cdot 2 \cdot \overset{1}{\cancel{3}} \cdot 3}{\underset{1}{\cancel{3}} \cdot 53} = \dfrac{12}{53}$

73. $\dfrac{413}{5} \div \dfrac{191}{10} = \dfrac{413}{5} \times \dfrac{10}{191} = \dfrac{7 \cdot 59 \cdot 2 \cdot \overset{1}{\cancel{5}}}{\underset{1}{\cancel{5}} \cdot 191}$

$= \dfrac{826}{191} = 4\dfrac{62}{191}$

75. $\dfrac{102}{1} \div \dfrac{3}{2} = \dfrac{102}{1} \times \dfrac{2}{3} = \dfrac{2 \cdot \overset{1}{\cancel{3}} \cdot 17 \cdot 2}{\underset{1}{\cancel{3}}} = 68$

77. $\dfrac{58}{7} \div 1 = \dfrac{58}{7} \times 1 = \dfrac{58}{7} = 8\dfrac{2}{7}$

79. $\dfrac{80}{9} \div \dfrac{49}{18} = \dfrac{80}{9} \times \dfrac{18}{49} = \dfrac{2 \cdot 2 \cdot 2 \cdot 2 \cdot 5 \cdot 2 \cdot \overset{1}{\cancel{3}} \cdot \overset{1}{\cancel{3}}}{\underset{1}{\cancel{3}} \cdot \underset{1}{\cancel{3}} \cdot 7 \cdot 7}$

$\qquad\qquad = \dfrac{160}{49} = 3\dfrac{13}{49}$

81. $\dfrac{59}{8} \div \dfrac{59}{32} = \dfrac{59}{8} \times \dfrac{32}{59} = \dfrac{\overset{1}{\cancel{59}} \cdot \overset{1}{\cancel{2}} \cdot \overset{1}{\cancel{2}} \cdot \overset{1}{\cancel{2}} \cdot 2}{\underset{1}{\cancel{2}} \cdot \underset{1}{\cancel{2}} \cdot \underset{1}{\cancel{2}} \cdot \underset{1}{\cancel{59}}} = 4$

83. $2\dfrac{3}{4} \div 1\dfrac{23}{32} = \dfrac{11}{4} \div \dfrac{55}{32} = \dfrac{11}{4} \times \dfrac{32}{55} = \dfrac{\overset{1}{\cancel{11}} \cdot \overset{1}{\cancel{2}} \cdot \overset{1}{\cancel{2}} \cdot 2 \cdot 2}{\underset{1}{\cancel{2}} \cdot \underset{1}{\cancel{2}} \cdot 5 \cdot \underset{1}{\cancel{11}}}$

$\qquad\qquad\qquad = \dfrac{8}{5} = 1\dfrac{3}{5}$

85. $\dfrac{14}{77} \div 3\dfrac{1}{9} = \dfrac{14}{77} \div \dfrac{28}{9} = \dfrac{14}{77} \times \dfrac{9}{28} = \dfrac{\overset{1}{\cancel{2}} \cdot \overset{1}{\cancel{7}} \cdot 3 \cdot 3}{17 \cdot \underset{1}{\cancel{2}} \cdot 2 \cdot \underset{1}{\cancel{7}}} = \dfrac{9}{34}$

Objective C Exercises

87. Strategy To find the number of servings in 16 ounces of cereal, divide 16 by the amount in each serving $\left(1\dfrac{1}{3} \text{ ounces}\right)$.

Solution $16 \div 1\dfrac{1}{3} = 16 \div \dfrac{4}{3}$

$\qquad\qquad = 16 \times \dfrac{3}{4} = \dfrac{16 \cdot 3}{4}$

$\qquad\qquad = 12$

There are 12 servings in16 ounces of cereal.

89. Strategy To find the cost of each acre, divide the total cost ($200,000) by the number of acres $\left(8\dfrac{1}{3}\right)$.

Solution $\$200,000 \div 8\dfrac{1}{3} = 200,000 \div \dfrac{25}{3}$

$\qquad\qquad = 200,000 \times \dfrac{3}{25}$

$\qquad\qquad = \dfrac{200,000 \cdot 3}{25}$

$\qquad\qquad = \$24,000$

Each acre costs $24,000.

91. Strategy To find the number of turns, divide the distance for the nut to move $\left(1\dfrac{7}{8} \text{ inches}\right)$ by the distance the nut moves for each turn $\left(\dfrac{5}{32} \text{ inch}\right)$.

Solution $1\dfrac{7}{8} \div \dfrac{5}{32} = \dfrac{15}{8} \div \dfrac{5}{32}$

$\qquad\qquad = \dfrac{15}{8} \times \dfrac{32}{5}$

$\qquad\qquad = \dfrac{3 \cdot \overset{1}{\cancel{5}} \cdot \overset{1}{\cancel{2}} \cdot \overset{1}{\cancel{2}} \cdot \overset{1}{\cancel{2}} \cdot 2 \cdot 2}{\underset{1}{\cancel{2}} \cdot \underset{1}{\cancel{2}} \cdot \underset{1}{\cancel{2}} \cdot \underset{1}{\cancel{5}}}$

$\qquad\qquad = 12$

The nut will make 12 turns in moving $1\dfrac{7}{8}$ inches.

93a. Strategy To find the total weight of the fat and bone, subtract the weight after trimming $\left(9\dfrac{1}{3} \text{ pounds}\right)$ from the original weight $\left(10\dfrac{3}{4} \text{ pounds}\right)$.

Solution
$$10\dfrac{3}{4} = 10\dfrac{9}{12}$$
$$\underline{-9\dfrac{1}{3} = \ 9\dfrac{4}{12}}$$
$$1\dfrac{5}{12}$$

The total weight of the fat and bone was $1\dfrac{5}{12}$ pounds.

b. Strategy To find the number of servings, divide the weight after trimming $\left(9\dfrac{1}{3} \text{ pounds}\right)$ by the weight of one serving $\left(\dfrac{1}{3} \text{ pound}\right)$.

Solution $9\dfrac{1}{3} \div \dfrac{1}{3} = \dfrac{28}{3} \div \dfrac{1}{3} = \dfrac{28}{3} \times \dfrac{3}{1}$

$\qquad\qquad = \dfrac{28 \cdot \overset{1}{\cancel{3}}}{\underset{1}{\cancel{3}} \cdot 1} = 28$

The chef can cut 28 servings from the roast.

95. $6\dfrac{1}{4} \div \dfrac{1}{2} = \dfrac{25}{4} \times \dfrac{2}{1} = \dfrac{5 \cdot 5 \cdot \overset{1}{\cancel{2}}}{\underset{1}{\cancel{2}} \cdot 2} = \dfrac{25}{2} = 12\dfrac{1}{2}$

The actual length of wall a is $12\dfrac{1}{2}$ feet.

$9 \div \dfrac{1}{2} = \dfrac{9}{1} \times \dfrac{2}{1} = 18$

The actual length of wall b is 18 feet.

$7\dfrac{7}{8} \div \dfrac{1}{2} = \dfrac{63}{8} \times \dfrac{2}{1} = \dfrac{3 \cdot 3 \cdot 7 \cdot \overset{1}{\cancel{2}}}{\underset{1}{\cancel{2}} \cdot 2 \cdot 2} = \dfrac{63}{4} = 15\dfrac{3}{4}$

The actual length of wall c is $15\dfrac{3}{4}$ feet.

Applying the Concepts

97. Strategy To find the fractional part of money borrowed on home-equity loans that is spent on home improvement, cars, and tuition, add the fraction spent on home improvement $\left(\dfrac{6}{25}\right)$, the fraction spent on cars $\left(\dfrac{1}{20}\right)$, and the fraction spent on tuition $\left(\dfrac{1}{20}\right)$.

Solution $\dfrac{6}{25} + \dfrac{1}{20} + \dfrac{1}{20} = \dfrac{24}{100} + \dfrac{5}{100} + \dfrac{5}{100}$

$= \dfrac{34}{100} = \dfrac{17}{50}$

$\dfrac{17}{50}$ of the money borrowed is spent on home improvement, cars, and tuition.

99. Strategy To find the capacity of the music center, divide the number of people attending (1200) by $\dfrac{2}{3}$.

Solution $1200 \div \dfrac{2}{3} = 1200 \cdot \dfrac{3}{2} = \dfrac{3600}{2} = 1800$

The capacity of the music center is 1800 people.

101. Strategy To find the fraction of the puzzle left to complete:
● Add the fraction completed yesterday $\left(\dfrac{1}{3}\right)$ to the fraction completed today $\left(\dfrac{1}{2}\right)$.
● Subtract that from 1.

Solution $\dfrac{1}{3} + \dfrac{1}{2} = \dfrac{2}{6} + \dfrac{3}{6} = \dfrac{5}{6}$ $1 - \dfrac{5}{6} = \dfrac{1}{6}$

$\dfrac{1}{6}$ of the puzzle is left to complete.

103. Strategy To find your total earnings for the week, add the four numbers representing the hours you worked and multiply the sum by your rate of pay ($9 per hour).

Solution $5 = 5$

$3\dfrac{3}{4} = 3\dfrac{9}{12}$

$1\dfrac{1}{4} = 1\dfrac{3}{12}$

$2\dfrac{1}{3} = 2\dfrac{4}{12}$

$11\dfrac{16}{12} = 12\dfrac{4}{12} = 12\dfrac{1}{3}$ hours

$12\dfrac{1}{3} \times 9 = \dfrac{37}{3} \cdot 9 = 3.8 = \$1$

Your total earnings for last week's work are $111.

105. Strategy To find the number of calories an average teenage boy consumes each week in soda:
● Multiply $3\dfrac{1}{3}$ by 7 to determine the number of cans of soda he drinks.
● Multiply that number by the number of calories per can (150).

Solution $3\dfrac{1}{3} \times 7 = \dfrac{10}{3} \times 7 = \dfrac{70}{3}$

$= 23\dfrac{1}{3}$ cans per week

$23\dfrac{1}{3} \times 150 = \dfrac{70}{3} \times 150 = 70 \times 50$

$= 3500$ calories

The average teenage boy consumes 3500 calories each week in soda.

107. Strategy To find the number of miles:
● Find out how many units of $\dfrac{3}{8}$ inch there are in $4\dfrac{5}{8}$ inches.
● Multiply by 60.

Solution $4\dfrac{5}{8} \div \dfrac{3}{8} = \dfrac{37}{8} \times \dfrac{8}{3} = \dfrac{37}{3}$ units

Then, because each unit represents 60 miles, $\dfrac{37}{3} \times \dfrac{60}{1} = 740$

The distance is 740 miles.

109a. $\dfrac{1}{2} \div \dfrac{3}{4} = \dfrac{1}{2} \times \dfrac{4}{3} = \dfrac{2}{3}$

Factor 3 factor = product

In order to find a factor, divide the product by the known factor.

b. $1\dfrac{3}{4} \div \dfrac{2}{3} = \dfrac{7}{4} \times \dfrac{3}{2} = \dfrac{21}{8} = 2\dfrac{5}{8}$

111. The quotient. To divide by a proper fraction is to multiply by an improper fraction, which is greater than 1.

Examples:

$$5 \times \frac{2}{3} = \frac{5}{1} \times \frac{2}{3} = \frac{10}{3} = 3\frac{1}{3}$$

$$5 \div \frac{2}{3} = \frac{5}{1} \times \frac{3}{2} = \frac{15}{2} = 7\frac{1}{2}$$

Section 2.8

Objective A Exercises

1. $\dfrac{11}{40} < \dfrac{19}{40}$

3. $\dfrac{2}{3} = \dfrac{14}{21}, \dfrac{5}{7} = \dfrac{15}{21}, \dfrac{2}{3} < \dfrac{5}{7}$

5. $\dfrac{5}{8} = \dfrac{15}{24}, \dfrac{7}{12} = \dfrac{14}{24}, \dfrac{5}{8} > \dfrac{7}{12}$

7. $\dfrac{7}{9} = \dfrac{28}{36}, \dfrac{11}{12} = \dfrac{33}{36}, \dfrac{7}{9} < \dfrac{11}{12}$

9. $\dfrac{13}{14} = \dfrac{39}{42}, \dfrac{19}{21} = \dfrac{38}{42}, \dfrac{13}{14} > \dfrac{19}{21}$

11. $\dfrac{7}{24} = \dfrac{35}{120}, \dfrac{11}{30} = \dfrac{44}{120}, \dfrac{7}{24} < \dfrac{11}{30}$

Objective B Exercises

13. $\left(\dfrac{3}{8}\right)^2 = \dfrac{3}{8} \cdot \dfrac{3}{8} = \dfrac{9}{64}$

15. $\left(\dfrac{2}{9}\right)^3 = \dfrac{2}{9} \cdot \dfrac{2}{9} \cdot \dfrac{2}{9} = \dfrac{8}{729}$

17. $\left(\dfrac{2}{3}\right) \cdot \left(\dfrac{1}{2}\right)^4 = \left(\dfrac{2}{3}\right) \cdot \left(\dfrac{1}{2} \cdot \dfrac{1}{2} \cdot \dfrac{1}{2} \cdot \dfrac{1}{2}\right)$

$$= \frac{\overset{1}{\cancel{2}} \cdot 1 \cdot 1 \cdot 1 \cdot 1}{3 \cdot \underset{1}{\cancel{2}} \cdot 2 \cdot 2 \cdot 2} = \frac{1}{24}$$

19. $\left(\dfrac{2}{5}\right)^3 \cdot \left(\dfrac{5}{7}\right)^2 = \left(\dfrac{2}{5} \cdot \dfrac{2}{5} \cdot \dfrac{2}{5}\right) \cdot \left(\dfrac{5}{7} \cdot \dfrac{5}{7}\right)$

$$= \frac{2 \cdot 2 \cdot 2 \cdot \overset{1}{\cancel{5}} \cdot \overset{1}{\cancel{5}}}{\underset{1}{\cancel{5}} \cdot \underset{1}{\cancel{5}} \cdot 5 \cdot 7 \cdot 7} = \frac{8}{245}$$

21. $\left(\dfrac{1}{3}\right)^4 \cdot \left(\dfrac{9}{11}\right)^2 = \left(\dfrac{1}{3} \cdot \dfrac{1}{3} \cdot \dfrac{1}{3} \cdot \dfrac{1}{3}\right) \cdot \left(\dfrac{9}{11} \cdot \dfrac{9}{11}\right)$

$$= \frac{1 \cdot 1 \cdot 1 \cdot 1 \cdot \overset{1}{\cancel{3}} \cdot \overset{1}{\cancel{3}} \cdot \overset{1}{\cancel{3}} \cdot \overset{1}{\cancel{3}}}{\underset{1}{\cancel{3}} \cdot \underset{1}{\cancel{3}} \cdot \underset{1}{\cancel{3}} \cdot \underset{1}{\cancel{3}} \cdot 11 \cdot 11} = \frac{1}{121}$$

23. $\left(\dfrac{2}{3}\right)^4 \cdot \left(\dfrac{81}{100}\right)^2 = \left(\dfrac{2}{3} \cdot \dfrac{2}{3} \cdot \dfrac{2}{3} \cdot \dfrac{2}{3}\right) \cdot \left(\dfrac{81}{100} \cdot \dfrac{81}{100}\right)$

$$= \frac{2 \cdot 2 \cdot 2 \cdot 2 \cdot \overset{1}{\cancel{3}} \cdot \overset{1}{\cancel{3}} \cdot \overset{1}{\cancel{3}} \cdot \overset{1}{\cancel{3}} \cdot 3 \cdot 3 \cdot 3 \cdot 3}{\underset{1}{\cancel{3}} \cdot \underset{1}{\cancel{3}} \cdot \underset{1}{\cancel{3}} \cdot \underset{1}{\cancel{3}} \cdot 2 \cdot 2 \cdot 5 \cdot 5 \cdot \underset{1}{\cancel{2}} \cdot \underset{1}{\cancel{2}} \cdot 5 \cdot 5} = \frac{81}{625}$$

25. $\left(\dfrac{2}{7}\right) \cdot \left(\dfrac{7}{8}\right)^2 \cdot \left(\dfrac{8}{9}\right) = \left(\dfrac{2}{7}\right) \cdot \left(\dfrac{7}{8} \cdot \dfrac{7}{8}\right) \cdot \left(\dfrac{8}{9}\right)$

$$= \frac{\overset{1}{\cancel{2}} \cdot \overset{1}{\cancel{7}} \cdot 7 \cdot \overset{1}{\cancel{2}} \cdot \overset{1}{\cancel{2}} \cdot \overset{1}{\cancel{2}}}{\underset{1}{\cancel{7}} \cdot \underset{1}{\cancel{2}} \cdot \underset{1}{\cancel{2}} \cdot \underset{1}{\cancel{2}} \cdot 2 \cdot 2 \cdot 3 \cdot 3} = \frac{7}{36}$$

27. $4 \cdot \left(\dfrac{3}{4}\right)^3 \cdot \left(\dfrac{4}{7}\right)^2 = \left(\dfrac{4}{1}\right) \cdot \left(\dfrac{3}{4} \cdot \dfrac{3}{4} \cdot \dfrac{3}{4}\right) \cdot \left(\dfrac{4}{7} \cdot \dfrac{4}{7}\right)$

$$= \frac{\overset{1}{\cancel{2}} \cdot \overset{1}{\cancel{2}} \cdot 3 \cdot 3 \cdot 3 \cdot \overset{1}{\cancel{2}} \cdot \overset{1}{\cancel{2}} \cdot \overset{1}{\cancel{2}} \cdot \overset{1}{\cancel{2}}}{\underset{1}{\cancel{2}} \cdot \underset{1}{\cancel{2}} \cdot \underset{1}{\cancel{2}} \cdot \underset{1}{\cancel{2}} \cdot \underset{1}{\cancel{2}} \cdot \underset{1}{\cancel{2}} \cdot 7 \cdot 7} = \frac{27}{49}$$

Objective C Exercises

29. $\dfrac{1}{2} - \dfrac{1}{3} + \dfrac{2}{3} = \dfrac{3}{6} - \dfrac{2}{6} + \dfrac{2}{3}$

$$= \frac{1}{6} + \frac{2}{3}$$

$$= \frac{1}{6} + \frac{4}{6}$$

$$= \frac{5}{6}$$

31. $\dfrac{1}{3} \div \dfrac{1}{2} + \dfrac{3}{4} = \dfrac{1}{3} \cdot \dfrac{2}{1} + \dfrac{3}{4}$

$$= \frac{2}{3} + \frac{3}{4}$$

$$= \frac{8}{12} + \frac{9}{12}$$

$$= \frac{17}{12} = 1\frac{5}{12}$$

33. $\left(\dfrac{3}{4}\right)^2 - \dfrac{5}{12} = \dfrac{9}{16} - \dfrac{5}{12}$

$$= \frac{27}{48} - \frac{20}{48}$$

$$= \frac{7}{48}$$

35. $\dfrac{5}{6} \cdot \left(\dfrac{2}{3} - \dfrac{1}{6}\right) + \dfrac{7}{18} = \dfrac{5}{6} \cdot \left(\dfrac{4}{6} - \dfrac{1}{6}\right) + \dfrac{7}{18}$

$$= \frac{5}{6} \cdot \frac{3}{6} + \frac{7}{18}$$

$$= \frac{5}{12} + \frac{7}{18}$$

$$= \frac{15}{36} + \frac{14}{36}$$

$$= \frac{29}{36}$$

37. $\dfrac{7}{12}-\left(\dfrac{2}{3}\right)^2+\dfrac{5}{8}=\dfrac{7}{12}-\dfrac{4}{9}+\dfrac{5}{8}$

$\qquad\qquad\qquad\quad =\dfrac{21}{36}-\dfrac{16}{36}+\dfrac{5}{8}$

$\qquad\qquad\qquad\quad =\dfrac{5}{36}+\dfrac{5}{8}$

$\qquad\qquad\qquad\quad =\dfrac{10}{72}+\dfrac{45}{72}$

$\qquad\qquad\qquad\quad =\dfrac{55}{72}$

39. $\dfrac{3}{4}\cdot\left(\dfrac{4}{9}\right)^2+\dfrac{1}{2}=\dfrac{3}{4}\cdot\dfrac{16}{81}+\dfrac{1}{2}$

$\qquad\qquad\qquad\quad =\dfrac{4}{27}+\dfrac{1}{2}$

$\qquad\qquad\qquad\quad =\dfrac{8}{54}+\dfrac{27}{54}$

$\qquad\qquad\qquad\quad =\dfrac{35}{54}$

41. $\left(\dfrac{1}{2}+\dfrac{3}{4}\right)\div\dfrac{5}{8}=\left(\dfrac{2}{4}+\dfrac{3}{4}\right)\div\dfrac{5}{8}$

$\qquad\qquad\qquad\quad =\dfrac{5}{4}\cdot\dfrac{8}{5}$

$\qquad\qquad\qquad\quad =2$

43. $\dfrac{3}{8}\div\left(\dfrac{5}{12}+\dfrac{3}{8}\right)=\dfrac{3}{8}\div\left(\dfrac{10}{24}+\dfrac{9}{24}\right)$

$\qquad\qquad\qquad\quad\;\; =\dfrac{3}{8}\div\dfrac{19}{24}$

$\qquad\qquad\qquad\quad\;\; =\dfrac{3}{8}\cdot\dfrac{24}{19}$

$\qquad\qquad\qquad\quad\;\; =\dfrac{9}{19}$

45. $\left(\dfrac{3}{8}\right)^2\div\left(\dfrac{3}{7}+\dfrac{3}{14}\right)=\left(\dfrac{3}{8}\right)^2\div\left(\dfrac{6}{14}+\dfrac{3}{14}\right)$

$\qquad\qquad\qquad\qquad\quad =\left(\dfrac{3}{8}\right)^2\div\dfrac{9}{14}$

$\qquad\qquad\qquad\qquad\quad =\dfrac{9}{64}\cdot\dfrac{14}{9}$

$\qquad\qquad\qquad\qquad\quad =\dfrac{7}{32}$

47. $\dfrac{2}{5}\div\dfrac{3}{8}\cdot\dfrac{4}{5}=\dfrac{2}{5}\cdot\dfrac{8}{3}\cdot\dfrac{4}{5}$

$\qquad\qquad\quad =\dfrac{16}{15}\cdot\dfrac{4}{5}$

$\qquad\qquad\quad =\dfrac{64}{75}$

Applying the Concepts

49. The "puzzle" works as it does because the sum of the fractions $\dfrac{1}{2}$, $\dfrac{1}{3}$, and $\dfrac{1}{9}$ is $\dfrac{17}{18}$, not 1. As a result, the first child actually received $\dfrac{9}{17}$ of the horses, not $\dfrac{1}{2}$; the second child received $\dfrac{6}{17}$ of the horses, not $\dfrac{1}{3}$; and the third child got $\dfrac{2}{17}$ of the horses, not $\dfrac{1}{9}$.

Chapter 2 Review Exercises

1. $\dfrac{30}{45}=\dfrac{2\cdot\overset{1}{\cancel{3}}\cdot\overset{1}{\cancel{5}}}{3\cdot\underset{1}{\cancel{3}}\cdot\underset{1}{\cancel{5}}}=\dfrac{2}{3}$

2. $\left(\dfrac{3}{4}\right)^3\cdot\dfrac{20}{27}=\left(\dfrac{3}{4}\cdot\dfrac{3}{4}\cdot\dfrac{3}{4}\right)\left(\dfrac{20}{27}\right)$

$=\dfrac{\overset{1}{\cancel{3}}\cdot\overset{1}{\cancel{3}}\cdot\overset{1}{\cancel{3}}\cdot\overset{1}{\cancel{2}}\cdot\overset{1}{\cancel{2}}\cdot 5}{2\cdot 2\cdot 2\cdot 2\cdot\underset{1}{\cancel{2}}\cdot\underset{1}{\cancel{2}}\cdot\underset{1}{\cancel{3}}\cdot\underset{1}{\cancel{3}}\cdot\underset{1}{\cancel{3}}}=\dfrac{5}{16}$

3. $\dfrac{13}{4}$

4. $\begin{aligned}\dfrac{2}{3}&=\dfrac{12}{18}\\[2pt]\dfrac{5}{6}&=\dfrac{15}{18}\\[2pt]+\dfrac{2}{9}&=\dfrac{4}{18}\\[2pt]\hline &\dfrac{31}{18}=1\dfrac{13}{18}\end{aligned}$

5. $\dfrac{11}{18}=\dfrac{44}{72},\dfrac{17}{24}=\dfrac{51}{72},\dfrac{11}{18}<\dfrac{17}{24}$

6. $\begin{aligned}18\dfrac{1}{6}&=18\dfrac{7}{42}=17\dfrac{49}{42}\\[2pt]-3\dfrac{5}{7}&=3\dfrac{30}{42}=3\dfrac{30}{42}\\[2pt]\hline &\qquad\qquad\;\; 14\dfrac{19}{42}\end{aligned}$

7. $\dfrac{2}{7}\left[\dfrac{5}{8}-\dfrac{1}{3}\right]\div\dfrac{3}{5}=\dfrac{2}{7}\left[\dfrac{15}{24}-\dfrac{8}{24}\right]\div\dfrac{3}{5}$

$=\dfrac{2}{7}\left[\dfrac{7}{24}\right]\div\dfrac{3}{5}=\dfrac{2\cdot\overset{1}{\cancel{7}}}{7\cdot 24}\div\dfrac{3}{5}$

$=\dfrac{1}{12}\times\dfrac{5}{3}=\dfrac{5}{36}$

8. $2\dfrac{1}{3}\times 3\dfrac{7}{8}=\dfrac{7}{3}\times\dfrac{31}{8}=\dfrac{7\cdot 31}{3\cdot 8}=\dfrac{217}{24}=9\dfrac{1}{24}$

9. $1\dfrac{1}{3}\div\dfrac{2}{3}=\dfrac{4}{3}\div\dfrac{2}{3}=\dfrac{4}{3}\times\dfrac{3}{2}=\dfrac{4\cdot 3}{3\cdot 2}=\dfrac{2\cdot\overset{1}{\cancel{2}}\cdot\overset{1}{\cancel{3}}}{\underset{1}{\cancel{3}}\cdot\underset{1}{\cancel{2}}}=2$

10.
$$\frac{17}{24} = \frac{34}{48}$$
$$-\frac{3}{16} = \frac{9}{48}$$
$$\frac{25}{48}$$

11.
$$8\frac{2}{3} \div 2\frac{3}{5} = \frac{26}{3} \div \frac{13}{5} = \frac{26}{3} \times \frac{5}{13} = \frac{26 \cdot 5}{3 \cdot 13} = \frac{2 \cdot \overset{1}{\cancel{13}} \cdot 5}{3 \cdot \cancel{13}}$$
$$= \frac{10}{3} = 3\frac{1}{3}$$

12.
$$20 = \boxed{\begin{array}{c|c|c} 2 & 3 & 5 \\ \hline \boxed{2 \cdot 2} & & 5 \end{array}}$$
$$48 = \begin{array}{c|c|c} 2 \cdot 2 \cdot 2 \cdot 2 & 3 & \end{array}$$
$$\text{GCF} = 2 \cdot 2 = 4$$

13.
$$\frac{2 \cdot 12}{3 \cdot 12} = \frac{24}{36}$$

14.
$$\frac{15}{28} \div \frac{5}{7} = \frac{15}{28} \times \frac{7}{5} = \frac{15 \cdot 7}{28 \cdot 5} = \frac{3 \cdot \overset{1}{\cancel{5}} \cdot \overset{1}{\cancel{7}}}{2 \cdot 2 \cdot \underset{1}{\cancel{7}} \cdot \underset{1}{\cancel{5}}} = \frac{3}{4}$$

15.
$$\frac{8 \cdot 4}{11 \cdot 4} = \frac{32}{44}$$

16.
$$2\frac{1}{4} \times 7\frac{1}{3} = \frac{9}{4} \times \frac{22}{3} = \frac{9 \cdot 22}{4 \cdot 3} = \frac{3 \cdot \overset{1}{\cancel{3}} \cdot 2 \cdot 11}{\underset{1}{\cancel{2}} \cdot 2 \cdot \underset{1}{\cancel{3}}}$$
$$= \frac{33}{2} = 16\frac{1}{2}$$

17.
$$18 = \begin{array}{c|c} 2 & 3 \\ \hline 2 & 3 \cdot 3 \end{array}$$
$$12 = \begin{array}{c|c} \boxed{2 \cdot 2} & \boxed{3 \cdot 3} \end{array}$$
$$\text{LCM} = 2 \cdot 2 \cdot 3 \cdot 3 = 36$$

18.
$$\frac{16}{44} = \frac{\overset{1}{\cancel{2}} \cdot \overset{1}{\cancel{2}} \cdot 2 \cdot 2}{\underset{1}{\cancel{2}} \cdot \underset{1}{\cancel{2}} \cdot 11} = \frac{4}{11}$$

19.
$$\frac{3}{8}$$
$$\frac{5}{8}$$
$$+\frac{1}{8}$$
$$\frac{9}{8} = 1\frac{1}{8}$$

20.
$$16 = 15\frac{8}{8}$$
$$-5\frac{7}{8} = 5\frac{7}{8}$$
$$10\frac{1}{8}$$

21.
$$4\frac{4}{9} = 4\frac{24}{54}$$
$$2\frac{1}{6} = 2\frac{9}{54}$$
$$+11\frac{17}{27} = 11\frac{34}{54}$$
$$17\frac{67}{54} = 18\frac{13}{54}$$

22.
$$15 = \begin{array}{c|c} 3 & 5 \\ \hline 3 & \boxed{5} \end{array}$$
$$25 = \begin{array}{c|c} & 5 \cdot 5 \end{array}$$
$$\text{GCF} = 5$$

23.
$$\begin{array}{r} 3 \\ 5\overline{)17} \\ -15 \\ \hline 2 \end{array} \qquad \frac{17}{5} = 3\frac{2}{5}$$

24.
$$\left[\frac{4}{5} - \frac{2}{3}\right]^2 \div \frac{4}{15} = \left[\frac{12}{15} - \frac{10}{15}\right]^2 \div \frac{4}{15}$$
$$= \left(\frac{2}{15}\right)^2 \div \frac{4}{15} = \left(\frac{2}{15}\right)\left(\frac{2}{15}\right) \div \left(\frac{4}{15}\right)$$
$$= \frac{4}{225} \times \frac{15}{4} = \frac{4 \cdot 15}{225 \cdot 4} = \frac{1}{15}$$

25.
$$\frac{3}{8} = \frac{9}{24}$$
$$1\frac{2}{3} = 1\frac{16}{24}$$
$$+3\frac{5}{6} = 3\frac{20}{24}$$
$$4\frac{45}{24} = 5\frac{21}{24} = 5\frac{7}{8}$$

26.
$$18 = \begin{array}{c|c} 2 & 3 \\ \hline \boxed{2} & 3 \cdot 3 \end{array}$$
$$27 = \begin{array}{c|c} & \boxed{3 \cdot 3 \cdot 3} \end{array}$$
$$\text{LCM} = 2 \cdot 3 \cdot 3 \cdot 3 = 54$$

27.
$$\frac{11}{18}$$
$$-\frac{5}{18}$$
$$\frac{6}{18} = \frac{1}{3}$$

28.
$$2\frac{5}{7} = \frac{14 + 5}{7} = \frac{19}{7}$$

29.
$$\frac{5}{6} \div \frac{5}{12} = \frac{5}{6} \cdot \frac{12}{5} = \frac{5 \cdot 12}{6 \cdot 5} = \frac{\overset{1}{\cancel{5}} \cdot \overset{1}{\cancel{2}} \cdot 2 \cdot \overset{1}{\cancel{3}}}{\underset{1}{\cancel{2}} \cdot \underset{1}{\cancel{3}} \cdot \underset{1}{\cancel{5}}} = 2$$

30.
$$\frac{5}{12} \times \frac{4}{25} = \frac{5 \cdot 4}{12 \cdot 25} = \frac{\overset{1}{\cancel{5}} \cdot \overset{1}{\cancel{2}} \cdot 2}{\underset{1}{\cancel{2}} \cdot 2 \cdot 3 \cdot \underset{1}{\cancel{5}} \cdot 5} = \frac{1}{15}$$

31.
$$\frac{11}{50} \times \frac{25}{44} = \frac{11 \cdot 25}{50 \cdot 44} = \frac{\overset{1}{\cancel{11}} \cdot \overset{1}{\cancel{5}} \cdot \overset{1}{\cancel{5}}}{2 \cdot \underset{1}{\cancel{5}} \cdot \underset{1}{\cancel{5}} \cdot 2 \cdot 2 \cdot \underset{1}{\cancel{11}}} = \frac{1}{8}$$

32. $1\dfrac{7}{8}$

33. Strategy To find the total rainfall for the 3 months, add the amounts of rain from each month $\left(5\dfrac{7}{8}, 6\dfrac{2}{3}, \text{and } 8\dfrac{3}{4} \text{ inches}\right)$.

Solution

$$5\dfrac{7}{8} = 5\dfrac{21}{24}$$
$$6\dfrac{2}{3} = 6\dfrac{16}{24}$$
$$+8\dfrac{3}{4} = 8\dfrac{18}{24}$$
$$\overline{\phantom{+8\dfrac{3}{4} =} 19\dfrac{55}{24} = 21\dfrac{7}{24}}$$

The total rainfall for the 3 months was $21\dfrac{7}{24}$ inches.

34. Strategy To find the cost of each acre, divide the total cost ($168,000) by the number of acres $\left(4\dfrac{2}{3}\right)$.

Solution

$$\$168,000 \div 4\dfrac{2}{3} = 168,000 \div \dfrac{14}{3}$$
$$= 168,000 \times \dfrac{3}{14}$$
$$= \$36,000$$

The cost per acre was $36,000.

35. Strategy To find how many miles the second checkpoint is from the finish line:
• Add the distance to the first checkpoint $\left(4\dfrac{1}{2} \text{ miles}\right)$ to the distance between the first checkpoint and the second checkpoint $\left(5\dfrac{3}{4} \text{ miles}\right)$.
• Subtract the total distance to the second checkpoint from the entire length of the race (15 miles).

Solution

$$4\dfrac{1}{2} = 4\dfrac{2}{4} \qquad\qquad 15 = 14\dfrac{4}{4}$$
$$+5\dfrac{3}{4} = 5\dfrac{3}{4} \qquad -10\dfrac{1}{4} = 10\dfrac{1}{4}$$
$$\overline{9\dfrac{5}{4} = 10\dfrac{1}{4}} \qquad \overline{\phantom{-10\dfrac{1}{4} =}4\dfrac{3}{4}}$$

The second checkpoint is $4\dfrac{3}{4}$ miles from the finish line.

36. Strategy To find how many miles the car can travel, multiply the number of miles the car can travel on 1 gallon (36) by the number of gallons used $\left(6\dfrac{3}{4}\right)$.

Solution $36 \times 6\dfrac{3}{4} = 36 \times \dfrac{27}{4} = \dfrac{36 \cdot 27}{4} = 243$

The car can travel 243 miles.

Chapter 2 Test

1. $\dfrac{9}{11} \times \dfrac{44}{81} = \dfrac{9 \cdot 44}{11 \cdot 81} = \dfrac{\overset{1}{\cancel{3}} \cdot \overset{1}{\cancel{3}} \cdot 2 \cdot 2 \cdot \overset{1}{\cancel{11}}}{\cancel{11} \cdot \underset{1}{\cancel{3}} \cdot \underset{1}{\cancel{3}} \cdot 3 \cdot 3} = \dfrac{4}{9}$

2.
$$24 = \boxed{2 \cdot 2 \cdot 2} \;\boxed{3}$$
$$80 = \boxed{2 \cdot 2 \cdot 2}\;\boxed{}\;\boxed{5}$$
$$\text{GCF} = 2 \cdot 2 \cdot 2 = 8$$

3. $\dfrac{5}{9} \div \dfrac{7}{18} = \dfrac{5}{9} \times \dfrac{18}{7} = \dfrac{5 \cdot 2 \cdot \overset{1}{\cancel{3}} \cdot \overset{1}{\cancel{3}}}{\underset{1}{\cancel{3}} \cdot \underset{1}{\cancel{3}} \cdot 7} = \dfrac{10}{7} = 1\dfrac{3}{7}$

4. $\left(\dfrac{3}{4}\right)^2 \div \left(\dfrac{2}{3} + \dfrac{5}{6}\right) - \dfrac{1}{12} = \left(\dfrac{3}{4} \cdot \dfrac{3}{4}\right) \div \left(\dfrac{4}{6} + \dfrac{5}{6}\right) - \dfrac{1}{12}$

$= \dfrac{9}{16} \div \left(\dfrac{9}{6}\right) - \dfrac{1}{12}$

$= \dfrac{9}{16} \div \dfrac{3}{2} - \dfrac{1}{12}$

$= \dfrac{9}{16} \times \dfrac{2}{3} - \dfrac{1}{12}$

$= \dfrac{\overset{1}{\cancel{3}} \cdot 3 \cdot \overset{1}{\cancel{2}}}{2 \cdot 2 \cdot 2 \cdot \underset{1}{\cancel{2}} \cdot \underset{1}{\cancel{3}}} - \dfrac{1}{12} = \dfrac{3}{8} - \dfrac{1}{12} = \dfrac{9}{24} - \dfrac{2}{24} = \dfrac{7}{24}$

5. $9\dfrac{4}{5} = \dfrac{45 + 4}{5} = \dfrac{49}{5}$

6. $5\dfrac{2}{3} \times 1\dfrac{7}{17} = \dfrac{17}{3} \times \dfrac{24}{17} = \dfrac{17 \cdot 24}{3 \cdot 17} = \dfrac{\overset{1}{\cancel{17}} \cdot 2 \cdot 2 \cdot 2 \cdot \overset{1}{\cancel{3}}}{\underset{1}{\cancel{3}} \cdot \underset{1}{\cancel{17}}} = 8$

7. $\dfrac{40}{64} = \dfrac{\overset{1}{\cancel{2}} \cdot \overset{1}{\cancel{2}} \cdot \overset{1}{\cancel{2}} \cdot 5}{\underset{1}{\cancel{2}} \cdot \underset{1}{\cancel{2}} \cdot \underset{1}{\cancel{2}} \cdot 2 \cdot 2} = \dfrac{5}{8}$

8. $\dfrac{3}{8} = \dfrac{9}{24}, \dfrac{5}{12} = \dfrac{10}{24}, \dfrac{3}{8} < \dfrac{5}{12}$

9. $\left(\dfrac{1}{4}\right)^3 \div \left(\dfrac{1}{8}\right)^2 - \dfrac{1}{6} = \left(\dfrac{1}{4} \cdot \dfrac{1}{4} \cdot \dfrac{1}{4}\right) \div \left(\dfrac{1}{8} \cdot \dfrac{1}{8}\right) - \dfrac{1}{6}$

$= \dfrac{1}{64} \div \dfrac{1}{64} - \dfrac{1}{6}$

$= \dfrac{1}{64} \times \dfrac{64}{1} - \dfrac{1}{6}$

$= 1 - \dfrac{1}{6}$

$= \dfrac{6}{6} - \dfrac{1}{6} = \dfrac{5}{6}$

10.

$$24 = \boxed{2 \cdot 2 \cdot 2} \quad \boxed{3}$$
$$40 = \boxed{2 \cdot 2 \cdot 2} \qquad \boxed{5}$$
$$\text{LCM} = 2 \cdot 2 \cdot 2 \cdot 3 \cdot 5 = 120$$

11.

$$\frac{17}{24}$$
$$-\frac{11}{24}$$
$$\overline{\frac{6}{24} = \frac{1}{4}}$$

12.

$$\begin{array}{r} 3 \\ 5{\overline{)18}} \\ \underline{-15} \\ 3 \end{array} \qquad \frac{18}{5} = 3\frac{3}{5}$$

13.

$$6\frac{2}{3} \div 3\frac{1}{6} = \frac{20}{3} \div \frac{19}{6} = \frac{20}{3} \times \frac{6}{19}$$
$$= \frac{2 \cdot 2 \cdot 5 \cdot 2 \cdot \overset{1}{\cancel{3}}}{\underset{1}{\cancel{3}} \cdot 19} = \frac{40}{19} = 2\frac{2}{19}$$

14. $\dfrac{5 \cdot 9}{8 \cdot 9} = \dfrac{45}{72}$

15.

$$\frac{5}{6} = \frac{75}{90}$$
$$\frac{7}{9} = \frac{70}{90}$$
$$+\frac{1}{15} = \frac{6}{90}$$
$$\overline{\frac{151}{90} = 1\frac{61}{90}}$$

16.

$$23\frac{1}{8} = 23\frac{11}{88} = 22\frac{99}{88}$$
$$-9\frac{9}{44} = 9\frac{18}{88} = 9\frac{18}{88}$$
$$\overline{\phantom{-9\frac{9}{44} = 9\frac{18}{88} = } 13\frac{81}{88}}$$

17.

$$\frac{9}{16} = \frac{27}{48}$$
$$-\frac{5}{12} = \frac{20}{48}$$
$$\overline{\frac{7}{48}}$$

18.

$$\left(\frac{2}{3}\right)^4 \left(\frac{27}{32}\right) = \left(\frac{2}{3} \cdot \frac{2}{3} \cdot \frac{2}{3} \cdot \frac{2}{3}\right)\left(\frac{27}{32}\right)$$
$$= \frac{\overset{1}{\cancel{2}} \cdot \overset{1}{\cancel{2}} \cdot \overset{1}{\cancel{2}} \cdot \overset{1}{\cancel{2}} \cdot \overset{1}{\cancel{3}} \cdot \overset{1}{\cancel{3}} \cdot \overset{1}{\cancel{3}}}{\underset{1}{\cancel{3}} \cdot \underset{1}{\cancel{3}} \cdot \underset{1}{\cancel{3}} \cdot \underset{1}{\cancel{3}} \cdot \underset{1}{\cancel{2}} \cdot \underset{1}{\cancel{2}} \cdot \underset{1}{\cancel{2}} \cdot \underset{1}{\cancel{2}} \cdot 2} = \frac{1}{6}$$

19.

$$\frac{7}{12}$$
$$\frac{11}{12}$$
$$+\frac{5}{12}$$
$$\overline{\frac{23}{12} = 1\frac{11}{12}}$$

20.

$$12\frac{5}{12} = 12\frac{25}{60}$$
$$+9\frac{17}{20} = 9\frac{51}{60}$$
$$\overline{21\frac{76}{60} = 22\frac{16}{60} = 22\frac{4}{15}}$$

21. $\dfrac{11}{4}$

22. **Strategy** To find the electrician's earnings, multiply daily earnings ($240) by the number of days worked $\left(3\frac{1}{2}\right)$.

Solution $\$240 \times 3\frac{1}{2} = 240 \times \frac{7}{2} = \frac{240 \cdot 7}{2} = \$840$
The electrician earns $840.

23. **Strategy** To find how many lots were available:
● Find how many acres were being developed by subtracting the amount set aside for the park $\left(1\frac{3}{4} \text{ acres}\right)$ from the total parcel $\left(7\frac{1}{4} \text{ acres}\right)$.
● Divide the amount being developed by the size of each lot $\left(\frac{1}{2} \text{ acre}\right)$.

Solution

$$7\frac{1}{4} = 6\frac{5}{4}$$
$$-1\frac{3}{4} = 1\frac{3}{4}$$
$$\overline{\phantom{-1\frac{3}{4} = } 5\frac{2}{4} = 5\frac{1}{2}}$$

$$5\frac{1}{2} \div \frac{1}{2} = \frac{11}{2} \times \frac{2}{1} = \frac{11 \cdot 2}{2} = 11$$
11 lots were available for sale.

24. **Strategy**   To find how many houses the developer plans to build:
● Subtract the amount of land set aside (3 acres) from the total purchased $\left(25\frac{1}{2}\ \text{acres}\right)$ to determine the amount of land that can be used for houses.
● Divide the amount of land available for houses by $\frac{3}{4}$ to determine the number of $\frac{3}{4}$-acre plots that are available for houses.

Solution $25\frac{1}{2} - 3 = 22\frac{1}{2}$ acres

$22\frac{1}{2} \div \frac{3}{4} = \frac{45}{2} \cdot \frac{4}{3} = 15 \cdot 2 = 30$

The developer plans to build 30 houses on the property.

25. **Strategy**   To find the total rainfall for the 3-month period, add the rainfall amounts for each of the months $\left(11\frac{1}{2}, 7\frac{5}{8},\ \text{and}\ 2\frac{1}{3}\ \text{inches}\right)$.

Solution
$11\frac{1}{2} = 11\frac{12}{24}$

$7\frac{5}{8} = 7\frac{15}{24}$

$\underline{+2\frac{1}{3} = 2\frac{8}{24}}$

$\qquad 20\frac{35}{24} = 21\frac{11}{24}$

The total rainfall for the 3-month period was $21\frac{11}{24}$ inches.

Cumulative Review Exercises

1. 290,000

2.
$$
\begin{array}{r}
{\scriptstyle 9\ \ \ \ 9\,13} \\
{\scriptstyle 8\,10\ 10\,3\,17} \\
3\cancel{9}\cancel{0},\cancel{0}\cancel{4}\cancel{7} \\
-\ \ 98,769 \\
\hline
291,278
\end{array}
$$

3.
$$
\begin{array}{r}
926 \\
\times\ \ 79 \\
\hline
8334 \\
6482\ \ \\
\hline
73,154
\end{array}
$$

4.
$$
\begin{array}{r}
540\ \text{r}12 \\
57\overline{)30{,}792} \\
-\ 285\ \ \ \ \\
\hline
229\ \ \\
-228\ \ \\
\hline
12\ \\
-0\ \\
\hline
12
\end{array}
$$

5. $4 \cdot (6 - 3) \div 6 - 1 = 4 \cdot 3 \div 6 - 1$
$= 12 \div 6 - 1$
$= 2 - 1$
$= 1$

6. $44 = 2 \cdot 2 \cdot 11$

$$
\begin{array}{r|r}
& 44 \\
\hline
2 & 22 \\
2 & 11 \\
11 & 1
\end{array}
$$

7.
$$
\begin{array}{c|c|c|c|c}
 & 2 & 3 & 5 & 7 \\
\hline
30 = & ② & ③ & ⑤ & \\
\hline
42 = & 2 & 3 & & ⑦ \\
\end{array}
$$
LCM $= 2 \cdot 3 \cdot 5 \cdot 7 = 210$

8.
$$
\begin{array}{c|c|c|c}
 & 2 & 3 & 5 \\
\hline
60 = & ②\cdot② & 3 & ⑤ \\
\hline
80 = & 2\cdot2\cdot2\cdot2 & & 5 \\
\end{array}
$$
GCF $= 2 \cdot 2 \cdot 5 = 20$

9. $7\frac{2}{3} = \frac{21 + 2}{3} = \frac{23}{3}$

10.
$$
\begin{array}{r}
6\ \text{r}1 \\
4\overline{)25} \\
-24\ \\
\hline
1
\end{array}
\qquad
\frac{25}{4} = 6\frac{1}{4}
$$

11. $\frac{5 \cdot 3}{16 \cdot 3} = \frac{15}{48}$

12. $\frac{24}{60} = \frac{2 \cdot \overset{1}{\cancel{2}} \cdot \overset{1}{\cancel{2}} \cdot \overset{1}{\cancel{3}}}{\underset{1}{\cancel{2}} \cdot \underset{1}{\cancel{2}} \cdot \underset{1}{\cancel{3}} \cdot 5} = \frac{2}{5}$

13.
$\frac{7}{12} = \frac{28}{48}$

$\underline{+\frac{9}{16} = \frac{27}{48}}$

$\qquad \frac{55}{48} = 1\frac{7}{48}$

14.
$3\frac{7}{8} = 3\frac{42}{48}$

$7\frac{5}{12} = 7\frac{20}{48}$

$\underline{+2\frac{15}{16} = 2\frac{45}{48}}$

$\qquad 12\frac{107}{48} = 14\frac{11}{48}$

15.

$$\dfrac{11}{12} = \dfrac{22}{24}$$
$$-\dfrac{3}{8} = \dfrac{9}{24}$$
$$\overline{\qquad \dfrac{13}{24}}$$

16.

$$5\dfrac{1}{6} = 5\dfrac{3}{18} = 4\dfrac{21}{18}$$
$$-3\dfrac{7}{18} = 3\dfrac{7}{18} = 3\dfrac{7}{18}$$
$$\overline{\qquad\qquad 1\dfrac{14}{18} = 1\dfrac{7}{9}}$$

17. $\dfrac{3}{8} \times \dfrac{14}{15} = \dfrac{3 \cdot 14}{8 \cdot 15} = \dfrac{\overset{1}{\cancel{3}} \cdot 2 \cdot 7}{2 \cdot 2 \cdot 2 \cdot \underset{1}{\cancel{3}} \cdot 5} = \dfrac{7}{20}$

18. $3\dfrac{1}{8} \times 2\dfrac{2}{5} = \dfrac{25}{8} \times \dfrac{12}{5} = \dfrac{25 \cdot 12}{8 \cdot 5}$

$= \dfrac{5 \cdot \overset{1}{\cancel{5}} \cdot \overset{1}{\cancel{2}} \cdot \overset{1}{\cancel{2}} \cdot 3}{2 \cdot \underset{1}{\cancel{2}} \cdot \underset{1}{\cancel{2}} \cdot \underset{1}{\cancel{5}}} = \dfrac{15}{2} = 7\dfrac{1}{2}$

19. $\dfrac{7}{16} \div \dfrac{5}{12} = \dfrac{7}{16} \times \dfrac{12}{5} = \dfrac{7 \cdot 12}{16 \cdot 5}$

$= \dfrac{7 \cdot \overset{1}{\cancel{2}} \cdot \overset{1}{\cancel{2}} \cdot 3}{\underset{1}{\cancel{2}} \cdot \underset{1}{\cancel{2}} \cdot 2 \cdot 2 \cdot 5} = \dfrac{21}{20} = 1\dfrac{1}{20}$

20. $6\dfrac{1}{8} \div 2\dfrac{1}{3} = \dfrac{49}{8} \div \dfrac{7}{3} = \dfrac{49}{8} \times \dfrac{3}{7} = \dfrac{49 \cdot 3}{8 \cdot 7}$

$= \dfrac{7 \cdot \overset{1}{\cancel{7}} \cdot 3}{2 \cdot 2 \cdot 2 \cdot \underset{1}{\cancel{7}}} = \dfrac{21}{8} = 2\dfrac{5}{8}$

21. $\left(\dfrac{1}{2}\right)^3 \cdot \dfrac{8}{9} = \left(\dfrac{1}{2} \cdot \dfrac{1}{2} \cdot \dfrac{1}{2}\right) \cdot \dfrac{8}{9} = \dfrac{1}{8} \cdot \dfrac{8}{9} = \dfrac{1}{9}$

22. $\left(\dfrac{1}{2} + \dfrac{1}{3}\right) \div \left(\dfrac{2}{5}\right)^2 = \left(\dfrac{3}{6} + \dfrac{2}{6}\right) \div \left(\dfrac{2}{5} \cdot \dfrac{2}{5}\right)$

$= \dfrac{5}{6} \div \dfrac{4}{25} = \dfrac{5}{6} \times \dfrac{25}{4} = \dfrac{5 \cdot 25}{6 \cdot 4} = \dfrac{125}{24} = 5\dfrac{5}{24}$

23. **Strategy**   To find the amount in the checking account:
- Find the total of the checks written by adding the check amounts ($128, $54, and $315).
- Subtract the total of the checks written from the original balance in the checking account ($1359).

Solution

$128	$1359
54	− 497
+ 315	$862
$497	

The amount in the checking account at the end of the week was $862.

24. **Strategy**   To find the total income from the sale of the tickets:
- Find the income from the adult tickets by multiplying the ticket price ($10) by the number of tickets sold (87).
- Find the income from the student tickets by multiplying the ticket price ($4) by the number of tickets sold (135).
- Find the total income by adding the income from the adult tickets to the income from the student tickets.

Solution

87	135	$870
×$10	× $4	+540
$870	$540	$1410

The total income from the tickets was $1410.

25. **Strategy**   To find the total weight, add the three weights $\left(1\dfrac{1}{2}, 7\dfrac{7}{8}, \text{and } 2\dfrac{2}{3} \text{ pounds}\right)$.

Solution

$$1\dfrac{1}{2} = 1\dfrac{12}{24}$$
$$7\dfrac{7}{8} = 7\dfrac{21}{24}$$
$$+2\dfrac{2}{3} = 2\dfrac{16}{24}$$
$$\overline{\qquad 10\dfrac{49}{24} = 12\dfrac{1}{24}}$$

The total weight is $12\dfrac{1}{24}$ pounds.

26. **Strategy**   To find the length of the remaining piece, subtract the length of the cut piece $\left(2\dfrac{5}{8} \text{ feet}\right)$ from the original length of the board $\left(7\dfrac{1}{3} \text{ feet}\right)$.

Solution

$$7\dfrac{1}{3} = 7\dfrac{8}{24} = 6\dfrac{32}{24}$$
$$-2\dfrac{5}{8} = 2\dfrac{15}{24} = 2\dfrac{15}{24}$$
$$\overline{\qquad\qquad\quad 4\dfrac{17}{24}}$$

The length of the remaining piece is $4\dfrac{17}{24}$ feet.

27. Strategy To find how many miles the car can travel, multiply the number of gallons used $\left(8\frac{1}{3}\right)$ by the number of miles that the car travels on each gallon (27).

Solution $27 \times 8\frac{1}{3} = 27 \times \frac{25}{3} = 225$

The car travels 225 miles on $8\frac{1}{3}$ gallons of gas.

28. Strategy To find how many parcels can be sold:
- Find the amount of land that can be developed by subtracting the land donated for a park (2 acres) from the total amount of land purchased $\left(10\frac{1}{3}\text{ acres}\right)$.
- Divide the amount of land that can be developed by the size of each parcel $\left(\frac{1}{3}\text{ acre}\right)$.

Solution

$$10\frac{1}{3}$$
$$\underline{-2}$$
$$8\frac{1}{3}$$

$8\frac{1}{3} \div \frac{1}{3} = \frac{25}{3} \div \frac{1}{3} = \frac{25}{3} \times \frac{3}{1} = 25$

25 parcels can be sold from the remaining land.

Chapter 3: Decimals

Prep Test

1. $\dfrac{3}{10}$

2. 36,900

3. Four thousand seven hundred ninety-one

4. 6842

5. 9394

6. 1638

7.
```
    844
  × 91
    844
  7596
 76,804
```

8.
```
       278 r18
   23)6412
     −46
      181
     −161
      202
     −184
       18
```

Go Figure

There are 7 children in the family.
Maria has twice as many brothers as sisters. She
has 4 brothers (including Pedro) and 2 sisters.
Pedro has as many brothers as sisters. He has 3
sisters (including Maria) and 3 brothers.

Section 3.1

Objective A Exercises

1. The digit 5 is in the thousandths place.

3. The digit 5 is in the ten-thousandths place.

5. The digit 5 is in the hundredths place.

7. $\dfrac{3}{10} = 0.3$ (three tenths)

9. $\dfrac{21}{100} = 0.21$ (twenty-one hundredths)

11. $\dfrac{461}{1000} = 0.461$ (four hundred sixty-one thousandths)

13. $\dfrac{93}{1000} = 0.093$ (ninety-three thousandths)

15. $0.1 = \dfrac{1}{10}$ (one-tenth)

17. $0.47 = \dfrac{47}{100}$ (forty-seven-hundredths)

19. $0.289 = \dfrac{289}{1000}$ (two hundred eighty-nine-thousandths)

21. $0.09 = \dfrac{9}{100}$ (nine-hundredths)

23. Thirty-seven-hundredths

25. Nine and four-tenths

27. Fifty-three-ten-thousandths

29. Forty-five-thousandths

31. Twenty-six and four-hundredths

33. 3.0806

35. 407.03

37. 246.024

39. 73.02684

Objective B Exercises

41. ┌─── Given place value
5.398
└─9 > 5
5.398 rounded to the nearest tenth is 5.4.

43. ┌─── Given place value
30.0092
└─ 0 < 5
30.0092 rounded to the nearest tenth is 30.0.

45. ┌─── Given place value
413.5972
└─ 7 > 5
413.5972 rounded to the nearest hundredth is 413.60.

47. ┌─── Given place value
6.061745
└─7 > 5
6.061745 rounded to the nearest thousandth is 6.062.

49. ┌─── Given place value
96.8027
└─ 8 > 5
96.8027 rounded to the nearest whole number is 97.

51. ┌─── Given place value
5439.83
└─ 8 > 5
5439.83 rounded to the nearest whole number is 5440.

53. $\overline{}$ *Given place value*
0.023591
$\underline{}$ 9 > 5
0.023591 rounded to the nearest ten-thousandth is 0.0236.

55. 0.1763668 rounded to the nearest hundredth is 0.18. The weight of a nickel to the nearest hundredth is 0.18 ounce.

57. 26.21875 rounded to the nearest tenth is 26.2. To the nearest tenth, the Boston Marathon is 26.2 miles.

Applying the Concepts

59a. Answers will vary. For example, 0.11, 0.12, 0.13, 0.14, 0.15, 0.16, 0.17, 0.18, and 0.19 are numbers between 0.1 and 0.2. But any number of digits can be attached to 0.1, and the number will be between 0.1 and 0.2. For example, 0.123456789 is a number between 0.1 and 0.2.

b. Answers will vary. For example, 1.01, 1.02, 1.03, 1.04, 1.05, 1.06, 1.07, 1.08, and 1.09 are numbers between 1 and 1.1. But any number of digits can be attached to 1.0, and the number will be between 1 and 1.1. For example, 1.0123456789 is a number between 1 and 1.1.

c. Answers will vary. For example, 0.001, 0.002, 0.003 and 0.004 are numbers between 0 and 0.005. But any number of digits can be attached to 0.001, 0.002, 0.003, or 0.004, and the number will be between 0 and 0.005. For example, 0.00123456789 is a number between 0 and 0.005.

Section 3.2

Objective A Exercises

1.
$^{1}^{11}$
16.008
2.0385
+ 132.06
150.1065

3.
$^{1}^{1}$
1.792
67.
+ 27.0526
95.8446

5.
^{1}
3.02
62.7
+ 3.924
69.644

7.
$^{1\ 1}^{1}$
82.006
9.95
+ 0.927
92.883

9.
$^{2\ 1\ 11}$
4.307
99.82
+ 9.078
113.205

11.
0.29
+ 0.4
0.69

13.
7.3
+ 9.005
16.305

15.
$^{2\ 1\ 1}$
8.72
99.073
+ 2.9736
110.7666

17.
^{2}
8.
89.43
+ 7.0659
104.4959

19.

219.9	≈	220
0.872	≈	1
+ 13.42	≈	+ 13
Cal.: 234.192		Est.: 234

21.

678.92	≈	679
97.6	≈	98
+ 5.423	≈	+ 5
Cal.: 781.943		Est.: 782

Objective B Exercises

23. **Strategy** To find the length of the shaft, add the three measures on the shaft (1.87, 1.63, and 2.15 inches).

Solution 1.87
1.63
+ 2.15
5.65

The total length of the shaft is 5.65 inches.

25. **Strategy** To find the amount in your checking account:
- Find the total amount of the four deposits ($210.98 + $45.32 + $1236.34 +$27.99).
- Add the total of the deposits to the previous balance ($2143.57).

Solution

```
$ 210.98      $1520.63
   45.32      + 2143.57
 1236.34      $3664.20
+  27.99
$1520.63
```

The amount in the checking account is $3664.20.

27. **Strategy** To find total projected populations of Asia and Africa in 2050, add the expected population of Asia (5.3 billion) to the expected population of Africa (1.8 billion).

Solution

```
  5.3
+ 1.8
  7.1
```

The combined populations of Asia and Africa in 2050 are expected to be 7.1 billion people.

29. **Strategy** To find how many self-employed people earn more than $5000, add the numbers of people from the chart who make more than $5000 (3.9, 1.6, and 1.0 million).

Solution

```
  3.9    $5000–$24,999
  1.6    $25,000–$49,999
+ 1.0    $50,000 or more
  6.5
```

The number of self-employed people who earn more than $5000 is 6.5 million.

Applying the Concepts

31.
```
$ 3.29    Raisin bran
  1.49    Bread
  2.59    Milk
+ 2.79    Butter
$10.16
```
No, $10 is not enough.

33.
```
  1.4
×   4
  5.6
```
No, a 4-foot rope cannot be wrapped around the box.

Section 3.3

Objective A Exercises

1.
$$
\begin{array}{r}
24.037 \\
-18.41 \\
\hline
5.627
\end{array}
$$

3.
$$
\begin{array}{r}
123.0700 \\
-\ \ \ 9.4273 \\
\hline
113.6427
\end{array}
$$

5.
$$
\begin{array}{r}
16.5000 \\
-\ 9.7902 \\
\hline
6.7098
\end{array}
$$

7.
$$
\begin{array}{r}
235.790 \\
-\ 20.093 \\
\hline
215.697
\end{array}
$$

9.
$$
\begin{array}{r}
63.0050 \\
-\ 9.1274 \\
\hline
53.8776
\end{array}
$$

11.
$$
\begin{array}{r}
92.0000 \\
-19.2909 \\
\hline
72.7091
\end{array}
$$

13.
$$
\begin{array}{r}
0.3200 \\
-0.0058 \\
\hline
0.3142
\end{array}
$$

15.
$$
\begin{array}{r}
3.005 \\
-1.982 \\
\hline
1.023
\end{array}
$$

17.
$$
\begin{array}{r}
352.160 \\
-\ 90.994 \\
\hline
261.166
\end{array}
$$

19.
$$
\begin{array}{r}
724.32 \\
-\ 69. \\
\hline
655.32
\end{array}
$$

21.
$$
\begin{array}{r}
362.3940 \\
-\ 19.4672 \\
\hline
342.9268
\end{array}
$$

23.

$$\begin{array}{r}
\overset{\scriptscriptstyle 9\ \ \ \ 9}{\underset{8\ 101010}{}}\\
1\overset{}{9}.\overset{}{0}\overset{}{0}\overset{}{0}\\
-10.372\\
\hline
8.628
\end{array}$$

25.

$$\begin{array}{r}
\overset{\scriptscriptstyle 9\ \ 10}{\underset{6\ 100 10}{}}\\
7.0\overset{}{1}\overset{}{0}\\
-2.325\\
\hline
4.685
\end{array}$$

27.

$$\begin{array}{r}
\overset{\scriptscriptstyle 1214}{\underset{8\ 2410}{}}\\
19.350\\
-\ 8.967\\
\hline
10.383
\end{array}$$

29.
$$\begin{array}{rcr}
3.7529 & \approx & 4\\
-1.00784 & \approx & -1\\
\end{array}$$
Cal.: 2.74506   Est.: 3

31.
$$\begin{array}{rcr}
9.07325 & \approx & 9\\
-1.924 & \approx & -2\\
\end{array}$$
Cal.: 7.14925   Est.: 7

Objective B Exercises

33. Strategy To find the missing dimension, subtract 6.79 from 14.34.

Solution
$$\begin{array}{r}14.34\\-\ 6.79\\\hline 7.55\end{array}$$
The missing dimension is 7.55 inches.

35a. Strategy To find the total amount of the checks written, add the three amounts ($67.92, $43.10, and $496.34).

Solution
$$\begin{array}{r}\$\ 67.92\\43.10\\+\ 496.34\\\hline \$607.36\end{array}$$
The total amount of the checks is $607.36.

35b. Strategy To find the new balance in your checking account, subtract the total amount of the checks written ($607.36) from the original balance ($1029.74).

Solution
$$\begin{array}{r}\$1029.74\\-\ 607.36\\\hline \$\ 422.38\end{array}$$
Your new balance is $422.38.

37. Strategy To find the projected increase in the annual number of births from 2003 to 2012, subtract the number of projected births in 2003 (3.98 million) from the number of projected births in 2012 (4.37 million).

Solution
$$\begin{array}{r}4.37\ \ \text{million}\\-3.98\ \ \text{million}\\\hline 0.39\ \ \text{million}\end{array}$$
The increase is 0.39 million births.

39. Strategy To find the growth in online shopping, subtract the number of households shopping online in 2000 (17.7 million) from the number of households shopping online in 2003 (40.3 million).

Solution
$$\begin{array}{r}40.3\\-17.7\\\hline 22.6\end{array}$$
The growth in online shopping from 2000 to 2003 is 22.6 million households.

Applying the Concepts

41a. Rounding to tenths, the largest difference between a decimal and the decimal rounded to tenths is 0.05. Example: For numbers between 3.7 and 3.8, (1) Any number between 3.7 and 3.75 (not including 3.75) is rounded to 3.7, so the largest difference is *less than* 0.05. (2) Any number between 3.75 and 3.8 (including 3.75) is rounded to 3.8, so the largest difference is *equal to* 0.05. Therefore, rounding to tenths, the largest amount by which the estimate of the sum of two decimals could differ from the exact sum is the sum of the largest differences for each decimal. $0.05 + 0.05 = 0.1$.

b. For hundredths, $0.005 + 0.005 = 0.01$.

c. For thousandths, $0.0005 + 0.0005 = 0.001$.

Section 3.4

Objective A Exercises

1.
$$\begin{array}{r}0.9\\\times\ 0.4\\\hline 0.36\end{array}$$

3.
$$\begin{array}{r}0.5\\\times\ 0.6\\\hline 0.30\end{array}$$

5.
$$\begin{array}{r}0.5\\\times\ 0.5\\\hline 0.25\end{array}$$

7.
$$\begin{array}{r}0.9\\\times\ 0.5\\\hline 0.45\end{array}$$

9.
$$\begin{array}{r}7.7\\\times\ 0.9\\\hline 6.93\end{array}$$

11. $\begin{array}{r} 9.2 \\ \times\ 0.2 \\ \hline 1.84 \end{array}$

13. $\begin{array}{r} 7.2 \\ \times\ 0.6 \\ \hline 4.32 \end{array}$

15. $\begin{array}{r} 7.4 \\ \times\ 0.1 \\ \hline 0.74 \end{array}$

17. $\begin{array}{r} 7.9 \\ \times\ 5 \\ \hline 39.5 \end{array}$

19. $\begin{array}{r} 0.68 \\ \times\ 4 \\ \hline 2.72 \end{array}$

21. $\begin{array}{r} 0.67 \\ \times\ 0.9 \\ \hline 0.603 \end{array}$

23. $\begin{array}{r} 0.16 \\ \times\ 0.6 \\ \hline 0.096 \end{array}$

25. $\begin{array}{r} 2.5 \\ \times\ 5.4 \\ \hline 13.50 \end{array}$

27. $\begin{array}{r} 8.4 \\ \times 9.5 \\ \hline 79.80 \end{array}$

29. $\begin{array}{r} 0.83 \\ \times\ 5.2 \\ \hline 166 \\ 415 \\ \hline 4.316 \end{array}$

31. $\begin{array}{r} 0.46 \\ \times\ 3.9 \\ \hline 414 \\ 138 \\ \hline 1.794 \end{array}$

33. $\begin{array}{r} 0.2 \\ \times\ 0.3 \\ \hline 0.06 \end{array}$

35. $\begin{array}{r} 0.24 \\ \times\ 0.3 \\ \hline 0.072 \end{array}$

37. $\begin{array}{r} 1.47 \\ \times\ 0.09 \\ \hline 0.1323 \end{array}$

39. $\begin{array}{r} 8.92 \\ \times\ 0.004 \\ \hline 0.03568 \end{array}$

41. $\begin{array}{r} 0.49 \\ \times\ 0.16 \\ \hline 294 \\ 49 \\ \hline 0.0784 \end{array}$

43. $\begin{array}{r} 7.6 \\ \times\ 0.01 \\ \hline 0.076 \end{array}$

45. $\begin{array}{r} 8.62 \\ \times\ 4 \\ \hline 34.48 \end{array}$

47. $\begin{array}{r} 64.5 \\ \times\ 9 \\ \hline 580.5 \end{array}$

49. $\begin{array}{r} 2.19 \\ \times\ 9.2 \\ \hline 438 \\ 1971 \\ \hline 20.148 \end{array}$

51. $\begin{array}{r} 1.85 \\ \times\ 0.023 \\ \hline 555 \\ 370 \\ \hline 0.04255 \end{array}$

53. $\begin{array}{r} 0.478 \\ \times\ 0.37 \\ \hline 3346 \\ 1434 \\ \hline 0.17686 \end{array}$

55. $\begin{array}{r} 48.3 \\ \times\ 0.0041 \\ \hline 483 \\ 1932 \\ \hline 0.19803 \end{array}$

57. $\begin{array}{r} 2.437 \\ \times\ 6.1 \\ \hline 2437 \\ 14622 \\ \hline 14.8657 \end{array}$

59. $\begin{array}{r} 0.413 \\ \times\ 0.0016 \\ \hline 2478 \\ 413 \\ \hline 0.0006608 \end{array}$

61. $\begin{array}{r} 94.73 \\ \times\ 0.57 \\ \hline 66311 \\ 47365 \\ \hline 53.9961 \end{array}$

63. $\begin{array}{r} 8.005 \\ \times\ 0.067 \\ \hline 56035 \\ 48030 \\ \hline 0.536335 \end{array}$

65. $\begin{array}{r} 4.29 \\ \times\ 0.1 \\ \hline 0.429 \end{array}$

67. $\begin{array}{r} 5.29 \\ \times\ 0.4 \\ \hline 2.116 \end{array}$

69. $\begin{array}{r} 0.68 \\ \times\ 0.7 \\ \hline 0.476 \end{array}$

71. $\begin{array}{r} 1.4 \\ \times\ 0.73 \\ \hline 42 \\ 98 \\ \hline 1.022 \end{array}$

73. $\begin{array}{r} 5.2 \\ \times\ 7.3 \\ \hline 156 \\ 364 \\ \hline 37.96 \end{array}$

75. $\begin{array}{r} 3.8 \\ \times\ 0.61 \\ \hline 38 \\ 228 \\ \hline 2.318 \end{array}$

77. $0.32 \times 10 = 3.2$

79. $0.065 \times 100 = 6.5$

81. $6.2856 \times 1000 = 6285.6$

83. $3.2 \times 1000 = 3200$

85. $3.57 \times 10{,}000 = 35{,}700$

87. $0.63 \times 10^1 = 6.3$

89. $0.039 \times 10^2 = 3.9$

91. $4.9 \times 10^4 = 49{,}000$

93. $0.067 \times 10^2 = 6.7$

95. $\begin{array}{r} 3.45 \\ \times\ 0.0035 \\ \hline 1725 \\ 1035 \\ \hline 0.012075 \end{array}$

97. $\begin{array}{r} 0.00392 \\ \times\ 3.005 \\ \hline 1960 \\ 1176 \\ \hline 0.01177960 \end{array}$
     or 0.0117796

99.
```
      1.348
   ×  0.23
      4044
     2696
   0.31004
```

101.
```
     23.67
   × 0.0035
     11835
     7101
   0.082845
```

103.
```
   0.45      2.25
   ×5       ×2.3
   2.25      675
            450
           5.175
```

105.
```
    28.5   ≈        30
  × 3.2   ≈       × 3
   Cal.: 91.2      Est.: 90
```

107.
```
     2.38   ≈         2
   × 0.44  ≈       × 0.4
   Cal.: 1.0472      Est.: 0.8
```

109.
```
    0.866   ≈        0.9
  ×  4.5   ≈       ×  5
   Cal.: 3.897      Est.: 4.5
```

111.
```
    4.34   ≈          4
  × 2.59  ≈        × 3
   Cal.: 11.2406     Est.: 12
```

113.
```
    8.434   ≈          8
  × 0.044  ≈       × 0.04
   Cal.: 0.371096    Est.: 0.32
```

115.
```
    28.44   ≈         30
  ×  1.12  ≈       × 1
   Cal.: 31.8528     Est.: 30
```

117.
```
    49.6854   ≈          50
  × 39.0672  ≈        × 40
   Cal.: 1941.069459    Est.: 2000
```

119.
```
     0.00456   ≈        0.005
  ×  0.009542  ≈      × 0.01
   Cal.: 0.0000435152   Est.: 0.00005
```

Objective B Exercises

121. **Strategy** To find the cost of operating the electric motor, multiply the hourly cost ($.027) by the number of hours it is used (56).

Solution
```
   $.027
   × 56
    162
    135
   $1.512 ≈ $1.51
```
The motor costs $1.51 to operate.

123a. **Strategy** To estimate the amount received, round each number so that all the digits are zero except the first digit, and then multiply.

Solution
```
    520    ≈       500
  × $.045  ≈     $ .05
                 $25.00
```
The estimated amount received is $25.00.

b. **Strategy** To find the payment for recycling the newspapers, multiply the number of pounds (520) by the payment per pound ($.045).

Solution
```
     520
   × $.045
    2600
   2080
   $23.400
```
The amount received was $23.40.

125. **Strategy** To find the area of a square, multiply the length (6.75 feet) by the width (3.5 feet).

Solution
```
    6.75
   × 3.5
   3375
   2025
   23.625
```
The area is 23.625 square feet.

127a. **Strategy** To find the amount of overtime pay, multiply the overtime rate ($43.35) by the number of hours worked (15).

Solution
```
   $43.35
   ×  15
   21675
   4335
   $650.25
```
The amount of overtime pay is $650.25.

b. **Strategy** To find the nurse's total income for the week, add the overtime pay ($650.25) to the salary ($1156).

Solution
```
   $ 650.25
   + 1156.00
   $1806.25
```
The nurse's total income is $1806.25.

129. **Strategy** To find cost, multiply the rate for post office to addressee for up to $\frac{1}{2}$ pound ($13.65) by the number of packages (25).

Solution
$$\begin{array}{r} \$13.65 \\ \times\ \ 25 \\ \hline 6825 \\ 2730 \\ \hline \$341.25 \end{array}$$
The cost is $341.25.

131a.
$$\begin{array}{r} 2.2 \\ \times 8 \\ \hline 17.6 \\ \times 1.2 \\ \hline 352 \\ 176 \\ \hline 21.12 \end{array}$$
The total cost of grade 1 is $21.12.

b.
$$\begin{array}{r} 3.4 \\ \times\ \ 6.5 \\ \hline 170 \\ 204 \\ \hline 22.10 \\ \times\ 1.35 \\ \hline 11050 \\ 6630 \\ 2210 \\ \hline 29.8350 \end{array}$$
The total cost of grade 2 is $29.84.

c.
$$\begin{array}{r} 6.75 \\ \times\ 15.4 \\ \hline 2700 \\ 3375 \\ 675 \\ \hline 103.950 \\ \times\ 1.94 \\ \hline 415800 \\ 935550 \\ 103950 \\ \hline 201.66300 \end{array}$$
The total cost of grade 3 is $201.66.

d.
Grade 1	$ 21.12
Grade 2	29.84
Grade 3	+ 201.66
Total :	$252.62

The total cost is $252.62.

133. Car 1 $\dfrac{360}{367,921} < 0.001$

Car 2 $\dfrac{420}{401,346} > 0.001$

Car 3 $\dfrac{210}{298,773} < 0.001$

Car 4 $\dfrac{320}{330,045} < 0.001$

Car 5 $\dfrac{450}{432,989} > 0.001$

Cars 2 and 5 would fail the test.

Applying the Concepts

135. When a number is multiplied by 10, 100, 1000, 10,000, etc., the decimal point is moved as many places to the right as there are zeros in the multiple of 10. For example, the decimal point in a number multiplied by 1000 would be moved three places to the right.

137. $1.3 = 1\dfrac{3}{10}$

$2.31 = 2\dfrac{31}{100}$

$1\dfrac{3}{10} \times 2\dfrac{31}{100} = \dfrac{13}{10} \times \dfrac{231}{100} = \dfrac{3003}{1000} = 3\dfrac{3}{1000} = 3.003$

Section 3.5

Objective A Exercises

1.
$$\begin{array}{r} 0.82 \\ 3\overline{)2.46} \\ -24 \\ \hline 06 \\ -\ 06 \\ \hline 0 \end{array}$$

3.
$$\begin{array}{r} 4.8 \\ 0.8.\overline{)3.8.4} \\ -32 \\ \hline 64 \\ -\ 64 \\ \hline 0 \end{array}$$

5.
$$\begin{array}{r} 89. \\ 0.7.\overline{)62.3.} \\ -56 \\ \hline 63 \\ -\ 63 \\ \hline 0 \end{array}$$

7.
$$\begin{array}{r} 60. \\ 0.4.\overline{)24.0.} \\ -24 \\ \hline 00 \\ -\ 0 \\ \hline 0 \end{array}$$

9.
```
        84.3
0.7.)59.0.1
     -56
      30
     -28
      21
     -21
       0
```

11.
```
        32.3
0.5.)16.1.5
     -15
      11
     -10
      15
     -15
       0
```

13.
```
        5.06
0.7.)3.5.42
     -35
      04
     - 0
      42
     -42
       0
```

15.
```
        1.3
6.3.)8.1.9
     -63
     189
    -189
       0
```

17.
```
        0.11
3.6.)0.3.96
     - 36
       36
      -36
        0
```

19.
```
        3.8
6.9.)26.2.2
     -207
      552
     -552
        0
```

21.
```
         6.32 ≈ 6.3
8.8.)55.6.20
     -528
      282
     -264
      180
     -176
        4
```

23.
```
        0.57 ≈ 0.6
9.5.)5.4.27
     -475
      677
     -665
       12
```

25.
```
         2.52 ≈ 2.5
7.3.)18.4.00
     -146
      380
     -365
      150
     -146
        4
```

27.
```
          1.07 ≈ 1.1
0.17.)0.18.30
      -17
       13
      - 0
      130
     -119
       11
```

29.
```
           130.64 ≈ 130.6
0.053.)6.924.00
       -53
       162
      -159
        34
       -0
       340
      -318
       220
      -212
         8
```

31.
```
       0.808 ≈ 0.81
8)6.467
  -64
   06
  - 0
   67
  -64
    3
```

33.
```
          0.089 ≈ 0.09
0.72.)0.06.470
      - 576
        710
      - 648
         62
```

35.
```
         40.70
0.95.)38.66.5
      -380
       66
      - 0
      665
     -665
        0
```

37.
```
          0.456 ≈ 0.46
60.8.)27.7.380
      - 2432
        3418
       -3040
        3780
       -3648
         132
```

39.
```
        0.0190 ≈ 0.019
54)1.0280
  - 54
    488
   -486
     20
    - 0
      0
```

41.
```
          0.0874 ≈ 0.087
0.5.)0.0.4370
     - 40
       37
      -35
       20
      -20
        0
```

43.
```
            0.3600 ≈ 0.360
95.3.)34.3.1000
      -2859
       5720
      -5718
         20
        - 0
        200
        - 0
        200
```

45.
```
           0.1031 ≈ 0.103
4.72.)0.48.7100
      - 472
        151
       - 0
       1510
      -1416
        940
       -472
        468
```

47.
$$
\begin{array}{r}
0.0086 \approx 0.009 \\
26.7.\overline{)0.2.3070} \\
-2136 \\
\hline
1710 \\
-1602 \\
\hline
108
\end{array}
$$

49.
$$
\begin{array}{r}
0.9 \approx 1 \\
90\overline{)89.76} \\
-810 \\
\hline
87
\end{array}
$$

51.
$$
\begin{array}{r}
2.5 \approx 3 \\
0.413.\overline{)1.047.8} \\
-826 \\
\hline
2218 \\
-2065 \\
\hline
153
\end{array}
$$

53.
$$
\begin{array}{r}
1.0 \approx 1 \\
0.778.\overline{)0.790.0} \\
-778 \\
\hline
120
\end{array}
$$

55.
$$
\begin{array}{r}
56.8 \approx 57 \\
6.9.\overline{)392.0.0} \\
-345 \\
\hline
470 \\
-414 \\
\hline
560 \\
-552 \\
\hline
8
\end{array}
$$

57. $4.07 \div 10 = 0.407$

59. $42.67 \div 10 = 4.267$

61. $1.037 \div 100 = 0.01037$

63. $8.295 \div 1000 = 0.008295$

65. $825.37 \div 1000 = 0.82537$

67. $0.32 \div 10^1 = 0.032$

69. $23.627 \div 10^2 = 0.23627$

71. $0.0053 \div 10^2 = 0.000053$

73. $1.8932 \div 10^3 = 0.0018932$

75.
$$
\begin{array}{r}
18.42 \\
2.4.\overline{)44.2.08} \\
-24 \\
\hline
202 \\
-192 \\
\hline
100 \\
-96 \\
\hline
48 \\
-48 \\
\hline
0
\end{array}
$$

77.
$$
\begin{array}{r}
16.07 \\
45\overline{)723.15} \\
-45 \\
\hline
273 \\
-270 \\
\hline
31 \\
-0 \\
\hline
315 \\
-315 \\
\hline
0
\end{array}
$$

79. $13.5 \div 10^3 = 0.0135$

81. $23.678 \div 1000 = 0.023678$

83.
$$
\begin{array}{r}
0.112 \\
0.05.\overline{)0.00.560} \\
-5 \\
\hline
06 \\
-5 \\
\hline
10 \\
-10 \\
\hline
0
\end{array}
$$

85. Cal.: $42.42 \div 3.8 = 11.1632$
Est.: $40 \div 4 = 10$

87. Cal.: $389 \div 0.44 = 884.0909$
Est.: $400 \div 0.4 = 1000$

89. Cal.: $6.394 \div 3.5 = 1.8269$
Est.: $6 \div 4 = 1.5$

91. Cal.: $1.235 \div 0.021 = 58.8095$
Est.: $1 \div 0.02 = 50$

93. Cal.: $95.443 \div 1.32 = 72.3053$
Est.: $100 \div 1 = 100$

95. Cal.: $1.000523 \div 429.07 = 0.0023$
Est.: $1 \div 400 = 0.0025$

Objective B Exercises

97. **Strategy** To find the number of yards per carry, divide the total number of yards (162) by the number of carries (26).

Solution
$$
\begin{array}{r}
6.230 \approx 6.23 \\
26\overline{)162.000} \\
-156 \\
\hline
60 \\
-52 \\
\hline
80 \\
-78 \\
\hline
20 \\
-0 \\
\hline
0
\end{array}
$$

6.23 yards are gained per carry.

99. **Strategy** To find the cost per can, divide the cost of a case ($6.79) by the number of cans in a case (24).

Solution

$$
\begin{array}{r}
.282 \approx \$.28 \\
24)\overline{\$6.79} \\
\underline{-48} \\
199 \\
\underline{-192} \\
70 \\
\underline{-48} \\
22
\end{array}
$$

The cost per can is $.28.

101. **Strategy** To find the cost per mile, divide the toll ($5.60) by the number of miles (136 miles).

Solution
$5.60 \div 136 = 0.041$
$0.041 \approx 0.04$
The cost per mile is $.04.

103. **Strategy** To find the monthly payment, divide the yearly premium ($703.80) by 12.

Solution

$$
\begin{array}{r}
\$58.65 \\
12)\overline{\$703.80} \\
\underline{-60} \\
103 \\
\underline{-96} \\
78 \\
\underline{-72} \\
60 \\
\underline{-60} \\
0
\end{array}
$$

The monthly payment is $58.65.

105. **Strategy** To find how many more women than men will be enrolled at institutions of higher learning in 2010, subtract the expected number of men (7.3 million) from the expected number of women (10.2 million).

Solution

$$
\begin{array}{r}
10.2 \quad \text{million} \\
-\ 7.3 \quad \text{million} \\
\hline
2.9 \quad \text{million}
\end{array}
$$

2.9 million more women are expected to be attending institutions of higher learning in 2010.

107. **Strategy** To find how many times greater the Army's advertising budget is than the Navy's advertising budget, divide the Army's budget ($85.3 million) by the Navy's budget ($20.5 million).

Solution
$85.3 \div 20.5 \approx 4.2$
The Army's advertising budget was 4.2 times greater than the Navy's.

109. **Strategy** To find how many times greater the population of 85 and over is expected to be in 2030 than in 2000, divide the expected population in 2030 (8.9 million) by the population in 2000 (4.2 million).

Solution
$8.9 \div 4.2 \approx 2.1$
The population of this segment is expected to be 2.1 times greater in 2030 than in 2000.

111. **Strategy** To find how many times greater the cigarette consumption was in 2000 then in 1960, divide the 2000 consumption (5.5 trillion) by the 1960 consumption (2.2 trillion).

Solution
$5.5 \div 2.2 = 2.5$
The cigarette consumption in 2000 was 2.5 times greater than in 1960.

113. **Strategy** To find the total number of acres burned in the 4 years, add the numbers given in the table.

Solution

$$
\begin{array}{r}
2.9 \quad \text{million} \\
1.5 \quad \text{million} \\
3.8 \quad \text{million} \\
\underline{1.5} \quad \text{million} \\
9.7 \quad \text{million}
\end{array}
$$

A total of 9.7 million acres was burned.

Applying the Concepts

115. When a number is divided by 10, 100, 1000, 10,000, etc., the decimal point is moved as many places to the left as there are zeros in the power of 10. For example, the decimal point in a number divided by 1000 would be moved three places to the left.

117. To determine where a decimal point is placed in a quotient, first move the decimal point in the divisor to make it a whole number. Then move the decimal point in the dividend the same number of places to the right. The decimal point in the quotient lines up vertically with the decimal point in the dividend.

119. $3.46 \times 0.24 = 0.8304$

121. $0.064 \times 1.6 = 0.1024$

123. $3.0381 \div 1.23 = 2.47$

125. 2.53

Section 3.6

Objective A Exercises

1.
$$\begin{array}{r} 0.625 \\ 8)\overline{5.000} \end{array}$$

3.
$$\begin{array}{r} 0.6666 \approx 0.667 \\ 3)\overline{2.0000} \end{array}$$

5.
$$\begin{array}{r} 0.1666 \approx 0.167 \\ 6)\overline{1.0000} \end{array}$$

7.
$$\begin{array}{r} 0.4166 \approx 0.417 \\ 12)\overline{5.0000} \end{array}$$

9.
$$\begin{array}{r} 1.750 \\ 4)\overline{7.000} \end{array}$$

11. $1\frac{1}{2} = \frac{3}{2};$
$$\begin{array}{r} 1.500 \\ 2)\overline{3.000} \end{array}$$

13.
$$\begin{array}{r} 4.000 \\ 4)\overline{16.000} \end{array}$$

15.
$$\begin{array}{r} 0.003 \\ 1000)\overline{3.000} \end{array}$$

17. $7\frac{2}{25} = \frac{177}{25};$
$$\begin{array}{r} 7.080 \\ 25)\overline{177.000} \end{array}$$

19. $37\frac{1}{2} = \frac{75}{2};$
$$\begin{array}{r} 37.500 \\ 2)\overline{75.000} \end{array}$$

21.
$$\begin{array}{r} 0.160 \\ 25)\overline{4.000} \end{array}$$

23. $8\frac{2}{5} = \frac{42}{5};$
$$\begin{array}{r} 8.400 \\ 5)\overline{42.000} \end{array}$$

Objective B Exercises

25. $0.8 = \frac{8}{10} = \frac{4}{5}$

27. $0.32 = \frac{32}{100} = \frac{8}{25}$

29. $0.125 = \frac{125}{1000} = \frac{1}{8}$

31. $1.25 = 1\frac{25}{100} = 1\frac{1}{4}$

33. $16.9 = 16\frac{9}{10}$

35. $8.4 = 8\frac{4}{10} = 8\frac{2}{5}$

37. $8.437 = 8\frac{437}{1000}$

39. $2.25 = 2\frac{25}{100} = 2\frac{1}{4}$

41. $0.15\frac{1}{3} = \frac{15\frac{1}{3}}{100} = 15\frac{1}{3} \div 100 = \frac{46}{3} \times \frac{1}{100}$
$= \frac{46}{300} = \frac{23}{150}$

43. $0.87\frac{7}{8} = \frac{87\frac{7}{8}}{100} = 87\frac{7}{8} \div 100 = \frac{703}{8} \times \frac{1}{100} = \frac{703}{800}$

45. $7.38 = 7\frac{38}{100} = 7\frac{19}{50}$

47. $0.57 = \frac{57}{100}$

49. $0.66\frac{2}{3} = \frac{66\frac{2}{3}}{100} = 66\frac{2}{3} \div 100 = \frac{200}{3} \times \frac{1}{100} = \frac{2}{3}$

Objective C Exercises

51. $0.6 > 0.45$

53. $3.89 < 3.98$

55. $0.025 < 0.105$

57. $\frac{4}{5} = 0.8$
$0.8 < 0.802$
$\frac{4}{5} < 0.802$

59. $\frac{7}{8} = 0.875$
$0.85 < 0.875$
$0.85 < \frac{7}{8}$

61. $\frac{7}{12} \approx 0.583$
$0.583 > 0.58$
$\frac{7}{12} > 0.58$

63. $\frac{11}{12} \approx 0.9167$
$0.9167 < 0.92$
$\frac{11}{12} < 0.92$

65. $0.623 > 0.6023$

67. $0.87 > 0.087$

69. $0.033 < 0.3$

Applying the Concepts

71. Strategy To find whether the population ages 0 to 19 is more or less than $\frac{1}{4}$ the total population, compare the population ages 0 to 19 (80.5 million) with $\frac{1}{4}$ of the sum of the population in all the classes.

Solution
80.5 million (0–19)
81.6 million (20–39)
73.6 million (40–59)
36.6 million (60–79)
<u>9.2</u> million (80–up)
281.5 million

$\frac{1}{4} \times 281.5$ million ≈ 70.4 million

80.5 million > 70.4 million
The population ages 0 to 19 is more than $\frac{1}{4}$ of the total population.

73. $\frac{7}{13} = 0.538461538461\ldots$

Yes, the digits 538461 repeat.

75. A terminating decimal ends, or stops. For example, 3.25 and 9.762104 are terminating decimals. A repeating decimal never ends. One or more digits to the right of the decimal point repeat without end. The decimals in the Optional Student Activity for Objective 3.6C are examples of repeating decimals: 0.111..., 0.222..., 0.333..., 0.444..., etc.
A nonrepeating decimal never ends nor does it have any digits to the right of the decimal point that repeat. For example, 1.20200200020000200000... is a nonrepeating decimal.

Chapter 3 Review Exercises

1.
$$\begin{array}{r} 54.5 \\ 0.067.\overline{)3.651.5} \\ -335 \\ \hline 301 \\ -268 \\ \hline 335 \\ -335 \\ \hline 0 \end{array}$$

2.
$$\begin{array}{r} {}^{2\,3\,1\ 1} \\ 369.41 \\ 88.3 \\ 9.774 \\ +366.474 \\ \hline 833.958 \end{array}$$

3. $0.055 < 0.1$

4. Twenty-two and ninety-two ten-thousandths

5. Given place value
0.05678235
$2 < 5$
0.05678

6. $2\frac{1}{3} = \frac{7}{3}$; $\begin{array}{r} 2.333 \approx 2.33 \\ 3\overline{)7.000} \end{array}$

7. $0.375 = \frac{375}{1000} = \frac{3}{8}$

8.
$$\begin{array}{r} {}^{1\ 1} \\ 3.42 \\ 0.794 \\ +32.5 \\ \hline 36.714 \end{array}$$

9. 34.025

10. $\frac{5}{8} = 0.625$, $\frac{5}{8} > 0.62$

11. $\begin{array}{r} 0.7777 \approx 0.778 \\ 9\overline{)7.0000} \end{array}$

12. $0.66\frac{2}{3} = \frac{66\frac{2}{3}}{100} = 66\frac{2}{3} \div 100 = \frac{200}{3} \div 100$
$= \frac{200}{3} \times \frac{1}{100} = \frac{2}{3}$

13.
$$\begin{array}{r} {}^{6\ \ 2\ 10\ 10\ 10} \\ 27.3100 \\ -4.4465 \\ \hline 22.8635 \end{array}$$

14. Given place value
7.93704
$7 > 5$
7.94

15.
$$\begin{array}{r} 3.08 \\ \times\ 2.9 \\ \hline 2772 \\ 616 \\ \hline 8.932 \end{array}$$

16. Three hundred forty-two and thirty-seven hundredths

17. 3.06753

18.
$$\begin{array}{r} 34.79 \\ \times\ 0.74 \\ \hline 13916 \\ 24353 \\ \hline 25.7446 \end{array}$$

19.

$$
\begin{array}{r}
6.594 \\
0.053\overline{)0.349.482} \\
-318 \\
\hline
314 \\
-265 \\
\hline
498 \\
-477 \\
\hline
212 \\
-212 \\
\hline
0
\end{array}
$$

20.

$$
\begin{array}{r}
{}^{6}\ {}^{17}8\ {}^{8}_{}{}^{15}_{}10 \\
7.7\,9\,6\,\cancel{0} \\
-2.9\,1\,7\,5 \\
\hline
4.8\,7\,8\,5
\end{array}
$$

21. Strategy To find the new balance in your checking account:
● Find the total amount of the checks by adding the check amounts ($145.72 and $88.45).
● Subtract the total check amounts from the original balance ($895.68).

Solution

$$
\begin{array}{r}
\$145.72 \\
+\ \ 88.45 \\
\hline
\$234.17
\end{array}
\qquad
\begin{array}{r}
\$895.68 \\
-\ \ 234.17 \\
\hline
\$661.51
\end{array}
$$

The new balance in your account is $661.51.

22. Strategy To find the total number of children, add the number of children in public school (46.353 million), in private school (5.863 million), and in home-schooling (1.23 million).

Solution

$$
\begin{array}{r}
46.353 \\
5.863 \\
+\ \ 1.23 \\
\hline
53.446
\end{array}
$$

There are 53.446 million children in grades K–12.

23. Strategy To find the difference in the number of children, subtract the number of children in private school (5.863 million) from the number of children in public school (46.353 million).

Solution

$$
\begin{array}{r}
46.353 \\
-\ \ 5.863 \\
\hline
40.490
\end{array}
$$

There are 40.49 million more children in public school than in private school.

24. Strategy To find the amount of milk served during a 5-day school week, multiply the amount of milk served daily (1.9 million gallons) by 5 days.

Solution

$$
\begin{array}{r}
1.9 \\
\times 5 \\
\hline
9.5
\end{array}
$$

During a 5-day school week, 9.5 million gallons of milk are served.

25. Strategy To find how many times greater the number who drove (30.6 million) was than the number who flew (4.8 million), divide the number who drove by the number who flew.

Solution $30.6 \div 4.8 \approx 6.4$
The number who drove is 6.4 times greater than the number who flew.

Chapter 3 Test

1. $0.66 < 0.666$

2.

$$
\begin{array}{r}
{}^{2}\ {}^{9}_{}{}^{10}12 \\
1\,\cancel{3}.\cancel{0}\,2\,7 \\
-\ 8.9\,4\,0 \\
\hline
4.0\,8\,7
\end{array}
$$

3. Forty-five and three hundred two ten-thousandths

4. $\dfrac{9}{13} = \begin{array}{r} 0.6923 \approx 0.692 \\ 13\overline{)9.0000} \end{array}$

5. $0.825 = \dfrac{825}{1000} = \dfrac{33}{40}$

6.

┌ *Given place value*
0.07395
　└ 5 = 5
0.0740

7.

$$
\begin{array}{r}
1.5378 \approx 1.538 \\
0.037\overline{)0.056.9000} \\
-37 \\
\hline
199 \\
-185 \\
\hline
140 \\
-111 \\
\hline
290 \\
-259 \\
\hline
310 \\
-296 \\
\hline
14
\end{array}
$$

8.

$$
\begin{array}{r}
{}^{16}\ {}^{9}\ {}^{9}_{}10\,10\ {}^{2}_{}10\ {}^{9}_{}10 \\
3\,7.\cancel{0}\,\cancel{0}\,\cancel{3}\,\cancel{0}\,\cancel{0} \\
-9.2\,3\,6\,7\,4 \\
\hline
2\,7.7\,6\,6\,2\,6
\end{array}
$$

9.

┌── *Given place value*
7.0954625
　└── 4 < 5
7.095

10.

```
            232.
0.006.)1.392.
       -12
         19
        -18
         12
        -12
          0
```

11.

```
 212 11
270.93
 97.
  1.976
+ 88.675
458.581
```

12. Strategy To find the missing dimension, subtract the given length (4.86 inches) from the total length (6.23 inches).

Solution
```
  6.23
- 4.86
  1.37
```
The missing dimension is 1.37 inches.

13.
```
    1.37
×  0.004
 0.00548
```

14.
```
  11
  62.3
   4.007
+189.65
 255.957
```

15. 209.07086

16. Strategy To find the amount of each payment:
• Find the total amount to be paid by subtracting the down payment ($2500) from the cost of the car ($16,734.40).
• Divide the amount remaining to be paid by the number of payments (36).

Solution
```
$16,734.40
-  2,500.00
$14,234.40
```

```
         $395.40
36)$14,234.40
   -108
    343
   -324
    194
   -180
    144
   -144
     00
    - 0
      0
```
Each payment is $395.40.

17. Strategy To find your total income, add the salary ($727.50), commission ($1909.64), and bonus ($450).

Solution
```
$ 727.50
 1909.64
+ 450.00
$3087.14
```
Your total income is $3087.14.

18. Strategy To find the cost of the 12-minute call:
• Find the number of additional minutes charged above the 3-minute base by subtracting the base (3 minutes) from the total call length (12 minutes).
• Multiply the number of additional minutes by the rate ($.42).
• Add the charge for additional minutes to the base rate ($.85).

Solution 12 − 3 = 9
```
 $.42      $3.78
 × 9       + .85
$3.78      $4.63
```
The cost of the call is $4.63.

19. Strategy To find average hours per year, multiply the weekly computer use by a 10th-grade student (6.7 hours) by 52 weeks.

Solution
```
   6.7
 × 52
  134
  335
 348.4
```
The yearly average computer use by a 10th-grade student is 348.4 hours.

20. Strategy To find how many more hours a 2nd-grade student uses a computer than a 5th-grade student:
• Subtract the number of hours the 5th-grade student uses the computer (4.2) from the number of hours the 2nd-grade student uses the computer (4.9).
• Multiply the difference by 52 weeks.

Solution
```
 4.9       52
-4.2      × 0.7
 0.7      36.4
```
On average a 2nd-grade student uses the computer 36.4 hours more per year than a 5th-grade student.

Cumulative Review Exercises

1.
$$
\begin{array}{r}
235 \text{ r}17 \\
89)\overline{20932} \\
-178 \\
\hline
313 \\
-267 \\
\hline
462 \\
-445 \\
\hline
17
\end{array}
$$

2. $2^3 \cdot 4^2 = 8 \cdot 16 = 128$

3. $2^2 - (7 - 3) \div 2 + 1$
$4 - 4 \div 2 + 1$
$4 - 2 + 1$
3

4.

$$
\begin{array}{c|c|c}
 & 2 & 3 \\
\hline
9 = & & \fbox{3 \cdot 3} \\
12 = & 2 \cdot 2 & 3 \\
24 = & \fbox{2 \cdot 2 \cdot 2} & 3 \\
\end{array}
$$

$\text{LCM} = 2 \cdot 2 \cdot 2 \cdot 3 \cdot 3 = 72$

5.
$$
\frac{22}{5} = \begin{array}{r} 4\text{ r}2 \\ 5)\overline{22} \\ -20 \\ \hline 2 \end{array} = 4\frac{2}{5}
$$

6. $4\frac{5}{8} = \frac{32 + 5}{8} = \frac{37}{8}$

7. $\frac{5 \cdot 5}{12 \cdot 5} = \frac{25}{60}$

8.
$$
\begin{array}{r}
\dfrac{3}{8} = \dfrac{18}{48} \\[6pt]
\dfrac{5}{12} = \dfrac{20}{48} \\[6pt]
+\dfrac{9}{16} = \dfrac{27}{48} \\[6pt]
\hline
\dfrac{65}{48} = 1\dfrac{17}{48}
\end{array}
$$

9.
$$
\begin{array}{r}
5\dfrac{7}{12} = 5\dfrac{21}{36} \\[6pt]
+3\dfrac{7}{18} = 3\dfrac{14}{36} \\[6pt]
\hline
8\dfrac{35}{36}
\end{array}
$$

10.
$$
\begin{array}{r}
9\dfrac{5}{9} = 9\dfrac{20}{36} = 8\dfrac{56}{36} \\[6pt]
-3\dfrac{11}{12} = 3\dfrac{33}{36} = 3\dfrac{33}{36} \\[6pt]
\hline
5\dfrac{23}{36}
\end{array}
$$

11. $\dfrac{9}{16} \times \dfrac{4}{27} = \dfrac{9 \times 4}{16 \times 27} = \dfrac{\overset{1}{\cancel{3}} \cdot \overset{1}{\cancel{3}} \cdot \overset{1}{\cancel{2}} \cdot \overset{1}{\cancel{2}}}{\underset{1}{\cancel{2}} \cdot \underset{1}{\cancel{2}} \cdot 2 \cdot 2 \cdot \underset{1}{\cancel{3}} \cdot \underset{1}{\cancel{3}} \cdot 3} = \dfrac{1}{12}$

12. $2\dfrac{1}{8} \times 4\dfrac{5}{17} = \dfrac{17}{8} \times \dfrac{73}{17} = \dfrac{17 \cdot 73}{8 \cdot 17} = \dfrac{73}{8} = 9\dfrac{1}{8}$

13. $\dfrac{11}{12} \div \dfrac{3}{4} = \dfrac{11}{12} \times \dfrac{4}{3} = \dfrac{11 \cdot 4}{12 \cdot 3} = \dfrac{11 \cdot \overset{1}{\cancel{2}} \cdot \overset{1}{\cancel{2}}}{\underset{1}{\cancel{2}} \cdot \underset{1}{\cancel{2}} \cdot 3 \cdot 3} = \dfrac{11}{9} = 1\dfrac{2}{9}$

14. $2\dfrac{3}{8} \div 2\dfrac{1}{2} = \dfrac{19}{8} \div \dfrac{5}{2} = \dfrac{19}{8} \times \dfrac{2}{5} = \dfrac{19 \cdot 2}{8 \cdot 5}$
$\qquad = \dfrac{19 \cdot \overset{1}{\cancel{2}}}{2 \cdot 2 \cdot \underset{1}{\cancel{2}} \cdot 5} = \dfrac{19}{20}$

15. $\left(\dfrac{2}{3}\right)^2 \left(\dfrac{3}{4}\right)^3 = \left(\dfrac{2}{3} \cdot \dfrac{2}{3}\right)\left(\dfrac{3}{4} \cdot \dfrac{3}{4} \cdot \dfrac{3}{4}\right)$
$\qquad = \dfrac{2 \cdot 2 \cdot \overset{1}{\cancel{3}} \cdot \overset{1}{\cancel{3}} \cdot \overset{1}{\cancel{3}}}{\underset{1}{\cancel{3}} \cdot \underset{1}{\cancel{3}} \cdot \underset{1}{\cancel{2}} \cdot 2 \cdot \underset{1}{\cancel{2}} \cdot 2 \cdot 2 \cdot 2} = \dfrac{3}{16}$

16. $\left(\dfrac{2}{3}\right)^2 - \left(\dfrac{2}{3} - \dfrac{1}{2}\right) + 2$
$= \left(\dfrac{2}{3} \cdot \dfrac{2}{3}\right) - \left(\dfrac{4}{6} - \dfrac{3}{6}\right) + 2$
$= \dfrac{4}{9} - \dfrac{1}{6} + 2$
$= \dfrac{8}{18} - \dfrac{3}{18} + \dfrac{36}{18}$
$= \dfrac{41}{18} = 2\dfrac{5}{18}$

17. Sixty-five and three hundred nine ten-thousandths

18.
$$
\begin{array}{r}
{}^{2\,3\,1}{}^{1\,1\,1} \\
379.006 \\
27.523 \\
9.8707 \\
+\;88.2994 \\
\hline
504.6991
\end{array}
$$

19.
$$
\begin{array}{r}
{}^{\;\;8}\,{}^{9}\!{}^{\;9}\!{}^{14} \\
2\cancel{9}.\cancel{0}\cancel{0}\cancel{5}\cancel{0} \\
-\;7.9\,2\,8\,6 \\
\hline
21.0\,7\,6\,4
\end{array}
$$

20.
$$
\begin{array}{r}
9.074 \\
\times\;\;\;6.09 \\
\hline
81666 \\
544440 \\
\hline
55.26066
\end{array}
$$

21.
$$
\begin{array}{r}
2.1544 \approx 2.154 \\
8.09)\overline{17.42.9630} \\
-1618 \\
\hline
1249 \\
-809 \\
\hline
4406 \\
-4045 \\
\hline
3613 \\
-3236 \\
\hline
3770 \\
-3236 \\
\hline
534
\end{array}
$$

22.
$$
\frac{11}{15} = \begin{array}{r} 0.7333 \approx 0.733 \\ 15)\overline{11.000} \end{array}
$$

23.
$$0.16\frac{2}{3} = \frac{16\frac{2}{3}}{100} = \frac{\frac{50}{3}}{100} = \frac{50}{3} \div 100 = \frac{50}{3} \times \frac{1}{100}$$
$$= \frac{50}{300} = \frac{1}{6}$$

24. $\frac{8}{9} \approx 0.89, \frac{8}{9} < 0.98$

25. Strategy To find how many more vacation days are mandated in Sweden than in Germany, subtract the number of days mandated in Germany (18) from the number of days mandated in Sweden (32).

Solution
$$\begin{array}{r} 32 \\ -18 \\ \hline 14 \end{array}$$

Sweden mandates 14 days more vacation than Germany.

26. Strategy To find the loss needed the third month:
- Add the losses for the first two months.
- Subtract this sum from the goal (24 pounds).

Solution
$$9\frac{1}{2} + 6\frac{3}{4} = 9\frac{2}{4} + 6\frac{3}{4} = 15\frac{5}{4} = 16\frac{1}{4}$$
pounds lost first two months
$$24 - 16\frac{1}{4} = 23\frac{4}{4} - 16\frac{1}{4} = 7\frac{3}{4} \text{ pounds}$$

The patient must lose $7\frac{3}{4}$ pounds the third month to achieve the goal.

27. Strategy To find your balance after you write the checks:
- Find the total of the checks written by adding the amounts of the checks ($42.98, $16.43, and $137.56).
- Subtract the total of the checks written from the original balance ($814.35).

Solution
$$\begin{array}{r} \$\ 42.98 \\ 16.43 \\ +137.56 \\ \hline \$196.97 \end{array} \qquad \begin{array}{r} \$814.35 \\ -196.97 \\ \hline \$617.38 \end{array}$$

Your checking account balance is $617.38.

28. Strategy To find the resulting thickness, subtract the amount removed (0.017 inch) from the original thickness (1.412 inches).

Solution
$$\begin{array}{r} 1.412 \\ -0.017 \\ \hline 1.395 \end{array}$$
The resulting thickness is 1.395 inches.

29. Strategy To find the amount of income tax you paid:
- Find the amount of tax paid on profit by multiplying the profit ($64,860) by the rate (0.08).
- Add the amount of tax paid on profit to the base tax ($820).

Solution
$$\begin{array}{r} \$64,860 \\ \times\ \ 0.08 \\ \hline \$5188.80 \end{array} \qquad \begin{array}{r} \$5188.80 \\ +\ 820.00 \\ \hline \$6008.80 \end{array}$$

You paid $6008.80 in income tax last year.

30. Strategy To find the amount of the monthly payment:
- Find the amount to be paid in payments by subtracting the down payment ($20) from the cost ($210.96).
- Divide the amount to be paid in payments by the number of payments (8).

Solution
$$\begin{array}{r} \$210.96 \\ -20.00 \\ \hline \$190.96 \end{array}$$

$$\begin{array}{r} \$23.87 \\ 8)\overline{\$190.96} \\ -16 \\ \hline 30 \\ -24 \\ \hline 69 \\ -64 \\ \hline 56 \\ -56 \\ \hline 0 \end{array}$$

The amount of each payment is $23.87.

Chapter 4: Ratio and Proportion

Prep Test

1. $\dfrac{8}{10} = \dfrac{\overset{1}{2} \cdot 2 \cdot 2}{\underset{1}{2} \cdot 5} = \dfrac{4}{5}$

2. $\dfrac{450}{650 + 250} = \dfrac{450}{900} = \dfrac{\overset{1}{\cancel{450}}}{2 \cdot \underset{1}{\cancel{450}}} = \dfrac{1}{2}$

3. $15\overline{)372.0} \;\; 24.8$

4. $4 \times 33 = 132$
   $62 \times 2 = 124$
   $132 > 124$
   4×33 is greater.

5. $5\overline{)20} \;\; 4$

Go Figure

From the third statement, we know that the order of three of the four men is either Luis, Kim, and Reggie or Reggie, Kim, and Luis. From the fourth statement, Luis is standing between Dave and Kim. So building from what we already know, the order is either Dave, Luis, Kim, and Reggie or Reggie, Kim, Luis, and Dave. From the second statement, we know that Dave is not first. Therefore, the order is Reggie, Kim, Luis, and Dave.

Section 4.1

Objective A Exercises

1. $\dfrac{3 \text{ pints}}{15 \text{ pints}} = \dfrac{3}{15} = \dfrac{1}{5}$
   3 pints:15 pints = 3:15 = 1:5
   3 pints to 15 pints = 3 to 15 = 1 to 5

3. $\dfrac{\$40}{\$20} = \dfrac{40}{20} = \dfrac{2}{1}$
   $40:$20 = 40:20 = 2:1
   $40 to $20 = 40 to 20 = 2 to 1

5. $\dfrac{3 \text{ miles}}{8 \text{ miles}} = \dfrac{3}{8}$
   3 miles:8 miles = 3:8
   3 miles to 8 miles = 3 to 8

7. $\dfrac{37 \text{ hours}}{24 \text{ hours}} = \dfrac{37}{24}$
   37 hours:24 hours = 37:24
   37 hours to 24 hours = 37 to 24

9. $\dfrac{6 \text{ minutes}}{6 \text{ minutes}} = \dfrac{6}{6} = \dfrac{1}{1}$
   6 minutes:6 minutes = 6:6 = 1:1
   6 minutes to 6 minutes = 6 to 6 = 1 to 1

11. $\dfrac{35 \text{ cents}}{50 \text{ cents}} = \dfrac{35}{50} = \dfrac{7}{10}$
    35 cents:50 cents = 35:50 = 7:10
    35 cents to 50 cents = 35 to 50 = 7 to 10

13. $\dfrac{30 \text{ minutes}}{60 \text{ minutes}} = \dfrac{30}{60} = \dfrac{1}{2}$
    30 minutes:60 minutes = 30:60 = 1:2
    30 minutes to 60 minutes = 30 to 60 = 1 to 2

15. $\dfrac{32 \text{ ounces}}{16 \text{ ounces}} = \dfrac{32}{16} = \dfrac{2}{1}$
    32 ounces:16 ounces = 32:16 = 2:1
    32 ounces to 16 ounces = 32 to 16 = 2 to 1

17. $\dfrac{3 \text{ cups}}{4 \text{ cups}} = \dfrac{3}{4}$
    3 cups:4 cups = 3:4
    3 cups to 4 cups = 3 to 4

19. $\dfrac{\$5}{\$3} = \dfrac{5}{3}$
    $5:$3 = 5:3
    $5 to $3 = 5 to 3

21. $\dfrac{12 \text{ quarts}}{18 \text{ quarts}} = \dfrac{12}{18} = \dfrac{2}{3}$
    12 quarts:18 quarts = 12:18 = 2:3
    12 quarts to 18 quarts = 12 to 18 = 2 to 3

23. $\dfrac{14 \text{ days}}{7 \text{ days}} = \dfrac{14}{7} = \dfrac{2}{1}$
    14 days:7 days = 14:7 = 2:1
    14 days to 7 days = 14 to 7 = 2 to 1

Objective B Exercises

25. **Strategy** To find the ratio, write the ratio of housing ($1600) to total expenses ($4800) in simplest form.

    Solution $\dfrac{\$1600}{\$4800} = \dfrac{1600}{4800} = \dfrac{1}{3}$
    The ratio is $\dfrac{1}{3}$.

27. **Strategy** To find the ratio, write the ratio of utilities ($300) to food ($800) in simplest form.

    Solution $\dfrac{\$300}{\$800} = \dfrac{300}{800} = \dfrac{3}{8}$
    The ratio is $\dfrac{3}{8}$.

29. Strategy To find the ratio, write in simplest form the number of college freshmen playing basketball over the number of high school seniors playing basketball.

Solution $\dfrac{4000}{154,000} = \dfrac{2}{77}$

The ratio is $\dfrac{2}{77}$.

31. Strategy To find the ratio, write the ratio of turns in the primary coil (40) to the number of turns in the secondary coil (480) in simplest form.

Solution $\dfrac{40}{480} = \dfrac{1}{12}$

The ratio is $\dfrac{1}{12}$.

33a. Strategy To find the amount of the increase, subtract the original value ($90,000) from the increased value ($110,000).

Solution
$\begin{array}{r} \$110,000 \\ -\ \ 90,000 \\ \hline \$20,000 \end{array}$

The amount of the increase is $20,000.

b. Strategy To find the ratio, write the ratio of the increase ($20,000) to the original value ($90,000) in simplest form.

Solution $\dfrac{\$20,000}{\$90,000} = \dfrac{20,000}{90,000} = \dfrac{2}{9}$

The ratio is $\dfrac{2}{9}$.

Applying the Concepts

35. Income = $5500 + $450 + $250 = $6200
Debts = $1200 + $300 + $450 + $250 = $2200
$\dfrac{\$2200}{\$6200} = \dfrac{11}{31}$

The ratio is $\dfrac{11}{31}$.

37. Income = $3400 + $83 + $650 + $34 = $4167
Debts = $1800 + $104 + $35 + $120 + $234 + $197 = $2490

No, the ratio $= \dfrac{\$2490}{\$4167} = \dfrac{830}{1389} \approx 0.5976$, which is greater than $\dfrac{2}{5}$ (0.4).

Section 4.2

Objective A Exercises

1. $\dfrac{3 \text{ pounds}}{4 \text{ people}}$

3. $\dfrac{\$80}{12 \text{ boards}} = \dfrac{\$20}{3 \text{ boards}}$

5. $\dfrac{300 \text{ miles}}{15 \text{ gallons}} = \dfrac{20 \text{ miles}}{1 \text{ gallon}}$

7. $\dfrac{20 \text{ children}}{8 \text{ families}} = \dfrac{5 \text{ children}}{2 \text{ families}}$

9. $\dfrac{16 \text{ gallons}}{2 \text{ hours}} = \dfrac{8 \text{ gallons}}{1 \text{ hour}}$

Objective B Exercises

11. $\dfrac{10 \text{ feet}}{4 \text{ seconds}} = 2.5 \text{ feet/second}$

13. $\dfrac{\$3900}{4 \text{ weeks}} = \$975/\text{week}$

15. $\dfrac{1100 \text{ trees}}{10 \text{ acres}} = 110 \text{ trees/acre}$

17. $\dfrac{\$131.88}{7 \text{ hours}} = \$18.84/\text{hour}$

19. $\dfrac{628.8 \text{ miles}}{12 \text{ hours}} = 52.4 \text{ miles/hour}$

21. $\dfrac{344.4 \text{ miles}}{12.3 \text{ gallons}} = 28 \text{ miles/gallon}$

23. $\dfrac{\$349.80}{212 \text{ pounds}} = \$1.65/\text{pound}$

Objective C Exercises

25. Strategy To find the number of miles driven per gallon of gas, divide the total number of miles (326.6) by the total number of gallons (11.5).

Solution $11.5\overline{)326.6} = 28.4$

The gas mileage was 28.4 miles/gallon.

27. Strategy To find how much fuel the rocket uses in 1 minute, divide the total fuel (534,000 gallons) by the number of minutes (2.5).

Solution $2.5\overline{)534,000} = 213,600$

The rocket uses 213,600 gallons/minute.

29a. Strategy To find how many pounds of beef were packaged, subtract the waste (75 pounds) from the original weight (250 pounds).

Solution
$\begin{array}{r} 250 \\ -\ 75 \\ \hline 175 \end{array}$

175 pounds of beef was packaged.

b. Strategy To find the cost per pound of the packaged beef, divide the total cost ($365.75) by the weight of the packaged beef (175 pounds).

Solution

$$\frac{\$2.09}{175)\$365.75}$$

The beef cost $2.09/pound.

31a. Strategy To find the rate per minute, multiply the rate (5.6 feet per second) by 60 seconds per minute.

Solution $\frac{5.6 \text{ ft}}{1 \text{ sec}} \times \frac{60 \text{ sec}}{1 \text{ min}} = 336 \text{ ft/min}$

The camera goes through film at the rate of 336 feet/minute.

b. Strategy To find how fast the camera uses a 500-foot roll, divide the length of the roll (500 feet) by the rate that it is used (5.6 feet per second).

Solution

$$\frac{89.2}{5.6)500}$$

$89.2 \approx 89$

The camera uses the film at a rate of 89 seconds/roll.

33a. Strategy To find the price of the computer hardware in euros, multiply the price ($120,000) by the euro exchange rate (0.9103 euros per U.S. dollar).

Solution $\frac{\$120,000}{1} \times \frac{0.9103 \text{ euros}}{\$1} = 109,236 \text{ euros}$

The price of the computer hardware would be 109,236 euros.

b. Strategy To find the price of a car in yen, multiply the price ($34,000) by the Japanese yen exchange rate (117 yen per U.S. dollar).

Solution $\frac{\$34,000}{1} \times \frac{117 \text{ yen}}{\$1} = 3,978,000 \text{ yen}$

The price of the car would be 3,978,000 yen.

Applying the Concepts

35. The price–earnings ratio of a company's stock is computed by dividing the current price per share of the stock by the annual earnings per share. For example, if the price–earnings ratio of a company's stock is 8.5, the price of the stock is 8.5 times the earnings per share of the stock.

Section 4.3

Objective A Exercises

1. $\frac{4}{8} \diagup\!\!\!\!\diagdown \frac{10}{20} \rightarrow 8 \times 10 = 80$
$\rightarrow 4 \times 20 = 80$
The proportion is true.

3. $\frac{7}{8} \diagup\!\!\!\!\diagdown \frac{11}{12} \rightarrow 8 \times 11 = 88$
$\rightarrow 7 \times 12 = 84$
The proportion is not true.

5. $\frac{27}{8} \diagup\!\!\!\!\diagdown \frac{9}{4} \rightarrow 8 \times 9 = 72$
$\rightarrow 27 \times 4 = 108$
The proportion is not true.

7. $\frac{45}{135} \diagup\!\!\!\!\diagdown \frac{3}{9} \rightarrow 135 \times 3 = 405$
$\rightarrow 45 \times 9 = 405$
The proportion is true.

9. $\frac{16}{3} \diagup\!\!\!\!\diagdown \frac{48}{9} \rightarrow 3 \times 48 = 144$
$\rightarrow 16 \times 9 = 144$
The proportion is true.

11. $\frac{7}{40} \diagup\!\!\!\!\diagdown \frac{7}{8} \rightarrow 40 \times 7 = 280$
$\rightarrow 7 \times 8 = 56$
The proportion is not true.

13. $\frac{50}{2} \diagup\!\!\!\!\diagdown \frac{25}{1} \rightarrow 2 \times 25 = 50$
$\rightarrow 50 \times 1 = 50$
The proportion is true.

15. $\frac{6}{5} \diagup\!\!\!\!\diagdown \frac{30}{25} \rightarrow 5 \times 30 = 150$
$\rightarrow 6 \times 25 = 150$
The proportion is true.

17. $\frac{15}{4} \diagup\!\!\!\!\diagdown \frac{45}{12} \rightarrow 4 \times 45 = 180$
$\rightarrow 15 \times 12 = 180$
The proportion is true.

19. $\frac{300}{4} \diagup\!\!\!\!\diagdown \frac{450}{7} \rightarrow 4 \times 450 = 1800$
$\rightarrow 300 \times 7 = 2100$
The proportion is not true.

21. $\frac{65}{5} \diagup\!\!\!\!\diagdown \frac{26}{2} \rightarrow 5 \times 26 = 130$
$\rightarrow 65 \times 2 = 130$
The proportion is true.

23. $\frac{7}{4} \diagup\!\!\!\!\diagdown \frac{42}{20} \rightarrow 4 \times 42 = 168$
$\rightarrow 7 \times 20 = 140$
The proportion is not true.

Objective B Exercises

25. $n \times 8 = 4 \times 6$
$n \times 8 = 24$
$n = 24 \div 8$
$n = 3$

27. $12 \times 9 = 18 \times n$
$108 = 18 \times n$
$108 \div 18 = n$
$6 = n$

29.
$$6 \times 36 = n \times 24$$
$$216 = n \times 24$$
$$216 \div 24 = n$$
$$9 = n$$

31.
$$n \times 135 = 45 \times 17$$
$$n \times 135 = 765$$
$$n = 765 \div 135$$
$$n \approx 5.67$$

33.
$$n \times 3 = 6 \times 2$$
$$n \times 3 = 12$$
$$n = 12 \div 3$$
$$n = 4$$

35.
$$n \times 8 = 5 \times 7$$
$$n \times 8 = 35$$
$$n = 35 \div 8$$
$$n \approx 4.38$$

37.
$$n \times 4 = 11 \times 32$$
$$n \times 4 = 352$$
$$n = 352 \div 4$$
$$n = 88$$

39.
$$5 \times 8 = 12 \times n$$
$$40 = 12 \times n$$
$$40 \div 12 = n$$
$$3.33 \approx n$$

41.
$$n \times 12 = 15 \times 21$$
$$n \times 12 = 315$$
$$n = 315 \div 12$$
$$n = 26.25$$

43.
$$32 \times 3 = n \times 1$$
$$96 = n \times 1$$
$$96 \div 1 = n$$
$$96 = n$$

45.
$$18 \times n = 11 \times 16$$
$$18 \times n = 176$$
$$n = 176 \div 18$$
$$n \approx 9.78$$

47.
$$28 \times n = 8 \times 12$$
$$28 \times n = 96$$
$$n = 96 \div 28$$
$$n \approx 3.43$$

49.
$$0.3 \times 25 = 5.6 \times n$$
$$7.5 = 5.6 \times n$$
$$7.5 \div 5.6 = n$$
$$1.34 \approx n$$

51.
$$0.7 \times n = 9.8 \times 3.6$$
$$0.7 \times n = 35.28$$
$$n = 35.28 \div 0.7$$
$$n = 50.4$$

Objective C Exercises

53. Strategy To find out how many calories are in a 0.5-ounce serving of cereal, write and solve a proportion using n to represent the calories.

Solution
$$\frac{6 \text{ ounces}}{600 \text{ calories}} = \frac{0.5 \text{ ounces}}{n \text{ calories}}$$
$$6 \times n = 600 \times 0.5$$
$$6 \times n = 300$$
$$n = 300 \div 6$$
$$n = 50$$
A 0.5-ounce serving contains 50 calories.

55. Strategy To find out how many pounds of fertilizer are used, write and solve a proportion using n to represent the pounds of fertilizer.

Solution
$$\frac{2 \text{ pounds}}{100 \text{ square feet}} = \frac{n \text{ pounds}}{3500 \text{ square feet}}$$
$$2 \times 3500 = 100 \times n$$
$$7000 = 100 \times n$$
$$7000 \div 100 = n$$
$$70 = n$$
Ron used 70 pounds of fertilizer.

57. Strategy To find the number of wooden bats produced, write and solve a proportion using n to represent the number of wooden bats.

Solution
$$\frac{4 \text{ alum. bats}}{15 \text{ wooden bats}} = \frac{100 \text{ alum. bats}}{n \text{ wooden bats}}$$
$$4 \times n = 15 \times 100$$
$$4 \times n = 1500$$
$$n = 1500 \div 4$$
$$n = 375$$
There were 375 wooden bats produced.

59. Strategy To find the distance between two cities that are 2 inches apart on the map, write and solve a proportion using n to represent the number of miles.

Solution
$$\frac{1.25 \text{ inches}}{10 \text{ miles}} = \frac{2 \text{ inches}}{n \text{ miles}}$$
$$1.25 \times n = 10 \times 2$$
$$1.25 \times n = 20$$
$$n = 20 \div 1.25$$
$$n = 16$$
The distance is 16 miles.

61. **Strategy**   To find the dosage for a person who weighs 150 pounds, write and solve a proportion using n to represent the number of ounces.

Solution
$$\frac{n}{150 \text{ pounds}} = \frac{\frac{1}{3} \text{ ounce}}{40 \text{ pounds}}$$
$$40 \times n = \frac{1}{3} \times 150$$
$$40 \times n = 50$$
$$n = 50 \div 40$$
$$n = 1.25$$
1.25 ounces are required.

63. **Strategy**   To find how many people in a county of 240,000 eligible voters would vote in the election, write and solve a proportion using n to represent the number of voters.

Solution
$$\frac{n}{240,000} = \frac{2}{3}$$
$$2 \times 240,000 = 3 \times n$$
$$480,000 = 3 \times n$$
$$480,000 \div 3 = n$$
$$160,000 = n$$
160,000 people would vote.

65. **Strategy**   To find the monthly payment, write and solve a proportion using n to represent the monthly payment.

Solution
$$\frac{\$35.35}{\$10,000} = \frac{n}{\$50,000}$$
$$35.35 \times 50,000 = 10,000 \times n$$
$$1,767,500 = 10,000 \times n$$
$$1,767,500 \div 10,000 = n$$
$$176.75 = n$$
The monthly payment is $176.75.

67. **Strategy**   To find how many shares of stock you own after a split, write and solve a proportion using n to represent the number of shares.

Solution
$$\frac{5}{3} = \frac{n}{240}$$
$$5 \times 240 = n \times 3$$
$$1200 = n \times 3$$
$$1200 \div 3 = n$$
$$400 = n$$
You will own 400 shares.

69. **Strategy**   To find how much a bowling ball weighs on the moon, write and solve a proportion using n to represent the weight on the moon.

Solution
$$\frac{1}{6} = \frac{n}{16}$$
$$1 \times 16 = n \times 6$$
$$16 = n \times 6$$
$$16 \div 6 = n$$
$$2.67 = n$$
The bowling ball would weigh 2.67 pounds on the moon.

71. **Strategy**   To find what dividend Carlos would receive after purchasing additional shares:
● Find the total number of shares owned by adding the original number (50) to the number purchased (300).
● Find the dividend by writing and solving a proportion using n to represent the dividend.

Solution
$$\begin{array}{r} 300 \\ +\ 50 \\ \hline 350 \text{ shares} \end{array}$$
$$\frac{n}{350 \text{ shares}} = \frac{\$153}{50 \text{ shares}}$$
$$153 \times 350 = n \times 50$$
$$53,550 = n \times 50$$
$$53,550 \div 50 = n$$
$$\$1071 = n$$
The dividend would be $1071.

Applying the Concepts

73. The fact that the number of workers per retiree is decreasing means that for each retiree drawing money out of Social Security, there are fewer and fewer workers paying into the Social Security system. In other words, fewer workers are supporting each retiree. Therefore, unless the amount paid into the system by each worker is increased or other radical changes are made, the funds to pay the Social Security benefits will be depleted.

75. No, it is not possible. The sum of the fractions is $\frac{2}{5} + \frac{3}{4} = \frac{23}{20} = 1\frac{3}{20}$, which is greater than 1. In order for the responses to be possible, the sum of the fractions must be 1.

Chapter 4 Review Exercises

1. $\dfrac{2}{9} \bowtie \dfrac{10}{45}$ $\rightarrow 9 \times 10 = 90$
$\rightarrow 2 \times 45 = 90$
The proportion is true.

2. $\dfrac{\$32}{\$80} = \dfrac{32}{80} = \dfrac{2}{5}$
$\$32{:}\$80 = 32{:}80 = 2{:}5$
$\$32 \text{ to } \$80 = 32 \text{ to } 80 = 2 \text{ to } 5$

3. $\dfrac{250 \text{ miles}}{4 \text{ hours}} = 62.5 \text{ miles/hour}$

4. $\dfrac{8}{15} \bowtie \dfrac{32}{60}$ $\rightarrow 15 \times 32 = 480$
$\rightarrow 8 \times 60 = 480$
The proportion is true.

5. $\dfrac{16}{n} = \dfrac{4}{17}$
$16 \times 17 = n \times 4$
$272 = n \times 4$
$272 \div 4 = n$
$68 = n$

6. $\dfrac{\$300}{40 \text{ hours}} = \$7.50/\text{hour}$

7. $\dfrac{\$8.75}{5 \text{ pounds}} = \$1.75/\text{pound}$

8. $\dfrac{8 \text{ feet}}{28 \text{ feet}} = \dfrac{8}{28} = \dfrac{2}{7}$
$8 \text{ feet}{:}28 \text{ feet} = 8{:}28 = 2{:}7$
$8 \text{ feet to } 28 \text{ feet} = 8 \text{ to } 28 = 2 \text{ to } 7$

9. $\dfrac{n}{8} = \dfrac{9}{2}$
$n \times 2 = 8 \times 9$
$n \times 2 = 72$
$n = 72 \div 2$
$n = 36$

10. $\dfrac{18}{35} = \dfrac{10}{n}$
$n \times 18 = 35 \times 10$
$n \times 18 = 350$
$n = 350 \div 18$
$n \approx 19.44$

11. $\dfrac{6 \text{ inches}}{15 \text{ inches}} = \dfrac{6}{15} = \dfrac{2}{5}$
$6 \text{ inches}{:}15 \text{ inches} = 6{:}15 = 2{:}5$
$6 \text{ inches to } 15 \text{ inches} = 6 \text{ to } 15 = 2 \text{ to } 5$

12. $\dfrac{3}{8} \bowtie \dfrac{10}{24}$ $\rightarrow 8 \times 10 = 80$
$\rightarrow 3 \times 24 = 72$
The proportion is not true.

13. $\dfrac{\$15}{4 \text{ hours}}$

14. $\dfrac{326.4 \text{ miles}}{12 \text{ gallons}} = 27.2 \text{ miles/gallon}$

15. $\dfrac{12 \text{ days}}{12 \text{ days}} = \dfrac{12}{12} = \dfrac{1}{1}$
$12 \text{ days}{:}12 \text{ days} = 12{:}12 = 1{:}1$
$12 \text{ days to } 12 \text{ days} = 12 \text{ to } 12 = 1 \text{ to } 1$

16. $\dfrac{5}{7} \bowtie \dfrac{25}{35}$ $\rightarrow 7 \times 25 = 175$
$\rightarrow 5 \times 35 = 175$
The proportion is true.

17. $\dfrac{24}{11} = \dfrac{n}{30}$
$24 \times 30 = n \times 11$
$720 = n \times 11$
$720 \div 11 = n$
$65.45 \approx n$

18. $\dfrac{100 \text{ miles}}{3 \text{ hours}}$

19. Strategy To find the ratio:
● Find the amount of the decrease by subtracting the current price ($24) from the original price ($40).
● Write the ratio between the decrease and the original price.

Solution
$\begin{array}{r} \$40 \\ -24 \\ \hline \$16 \end{array}$
$\dfrac{\$16}{\$40} = \dfrac{16}{40} = \dfrac{2}{5}$
The ratio is $\dfrac{2}{5}$.

20. Strategy To find the property tax on a home valued at $320,000, write and solve a proportion using n to represent the property tax.

Solution $\dfrac{n}{\$320,000} = \dfrac{\$4900}{\$245,000}$
$4900 \times 320,000 = 245,000 \times n$
$1,568,000,000 = 245,000 \times n$
$1,568,000,000 \div 245,000 = n$
$6400 = n$
The property tax is $6400.

21. Strategy To find the ratio, write the ratio of the high temperature (84 degrees) to the low temperature (42 degrees).

Solution $\dfrac{84 \text{ degrees}}{42 \text{ degrees}} = \dfrac{84}{42} = \dfrac{2}{1}$
The ratio is $\dfrac{2}{1}$.

22. **Strategy**   To find the cost per phone of the phones that did pass inspection:
• Find the number of phones that did pass inspection by subtracting the number that did not pass inspection (24) from the total (1000).
• Divide the total manufacturing cost ($36,600) by the number of phones that did pass inspection.

Solution
$$\begin{array}{r} 1000 \\ -\ 24 \\ \hline 976 \end{array}$$

$$\begin{array}{r} \$37.50 \\ 976\overline{)\$36,600} \end{array}$$
The cost per phone was $37.50.

23. **Strategy**   To find how many concrete blocks would be needed to build a wall 120 feet long, write and solve a proportion using n to represent the number of concrete blocks.

Solution
$$\frac{n}{120 \text{ feet}} = \frac{448 \text{ concrete blocks}}{40 \text{ feet}}$$
$$n \times 40 = 120 \times 448$$
$$n \times 40 = 53,760$$
$$n = 53,760 \div 40$$
$$n = 1344$$
1344 blocks would be needed.

24. **Strategy**   To find the ratio, write a ratio of radio advertising ($30,000) to newspaper advertising ($12,000).

Solution
$$\frac{\$30,000}{\$12,000} = \frac{30,000}{12,000} = \frac{5}{2}$$
The ratio is $\frac{5}{2}$.

25. **Strategy**   To find the cost per pound, divide the total cost ($13.95) by the number of pounds (15).

Solution
$$\begin{array}{r} \$.93 \\ 15\overline{)\$13.95} \end{array}$$
The turkey costs $.93/pound.

26. **Strategy**   To find the average number of miles driven per hour, divide the total number of miles driven (198.8) by the number of hours (3.5).

Solution
$$\begin{array}{r} 56.8 \\ 3.5\overline{)198.8} \end{array}$$
The average was 56.8 miles/hour.

27. **Strategy**   To find the cost of $50,000 of insurance, write and solve a proportion using n to represent the cost.

Solution
$$\frac{n}{\$50,000} = \frac{\$9.87}{\$1000}$$
$$n \times 1000 = 9.87 \times 50,000$$
$$n \times 1000 = 493,500$$
$$n = 493,500 \div 1000$$
$$n = 493.50$$
The cost is $493.50.

28. **Strategy**   To find the cost per share, divide the total cost ($3580) by the number of shares (80).

Solution
$$\begin{array}{r} \$44.75 \\ 80\overline{)\$3580} \end{array}$$
The cost is $44.75/share.

29. **Strategy**   To find how many pounds of fertilizer are used on a lawn that measures 3000 square feet, write and solve a proportion using n to represent the number of pounds of fertilizer.

Solution
$$\frac{n}{3000 \text{ square feet}} = \frac{1.5 \text{ pounds}}{200 \text{ square feet}}$$
$$n \times 200 = 1.5 \times 3000$$
$$n \times 200 = 4500$$
$$n = 4500 \div 200$$
$$n = 22.5$$
22.5 pounds of fertilizer will be used.

30. **Strategy**   To find the ratio:
• Find the amount of the increase by subtracting the original value ($80,000) from the increased value ($120,000).
• Write the ratio of the amount of the increase to the original value ($80,000).

Solution
$$\begin{array}{r} \$120,000 \\ -\ 80,000 \\ \hline \$40,000 \end{array}$$

$$\frac{\$40,000}{\$80,000} = \frac{40,000}{80,000} = \frac{1}{2}$$
The ratio is $\frac{1}{2}$.

Chapter 4 Test

1. $\dfrac{46,036.80}{12 \text{ months}} = \$3836.40/\text{month}$

2. $\dfrac{40 \text{ miles}}{240 \text{ miles}} = \dfrac{40}{240} = \dfrac{1}{6}$

40 miles:240 miles = 40:240 = 1:6
40 miles to 240 miles = 40 to 240 = 1 to 6

3. $\dfrac{18 \text{ supports}}{8 \text{ feet}} = \dfrac{9 \text{ supports}}{4 \text{ feet}}$

4. $\dfrac{40}{125} \diagdown\diagup \dfrac{5}{25} \rightarrow 125 \times 5 = 625$
$\rightarrow 40 \times 25 = 1000$
The proportion is not true.

5. $\dfrac{12 \text{ days}}{8 \text{ days}} = \dfrac{12}{8} = \dfrac{3}{2}$
12 days:8 days = 12:8 = 3:2
12 days to 8 days = 12 to 8 = 3 to 2

6. $\dfrac{5}{12} = \dfrac{60}{n}$
$n \times 5 = 12 \times 60$
$n \times 5 = 720$
$n = 720 \div 5$
$n = 144$

7. $\dfrac{256.2 \text{ miles}}{8.4 \text{ gallons}} = 30.5 \text{ miles/gallon}$

8. $\dfrac{\$27}{\$81} = \dfrac{27}{81} = \dfrac{1}{3}$
$27:\$81 = 27:81 = 1:3$
$27 to \$81 = 27 to 81 = 1 to 3

9. $\dfrac{5}{14} \diagdown\diagup \dfrac{25}{70} \rightarrow 14 \times 25 = 350$
$\rightarrow 5 \times 70 = 350$
The proportion is true.

10. $\dfrac{n}{18} = \dfrac{9}{4}$
$n \times 4 = 9 \times 18$
$n \times 4 = 162$
$n = 162 \div 4$
$n = 40.5$

11. $\dfrac{\$81}{12 \text{ boards}} = \dfrac{\$27}{4 \text{ boards}}$

12. $\dfrac{18 \text{ feet}}{30 \text{ feet}} = \dfrac{18}{30} = \dfrac{3}{5}$
18 feet:30 feet = 18:30 = 3:5
18 feet to 30 feet = 18 to 30 = 3 to 5

13. **Strategy** To find the dividend on 500 shares of the utility stock, write and solve a proportion using n to represent the dividend.

Solution $\dfrac{n}{500 \text{ shares}} = \dfrac{\$62.50}{50 \text{ shares}}$
$n \times 50 = 500 \times \$62.50$
$n \times 50 = 31{,}250$
$n = 31{,}250 \div 50$
$n = 625$
The dividend is $625.

14. **Strategy** To find the ratio, write the ratio of the city temperature (86°) to the desert temperature (112°).

Solution $\dfrac{86 \text{ degrees}}{112 \text{ degrees}} = \dfrac{86}{112} = \dfrac{43}{56}$
The ratio is $\dfrac{43}{56}$.

15. $\dfrac{2421 \text{ miles}}{4.5 \text{ hours}} = 538 \text{ miles/hour}$
The plane's speed is 538 miles/hour.

16. **Strategy** To estimate the number of pounds of water in a college student weighing 150 pounds, write and solve a proportion using n to represent the number of pounds of water.

Solution $\dfrac{88 \text{ pounds water}}{100 \text{ pounds body weight}}$
$= \dfrac{n}{150 \text{ pounds body weight}}$
$88 \times 150 = n \times 100$
$13{,}200 = n \times 100$
$13{,}200 \div 100 = n$
$132 = n$
The college student's body contains 132 pounds of water.

17. $\dfrac{\$69.20}{40 \text{ feet}} = \$1.73/\text{foot}$
The cost of the umber is $1.73/foot.

18. **Strategy** To find how many ounces of medication are required for a person who weighs 175 pounds, write and solve a proportion using n to represent the ounces of medication.

Solution $\dfrac{\frac{1}{4} \text{ ounce}}{50 \text{ pounds}} = \dfrac{n}{175 \text{ pounds}}$
$\dfrac{1}{4} \times 175 = n \times 50$
$43.75 = n \times 50$
$43.75 \div 50 = n$
$0.875 = n$
The amount of medication required is 0.875 ounce.

19. **Strategy** To find the ratio of the number of games won to the total number of games played, add the number of games won (20) to the number of games lost (5) to determine the number of games played. Then write the ratio of the number of games won to the number of games played.

Solution $20 + 5 = 25$ games played
$\dfrac{20}{25} = \dfrac{4}{5}$
The ratio of the number of games won to the total number of games played is $\dfrac{4}{5}$.

20. Strategy To find the number of defective hard drives in the production of 1200 hard drives, write and solve a proportion using n to represent the number of defective hard drives.

Solution

$$\frac{n}{1200} = \frac{3}{100}$$
$$n \times 100 = 1200 \times 3$$
$$n \times 100 = 3600$$
$$n = 3600 \div 100$$
$$n = 36$$

36 defective hard drives are expected to be found in the production of 1200 hard drives.

Cumulative Review Exercises

1.
$$\begin{array}{r} \overset{9}{\cancel{11}0}\ \overset{}{\cancel{10}}\overset{}{8}\overset{}{15} \\ 20{,}0\cancel{9}\cancel{5} \\ -10{,}937 \\ \hline 9{,}158 \end{array}$$

2. $2 \cdot 2 \cdot 2 \cdot 2 \cdot 3 \cdot 3 \cdot 3 = 2^4 \cdot 3^3$

3. $4 - (5 - 2)^2 \div 3 + 2 = 4 - (-3)^2 \div 3 + 2$
$$= 4 - 9 \div 3 + 2$$
$$= 4 - 3 + 2$$
$$= 1 + 2 = 3$$

4. $160 = 2 \cdot 2 \cdot 2 \cdot 2 \cdot 2 \cdot 5$

$$\begin{array}{c|c} \multicolumn{2}{c}{160} \\ \hline 2 & 80 \\ 2 & 40 \\ 2 & 20 \\ 2 & 10 \\ 2 & 5 \\ 5 & 1 \end{array}$$

5.

	2	3
9 =		3·3
12 =	(2·2)	3
18 =	2	(3·3)

LCM = $2 \cdot 2 \cdot 3 \cdot 3 = 36$

6.

	2	3	7
28 =	2·2		⑦
42 =	②	3	7

GCF = $2 \cdot 7 = 14$

7. $\dfrac{40}{64} = \dfrac{2 \cdot 2 \cdot 2 \cdot 5}{2 \cdot 2 \cdot 2 \cdot 2 \cdot 2 \cdot 2} = \dfrac{5}{8}$

8.
$$3\frac{5}{6} = 3\frac{25}{30}$$
$$+4\frac{7}{15} = 4\frac{14}{30}$$
$$\overline{7\frac{39}{30} = 8\frac{9}{30} = 8\frac{3}{10}}$$

9.
$$10\frac{1}{6} = 10\frac{3}{18} = 9\frac{21}{18}$$
$$-4\frac{5}{9} = 4\frac{10}{18} = 4\frac{10}{18}$$
$$\overline{\phantom{-4\frac{5}{9} = 4\frac{10}{18} = }5\frac{11}{18}}$$

10.
$$\frac{11}{12} \times 3\frac{1}{11} = \frac{11}{12} \times \frac{34}{11}$$
$$= \frac{11 \times 34}{12 \times 11}$$
$$= \frac{\overset{1}{\cancel{11}} \cdot 2 \cdot 17}{2 \cdot \underset{1}{\cancel{2}} \cdot 3 \cdot \underset{1}{\cancel{11}}} = \frac{17}{6} = 2\frac{5}{6}$$

11.
$$3\frac{1}{3} \div \frac{5}{7} = \frac{10}{3} \div \frac{5}{7}$$
$$= \frac{10}{3} \times \frac{7}{5}$$
$$= \frac{10 \cdot 7}{3 \cdot 5} = \frac{2 \cdot \overset{1}{\cancel{5}} \cdot 7}{3 \cdot \underset{1}{\cancel{5}}} = \frac{14}{3} = 4\frac{2}{3}$$

12.
$$\left(\frac{2}{5} + \frac{3}{4}\right) \div \frac{3}{2} = \left(\frac{8}{20} + \frac{15}{20}\right) \div \frac{3}{2}$$
$$= \frac{23}{20} \times \frac{2}{3}$$
$$= \frac{23 \times 2}{20 \times 3} = \frac{23 \cdot \overset{1}{\cancel{2}}}{2 \cdot \underset{1}{\cancel{2}} \cdot 5 \cdot 3} = \frac{23}{30}$$

13. Four and seven hundred nine ten-thousandths

14.
$$\overset{\text{Given place value}}{\underset{}{2.09762}}$$
$$7 > 5$$
$$2.10$$

15.
$$\begin{array}{r} 1.9898 \approx 1.990 \\ 8.09\overline{)16.09.7600} \\ -\ \underline{809} \\ 8007 \\ \underline{-7281} \\ 7266 \\ \underline{-6472} \\ 7940 \\ \underline{-7281} \\ 6590 \\ \underline{-6472} \\ 118 \end{array}$$

16.
$$0.06\frac{2}{3} = \frac{6\frac{2}{3}}{100} = 6\frac{2}{3} \div 100 = \frac{20}{3} \div 100$$
$$= \frac{20}{3} \times \frac{1}{100}$$
$$= \frac{20 \cdot 1}{3 \cdot 100} = \frac{1}{15}$$

17. $\dfrac{25 \text{ miles}}{200 \text{ miles}} = \dfrac{25}{200} = \dfrac{1}{8}$

18. $\dfrac{87¢}{6 \text{ pencils}} = \dfrac{29¢}{2 \text{ pencils}}$

19. $\dfrac{250.5 \text{ miles}}{7.5 \text{ gallons of gas}} = 33.4 \text{ miles/gallon}$

20.
$$\frac{40}{n} = \frac{160}{17}$$
$$40 \times 17 = n \times 160$$
$$680 = n \times 160$$
$$680 \div 160 = n$$
$$4.25 = n$$

21. $\dfrac{457.6 \text{ miles}}{8 \text{ hours}} = 57.2 \text{ miles/hour}$
The car's speed is 57.2 miles/hour.

22.
$$\frac{12}{5} = \frac{n}{15}$$
$$12 \times 15 = n \times 5$$
$$180 = n \times 5$$
$$180 \div 5 = n$$
$$36 = n$$

23. Strategy To find your new checking account balance:
• Find the total of the checks written by adding the two checks ($192 and $88).
• Subtract the total of the checks written from the original balance ($1024).

Solution
$$\begin{array}{r} \$192 \\ +\ 88 \\ \hline \$280 \end{array} \qquad \begin{array}{r} \$1024 \\ -\ 280 \\ \hline \$744 \end{array}$$
Your new balance is $744.

24. Strategy To find the monthly payment:
• Find the amount to be paid by subtracting the down payment ($5000) from the original cost ($32,360).
• Divide the amount remaining to be paid by the number of payments (48).

Solution
$$\begin{array}{r} \$32,360 \\ -\ 5,000 \\ \hline \$27,360 \end{array} \qquad \begin{array}{r} \$570 \\ 48)\overline{\$27,360} \end{array}$$

The monthly payment is $570.

25. Strategy To find how many pages remain to be read:
• Find the number read during vacation by multiplying the total (175 pages) by $\frac{2}{5}$.
• Subtract the number of pages read during vacation from the total (175 pages).

Solution $\dfrac{2}{5} \times 175 = \dfrac{2}{5} \times \dfrac{175}{1} = 70$
$175 - 70 = 105$
105 pages remain to be read.

26. Strategy To find the cost per acre, divide the total cost ($84,000) by the number of acres $\left(2\frac{1}{3}\right)$.

Solution $\$84,000 \div 2\dfrac{1}{3} = 84,000 \div \dfrac{7}{3}$
$$= 84,000 \times \dfrac{3}{7} = \$36,000$$
The cost per acre was $36,000.

27. Strategy To find the amount of change:
• Find the total amount of the purchases by adding the two purchases ($22.79 and $9.59).
• Subtract the total amount of the purchases from $50.

Solution
$$\begin{array}{r} \$22.79 \\ +\ 9.59 \\ \hline \$32.38 \end{array} \qquad \begin{array}{r} \$50.00 \\ -32.38 \\ \hline \$17.62 \end{array}$$
The change was $17.62.

28. Strategy To find your monthly salary, divide your annual salary ($41,691) by 12 months.

Solution
$$\begin{array}{r} 3468.25 \\ 12)\overline{41,619.00} \\ \underline{-36} \\ 56 \\ \underline{-48} \\ 81 \\ \underline{-72} \\ 99 \\ \underline{-96} \\ 30 \\ \underline{-24} \\ 60 \\ \underline{-60} \\ 0 \end{array}$$

Your monthly salary is $3468.25.

29. **Strategy** To find how many inches will be eroded in 50 months, write and solve a proportion using n to represent the number of inches.

Solution $\dfrac{3 \text{ inches}}{6 \text{ months}} = \dfrac{n}{50 \text{ months}}$

$3 \times 50 = n \times 6$

$150 = n \times 6$

$150 \div 6 = n$

$25 = n$

25 inches will erode in 50 months.

30. **Strategy** To find how many ounces of medication are required for a person who weighs 160 pounds, write and solve a proportion using n to represent the number of ounces.

Solution $\dfrac{n}{160} = \dfrac{\frac{1}{2} \text{ ounce}}{50 \text{ pounds}}$

$n \times 50 = \dfrac{1}{2} \times 160$

$n \times 50 = 80$

$n = 80 \div 50$

$n = 1.6$

1.6 ounces of medication are required.

Chapter 5: Percents

Prep Test

1. $\dfrac{19}{100}$

2. 0.23

3. 47

4. 2850

5.
$$\begin{array}{r} 4000. \\ 0.015.\overline{)60.000.} \\ \underline{-60} \\ 00 \\ \underline{-0} \\ 00 \\ \underline{-0} \\ 0 \end{array}$$

6. $8 \div \dfrac{1}{4} = \dfrac{8}{1} \times \dfrac{4}{1} = 32$

7. $\dfrac{5}{8} \times \dfrac{100}{1} = \dfrac{5 \cdot \overset{1}{\cancel{2}} \cdot \overset{1}{\cancel{2}} \cdot 5 \cdot 5}{\underset{1}{\cancel{2}} \cdot \underset{1}{\cancel{2}} \cdot 2} = \dfrac{125}{2} = 62\dfrac{1}{2} = 62.5$

8. $66\dfrac{2}{3}$

9.
$$\begin{array}{r} 1.75 \\ 16\overline{)28.00} \end{array}$$

Go Figure

a. The smallest three-digit palindrome is 101. However, any number that ends in 1 is not divisible by 2, and hence not by 6. So the numbers 111, 121, and 191 are also eliminated. The next smallest three-digit palindrome is 202, which is not divisible by 3, and neither is 212. However, 222 is divisible by 2 and 3, which means that it is a multiple of 6. So 222 is the smallest three-digit multiple of 6 that is a palindrome.

b. To use the process, add 874 and 478.
$$\begin{array}{r} 478 \\ + \ 874 \\ \hline 1352 \end{array}$$
Then add 1352 and 2531.
$$\begin{array}{r} 1352 \\ + \ 2531 \\ \hline 3883 \end{array}$$
The number 3883 is a palindrome.

Section 5.1

Objective A Exercises

1. $25\% = 25 \times \dfrac{1}{100} = \dfrac{25}{100} = \dfrac{1}{4}$
$25\% = 25 \times 0.01 = 0.25$

3. $130\% = 130 \times \dfrac{1}{100} = \dfrac{130}{100} = 1\dfrac{3}{10}$
$130\% = 130 \times 0.01 = 1.30$

5. $100\% = 100 \times \dfrac{1}{100} = \dfrac{100}{100} = 1$
$100\% = 100 \times 0.01 = 1.00$

7. $73\% = 73 \times \dfrac{1}{100} = \dfrac{73}{100}$
$73\% = 73 \times 0.01 = 0.73$

9. $383\% = 383 \times \dfrac{1}{100} = \dfrac{383}{100} = 3\dfrac{83}{100}$
$383\% = 383 \times 0.01 = 3.83$

11. $70\% = 70 \times \dfrac{1}{100} = \dfrac{70}{100} = \dfrac{7}{10}$
$70\% = 70 \times 0.01 = 0.70$

13. $88\% = 88 \times \dfrac{1}{100} = \dfrac{88}{100} = \dfrac{22}{25}$
$88\% = 88 \times 0.01 = 0.88$

15. $32\% = 32 \times \dfrac{1}{100} = \dfrac{32}{100} = \dfrac{8}{25}$
$32\% = 32 \times 0.01 = 0.32$

17. $66\dfrac{2}{3}\% = 66\dfrac{2}{3} \times \dfrac{1}{100} = \dfrac{200}{3} \times \dfrac{1}{100}$
$= \dfrac{200}{300} = \dfrac{2}{3}$

19. $83\dfrac{1}{3}\% = 83\dfrac{1}{3} \times \dfrac{1}{100} = \dfrac{250}{3} \times \dfrac{1}{100}$
$= \dfrac{250}{300} = \dfrac{5}{6}$

21. $11\dfrac{1}{9}\% = 11\dfrac{1}{9} \times \dfrac{1}{100} = \dfrac{100}{9} \times \dfrac{1}{100}$
$= \dfrac{100}{900} = \dfrac{1}{9}$

23. $45\dfrac{5}{11}\% = 45\dfrac{5}{11} \times \dfrac{1}{100} = \dfrac{500}{11} \times \dfrac{1}{100}$
$= \dfrac{500}{1100} = \dfrac{5}{11}$

25. $4\dfrac{2}{7}\% = 4\dfrac{2}{7} \times \dfrac{1}{100} = \dfrac{30}{7} \times \dfrac{1}{100}$
$= \dfrac{30}{700} = \dfrac{3}{70}$

27. $6\dfrac{2}{3}\% = 6\dfrac{2}{3} \times \dfrac{1}{100} = \dfrac{20}{3} \times \dfrac{1}{100} = \dfrac{20}{300} = \dfrac{1}{15}$

29. $6.5\% = 6.5 \times 0.01 = 0.065$

31. $12.3\% = 12.3 \times 0.01 = 0.123$

33. $0.55\% = 0.55 \times 0.01 = 0.0055$

35. $8.25\% = 8.25 \times 0.01 = 0.0825$

37. $5.05\% = 5.05 \times 0.01 = 0.0505$

39. $2\% = 2 \times 0.01 = 0.02$

41. $80.4\% = 80.4 \times 0.01 = 0.804$

43. $4.9\% = 4.9 \times 0.01 = 0.049$

Objective B Exercises

45. $0.73 = 0.73 \times 100\% = 73\%$

47. $0.01 = 0.01 \times 100\% = 1\%$

49. $2.94 = 2.94 \times 100\% = 294\%$

51. $0.006 = 0.006 \times 100\% = 0.6\%$

53. $3.106 = 3.106 \times 100\% = 310.6\%$

55. $0.70 = 0.70 \times 100\% = 70\%$

57. $\dfrac{37}{100} = \dfrac{37}{100} \times 100\% = \dfrac{3700}{100}\% = 37\%$

59. $\dfrac{2}{5} = \dfrac{2}{5} \times 100\% = \dfrac{200}{5}\% = 40\%$

61. $\dfrac{1}{8} = \dfrac{1}{8} \times 100\% = \dfrac{100}{8}\% = 12.5\%$

63. $1\dfrac{1}{2} = 1\dfrac{1}{2} \times 100\% = \dfrac{3}{2} \times 100\% = \dfrac{300}{2}\% = 150\%$

65. $1\dfrac{2}{3} = 1\dfrac{2}{3} \times 100\% = \dfrac{5}{3} \times 100\% = \dfrac{500}{3}\% \approx 166.7\%$

67. $\dfrac{7}{8} = \dfrac{7}{8} \times 100\% = \dfrac{700}{8}\% = 87.5\%$

69. $\dfrac{12}{25} = \dfrac{12}{25} \times 100\% = \dfrac{1200}{25}\% = 48\%$

71. $\dfrac{1}{3} = \dfrac{1}{3} \times 100\% = \dfrac{100}{3}\% = 33\dfrac{1}{3}\%$

73. $1\dfrac{2}{3} = 1\dfrac{2}{3} \times 100\% = \dfrac{5}{3} \times 100\% = \dfrac{500}{3}\% = 166\dfrac{2}{3}\%$

75. $\dfrac{7}{8} = \dfrac{7}{8} \times 100\% = \dfrac{700}{8}\% = 87\dfrac{1}{2}\%$

Applying the Concepts

77. Strategy To find the percent of those surveyed that did not name corn, cole slow, corn bread, or fries, add the percents representing these four side dishes and subtract the sum from 100%

Solution
38%	Corn on the Cob
35%	Cole slaw
11%	Corn bread
10%	Fries
94%	

$100\% - 94\% = 6\%$
6% of those surveyed named something other than corn on the cob, cole slaw, corn bread, or fries.

79. $50\% = \dfrac{50}{100} = \dfrac{1}{2}$; this represents $\dfrac{1}{2}$ off the regular price.

81. a. False **b.** For example, $200\% \times 4 = 2 \times 4 = 8$

Section 5.2

Objective A Exercises

1. $0.08 \times 100 = n$
$8 = n$

3. $0.27 \times 40 = n$
$10.8 = n$

5. $0.0005 \times 150 = n$
$0.075 = n$

7. $1.25 \times 64 = n$
$80 = n$

9. $0.107 \times 485 = n$
$51.895 = n$

11. $0.0025 \times 3000 = n$
$7.5 = n$

13. $0.80 \times 16.25 = n$
$13 = n$

15. $0.015 \times 250 = n$
$3.75 = n$

17. $\dfrac{1}{6} \times 120 = n$
$20 = n$

19. $\dfrac{1}{3} \times 630 = n$
$210 = n$

21. $0.05 \times 95 = n$ or $0.75 \times 6 = n$
$4.75 = n$ $4.5 = n$
Because $4.75 > 4.5$, 5% of 95 is larger.

23. $0.79 \times 16 = n$ or $0.20 \times 65 = n$
$12.64 = n$ $13 = n$
Because $12.64 < 13$, 79% of 16 is smaller.

25. $0.02 \times 1500 = n$ or $0.72 \times 40 = n$
$30 = n$ $28.8 = n$
Because $28.8 < 30$, 72% of 40 is smaller.

27. $0.31294 \times 82,460 = n$
$25,805.0324 = n$

Objective B Exercises

29. Strategy To find the number of people who do not have health insurance, write and solve the basic percent equation using n to represent the number of people between ages 18 and 24 who do not have health insurance. The percent is 30% and the base is 44.

Solution $30\% \times 44 = n$
$0.30 \times 44 = n$
$13.2 = n$
About 13.2 million people aged 18 to 24 do not have health insurance.

31. Strategy To find how many more faculty members described their political views as liberal than described their views as far left:
● Find the number that described their views as liberal by writing and solving the basic percent equation using n to represent the number with liberal views. The percent is 42.3% and the base is 32,840.
● Find the number that described their views as far left by writing and solving the basic percent equation using n to represent the number with far left views. The percent is 5.3% and the base is 32,840.
● Subtract the number with far left views from the number with liberal views.

Solution

Liberal	Far left
$42.3\% \times 32{,}840 = n$	$5.3\% \times 32{,}840 = n$
$0.423 \times 32{,}840 = n$	$0.053 \times 32{,}840 = n$
$13{,}891.32 = n$	$1740.52 = n$

$13{,}891.32 - 1740.52 = 12{,}150.8$
12,151 more faculty members described their political views as liberal than described their views as far left.

33a. Strategy To find the sales tax, write and solve the basic percent equation using n as the sales tax. The percent is 6% and the base is $29,500.

Solution $6\% \times \$29{,}500 = n$
$0.06 \times 29{,}500 = n$
$1770 = n$
The sales tax is $1770.

b. Strategy To find the total cost of the car, add the sales tax ($1770) to the purchase price of the car ($29,500).

Solution
$29,500
$\underline{+\ \ 1{,}770}$
$31,270
The total cost of the car is $31,270.

35. Strategy To find the number of respondents that did not answer yes to the question:
● Find the number that did answer yes by writing and solving the basic percent equation using n to represent the number that said yes. The percent is 29.8% and the base is 8878.
● Subtract the number of yes answers from the total number polled (8878).

Solution $29.8\% \times 8878 = n$
$0.298 \times 8878 = n$
$2646 \approx n$
$8878 - 2646 = 6232$
6232 respondents did not answer yes to the question.

Applying the Concepts

37. $43\% \times 112 = 48.16$
Employees spent 48.2 hours with family and friends.

39. Actual time: $112 \times 20\% = 22.4$
Preferred time: $112 \times 23\% = 25.76$
$25.76 - 22.4 = 3.36$
There are approximately 3.4 hours difference between the actual and preferred amounts of time the employees spent on self.

Section 5.3

Objective A Exercises

1. $n \times 75 = 24$
$n = 24 \div 75$
$n = 0.32$
$n = 32\%$

3. $n \times 90 = 15$
$n = 15 \div 90$
$n = 0.16\frac{2}{3}$
$n = 16\frac{2}{3}\%$

5. $n \times 12 = 24$
$n = 24 \div 12$
$n = 2$
$n = 200\%$

7. $n \times 16 = 6$
$n = 6 \div 16$
$n = 0.375$
$n = 37.5\%$

9. $n \times 100 = 18$
$n = 18 \div 100$
$n = 0.18$
$n = 18\%$

11. $n \times 2000 = 5$
$n = 5 \div 2000$
$n = 0.0025$
$n = 0.25\%$

13. $n \times 6 = 1.2$
$n = 1.2 \div 6$
$n = 0.2$
$n = 20\%$

15. $n \times 4.1 = 16.4$
$$n = 16.4 \div 4.1$$
$$n = 4$$
$$n = 400\%$$

17. $n \times 40 = 1$
$$n = 1 \div 40$$
$$n = 0.025$$
$$n = 2.5\%$$

19. $n \times 48 = 18$
$$n = 18 \div 48$$
$$n = 0.375$$
$$n = 37.5\%$$

21. $n \times 2800 = 7$
$$n = 7 \div 2800$$
$$n = 0.0025$$
$$n = 0.25\%$$

23. $n \times 175 = 4.2$
$$n = 4.2 \div 175$$
$$n = 0.024$$
$$n = 2.4\%$$

25. $n \times 86.5 = 8.304$
$$n = 8.304 \div 86.5$$
$$n = 0.096$$
$$n = 9.6\%$$

Objective B Exercises

27.   **Strategy**   To find what percent of couples disagree about financial matters, write and solve the basic percent equation using n to represent the unknown percent. The base is 10 and the amount is 7.

     Solution $n \times 10 = 7$
$$n = 7 \div 10$$
$$n = 0.70$$
70% of couples disagree about financial matters.

29.   **Strategy**   To find what percent of the vegetables was wasted, write and solve the basic percent equation using n to represent the unknown percent. The base is 63 billion and the amount is 16 billion.

     Solution $n \times 63$ billion $= 16$ billion
$$n = 16 \text{ billion} \div 63 \text{ billion}$$
$$n \approx 0.254$$
Approximately 25.4% of the vegetables were wasted.

31.   **Strategy**   To find what percent of the total amount spent on energy utilities is spent on lighting, write and solve the basic percent equation using n to represent the unknown percent. The base is $1355 and the amount is $81.30.

     Solution $n \times \$1355 = \$81.30$
$$n = 81.30 \div 1355$$
$$n = 0.06$$
The typical American household spends 6% of its total energy utilities on lighting.

33.   **Strategy**   To find what percent of food produced in the United States is wasted:
- Find the amount wasted by subtracting the amount not wasted (260 billion pounds) from the total (356 billion pounds).
- Find the percent by writing and solving the basic percent equation using n to represent the unknown percent. The base is 356 billion and the amount is (356 billion − 260 billion).

     Solution 356 billion $n \times 356$ billion $= 96$ billion
$$\underline{-260} \text{ billion} \quad n = 96 \text{ billion} \div 356 \text{ billion}$$
$$96 \text{ billion} \quad n \approx 0.27$$
27% of the food produced in United States is wasted.

Applying the Concepts

35.
$$
\begin{array}{r}
\$1,400 \\
1,200 \\
4,000 \\
3,900 \\
3,000 \\
+\ 1,100 \\
\hline
14,600
\end{array}
$$
$14,600 is the total amount spent.
$4000 is spent for food.

$$\frac{\$4,000}{\$14,600} \approx 0.274$$

Approximately 27.4% of the total expenses is spent for food.

37.
$1,400
1,200
4,000
3,900
3,000
+1,100
—————
$14,600
$14,600 is the total amount spent.

$14,600
−1,200
—————
$13,400 total spent on all categories except training.

$$\frac{\$13,400}{\$14,600} \approx 0.918$$

91.8% of the total is spent on all categories except training.

Section 5.4

Objective A Exercises

1. $0.12 \times n = 9$
$n = 9 \div 0.12$
$n = 75$

3. $0.16 \times n = 8$
$n = 8 \div 0.16$
$n = 50$

5. $0.10 \times n = 10$
$n = 10 \div 0.10$
$n = 100$

7. $0.30 \times n = 25.5$
$n = 25.5 \div 0.30$
$n = 85$

9. $0.025 \times n = 30$
$n = 30 \div 0.025$
$n = 1200$

11. $1.25 \times n = 24$
$n = 24 \div 1.25$
$n = 19.2$

13. $2.4 \times n = 18$
$n = 18 \div 2.4$
$n = 7.5$

15. $0.15 \times n = 4.8$
$n = 4.8 \div 0.15$
$n = 32$

17. $0.128 \times n = 25.6$
$n = 25.6 \div 0.128$
$n = 200$

19. $0.007 \times n = 0.56$
$n = 0.56 \div 0.007$
$n = 80$

21. $0.30 \times n = 2.7$
$n = 2.7 \div 0.30$
$n = 9$

23. $\frac{1}{6} \times n = 84$
$n = 84 \div \frac{1}{6}$
$n = 504$

25. $\frac{2}{3} \times n = 72$
$n = 72 \div \frac{2}{3}$
$n = 108$

Objective B Exercises

27. **Strategy** To find the number of travelers who allowed their children to miss school, write and solve the basic percent equation using n to represent the number of travelers. The percent is 11% and the amount is 1.738 million.

Solution $11\% \times n = 1.738$
$0.11 \times n = 1.738$
$n = 1.738 \div 0.11$
$n = 15.8$
There were 15.8 million travelers who allowed their children to miss school to go along on a trip.

29. **Strategy** To find how many people responded to the survey, write and solve the basic percent equation using n to represent the number of people that responded. The percent is 22% and the amount is 740 people.

Solution $22\% \times n = 740$
$0.22 \times n = 740$
$n = 740 \div 0.22$
$n \approx 3363.6$
3364 people responded to the survey.

31a. **Strategy** To find the number of computer boards tested, write and solve the basic percent equation using n to represent the number of computer boards tested. The percent is 0.8% and the amount is 24.

Solution $0.8\% \times n = 24$
$0.008 \times n = 24$
$n = 24 \div 0.008$
$n = 3000$
3000 boards were tested.

b. **Strategy** To find the number of boards that were tested as not defective, subtract the number of defective boards (24) from the total tested (3000).

Solution
$$\begin{array}{r} 3000 \\ - \ \ 24 \\ \hline 2976 \end{array}$$

2976 boards were tested as not defective.

Applying the Concepts

33. **Strategy** To find the number of people in the United States that were age 20 and older:
● Find the number that are under the age of 20 by using the basic percent equation using n to represent the number of people under 20. The percent is 28.6% and the base is 281,422,000.
● Subtract the number under 20 from the base (281,422,000).

Solution
$$28.6\% \times 281{,}422{,}000 = n$$
$$0.286 \times 281{,}422{,}000 = n$$
$$80{,}486{,}692 = n$$

$$\begin{array}{r} 281{,}422{,}000 \\ - \ 80{,}486{,}692 \\ \hline 200{,}935{,}308 \end{array}$$

200,935,308 people in the United States were age 20 and older in 2000.

35. $0.08 \div 0.04 = 2$ milligrams
The recommended daily amount of copper for an adult is 2 milligrams.

Section 5.5

Objective A Exercises

1. $\dfrac{26}{100} = \dfrac{n}{250}$
$$26 \times 250 = n \times 100$$
$$6500 = n \times 100$$
$$6500 \div 100 = n$$
$$65 = n$$

3. $\dfrac{n}{100} = \dfrac{37}{148}$
$$148 \times n = 37 \times 100$$
$$148 \times n = 3700$$
$$n = 3700 \div 148$$
$$n = 25$$
37 is 25% of 148.

5. $\dfrac{68}{100} = \dfrac{51}{n}$
$$68 \times n = 100 \times 51$$
$$68 \times n = 5100$$
$$n = 5100 \div 68$$
$$n = 75$$

7. $\dfrac{n}{100} = \dfrac{43}{344}$
$$n \times 344 = 100 \times 43$$
$$n \times 344 = 4300$$
$$n = 4300 \div 344$$
$$n = 12.5$$
12.5% of 344 is 43.

9. $\dfrac{20.5}{100} = \dfrac{82}{n}$
$$n \times 20.5 = 82 \times 100$$
$$n \times 20.5 = 8200$$
$$n = 8200 \div 20.5$$
$$n = 400$$

11. $\dfrac{6.5}{100} = \dfrac{n}{300}$
$$300 \times 6.5 = n \times 100$$
$$1950 = n \times 100$$
$$1950 \div 100 = n$$
$$19.5 = n$$

13. $\dfrac{n}{100} = \dfrac{7.4}{50}$
$$50 \times n = 7.4 \times 100$$
$$50 \times n = 740$$
$$n = 740 \div 50$$
$$n = 14.8$$
7.4 is 14.8% of 50.

15. $\dfrac{50.5}{100} = \dfrac{n}{124}$
$$50.5 \times 124 = n \times 100$$
$$6262 = n \times 100$$
$$6262 \div 100 = n$$
$$62.62 = n$$

17. $\dfrac{120}{100} = \dfrac{6}{n}$
$$120 \times n = 6 \times 100$$
$$120 \times n = 600$$
$$n = 600 \div 120$$
$$n = 5$$

19. $\dfrac{250}{100} = \dfrac{n}{18}$
$$250 \times 18 = n \times 100$$
$$4500 = n \times 100$$
$$4500 \div 100 = n$$
$$45 = n$$

21. $\dfrac{220}{100} = \dfrac{33}{n}$
$$n \times 220 = 33 \times 100$$
$$n \times 220 = 3300$$
$$n = 3300 \div 220$$
$$n = 15$$

Objective B Exercises

23. **Strategy** To find the total amount the charity organization collected, write and solve a proportion using n to represent the total collected (base). The percent is 12% and the amount is $2940.

Solution
$$\frac{12}{100} = \frac{2940}{n}$$
$12 \times n = 2940 \times 100$
$12 \times n = 294{,}000$
$n = 294{,}000 \div 12$
$n = 24{,}500$
The total amount collected was $24,500.

25. **Strategy** To find the total land area, write and solve a proportion using n to represent the total land area (base). The percent is 16% and the amount is 9,400,000 square miles.

Solution
$$\frac{16}{100} = \frac{9{,}400{,}000}{n}$$
$16 \times n = 9{,}400{,}000 \times 100$
$16 \times n = 940{,}000{,}000$
$n = 940{,}000{,}000 \div 16$
$n = 58{,}750{,}000$
The world's total land area is 58,750,000 square miles.

27. **Strategy** To find the number of hotels in the United States that are located along highways, write and solve a proportion using n to represent the number along highways. The percent is 42.2% and the base is 53,500.

Solution
$$\frac{42.2}{100} = \frac{n}{53{,}500}$$
$42.2 \times 53{,}500 = 100 \times n$
$2{,}257{,}700 = 100 \times n$
$2{,}257{,}700 \div 100 = n$
$22{,}577 = n$
22,577 hotels in the United States are located along highways.

29. **Strategy** To find the number of ounces of gold mined in the United States that year, write and solve a proportion using n to represent that the total number of ounces mined. The percent is 16%, and the amount is 2,240,000 ounces.

Solution
$$\frac{16}{100} = \frac{\$2{,}240{,}000}{n}$$
$16 \times n = 100 \times 2{,}240{,}000$
$16 \times n = 224{,}000{,}000$
$n = 224{,}000{,}000 \div 16$
$n = 14{,}000{,}000$
14,000,000 ounces of gold were mined in the United States that year.

31. **Strategy** To find the percent of the deaths due to traffic accidents:
• Find the total number of deaths.
• Write and solve a proportion using n to represent the percent. The base is the total deaths (156) and the amount is 73.

Solution
$$\begin{array}{r} 19 \\ 6 \\ 58 \\ + 73 \\ \hline 156 \end{array}$$
$$\frac{n}{100} = \frac{73}{156}$$
$n \times 156 = 100 \times 73$
$n \times 156 = 7300$
$n = 7300 \div 156$
$n \approx 46.8$
46.8% of the deaths were due to traffic accidents.

Applying the Concepts

33. 108th Senate   108th House of Representatives
$$\frac{51}{100} = \frac{n}{100} \qquad \frac{229}{435} = \frac{n}{100}$$
$n = 51\%$ $n \approx 52.6\%$ Republicans

The 108th House of Representatives had the larger percent of Republicans.

Chapter 5 Review Exercises

1. $0.30 \times 200 = n$
$60 = n$

2. $n \times 80 = 16$
$n = 16 \div 80$
$n = 0.2$
$n = 20\%$

3. $1\frac{3}{4} \times 100\% = 1.75 \times 100\% = 175\%$

4. $0.20 \times n = 15$
$n = 15 \div 0.20$
$n = 75$

5. $12\% = 12 \times \frac{1}{100} = \frac{12}{100} = \frac{3}{25}$

6. $0.22 \times 88 = n$
$19.36 = n$

7. $n \times 20 = 30$
$n = 30 \div 20$
$n = 1.5$
$n = 150\%$

8. $0.16\frac{2}{3} \times n = 84$

$\frac{1}{6} \times n = 84$

$n = 84 \div \frac{1}{6}$

$n = 84 \times 6$

$n = 504$

9. $42\% = 42 \times 0.01 = 0.42$

10. $0.075 \times 72 = n$

$5.4 = n$

11. $0.66\frac{2}{3} \times n = 105$

$\frac{2}{3} \times n = 105$

$n = 105 \div \frac{2}{3}$

$n = 105 \times \frac{3}{2}$

$n = 157.5$

12. $7.6\% = 7.6 \times 0.01 = 0.076$

13. $1.25 \times 62 = n$

$77.5 = n$

14. $16\frac{2}{3}\% = 16\frac{2}{3} \times \frac{1}{100} = \frac{50}{3} \times \frac{1}{100} = \frac{50}{300} = \frac{1}{6}$

15. $\frac{n}{100} = \frac{40}{25}$

$n \times 25 = 40 \times 100$

$n \times 25 = 4000$

$n = 4000 \div 25$

$n = 160$

160% of 25 is 40.

16. $\frac{20}{100} = \frac{15}{n}$

$20 \times n = 100 \times 15$

$20 \times n = 1,500$

$n = 1,500 \div 20$

$n = 75$

17. $0.38 \times 100\% = 38\%$

18. $0.78 \times n = 8.5$

$n = 8.5 \div 0.78$

$n \approx 10.89 \approx 10.9$

19. $n \times 30 = 2.2$

$n = 2.2 \div 30$

$n \approx 0.073$

$n \approx 7.3\%$

20. $n \times 15 = 92$

$n = 92 \div 15$

$n \approx 6.133$

$n \approx 613.3\%$

21. **Strategy** To find the percent of the questions answered correctly:
● Find the number of questions answered correctly by subtracting the number missed (9) from the total number of questions (60).
● Write and solve a proportion using n to represent the percent. The base is 60 and the amount is the number of questions answered correctly.

Solution $60 - 9 = 51$

$\frac{n}{100} = \frac{51}{60}$

$n \times 60 = 51 \times 100$

$n \times 60 = 5100$

$n = 5100 \div 60$

$n = 85$

The student answered 85% of the questions correctly.

22. **Strategy** To find how much of the budget was spent for newspaper advertising, write and solve the basic percent equation using n to represent the newspaper advertising. The percent is 7.5% and the base is $60,000.

Solution $7.5\% \times 60,000 = n$

$0.075 \times 60,000 = n$

$4500 = n$

The company spent $4500 for newspaper advertising.

23. **Strategy** To find what percent of total energy use is electricity:
● Find the total of the costs given on the graph. This sum is the base.
● Write and solve the basic percent equation using n as the unknown percent. The cost for electricity is the amount.

Solution
```
  910   Electricity
 1492   Motor gasoline
  383   Natural gas
+  83   Fuel oil, kerosene
─────
 2868
```
$n \times 2868 = 910$

$n = 910 \div 2868$

$n \approx 0.317$

31.7% of the cost is for electricity.

24. **Strategy** To find the total cost of the video camera:
● Find the amount of the sales tax by writing and solving the basic percent equation using n to represent the sales tax. The percent is 6.25% and the base is $980.
● Add the sales tax to the cost of the camera ($980).

Solution
$6.25\% \times \$980 = n$ $\$980.00$
$0.0625 \times 980 = n$ $\underline{+\ \ 61.25}$
$61.25 = n$ $\$1041.25$

The total cost of the video camera is $1041.25.

25. **Strategy** To find the percent of women who wore sunscreen often, write and solve the basic percent equation using n to represent the unknown percent. The base is 350 women and the amount is 275 women.

Solution
$n \times 350 = 275$
$n = 275 \div 350$
$n \approx 0.7857$
Approximately 78.6% of the women wore sunscreen often.

26. **Strategy** To find the world's population in 2000, write and solve the basic percent equation using n to represent the population in 2000. The percent is 149% and the amount is 9,100,000,000 people.

Solution
$149\% \times n = 9,100,000,000$
$1.49 \times n = 9,100,000,000$
$n = 9,100,000,000 \div 1.49$
$n \approx 6,100,000,000$
The world's population in 2000 was approximately 6,100,000,000 people.

27. **Strategy** To find the cost of the computer 4 years ago, write and solve a proportion using n to represent the cost 4 years ago. The percent is 60% and the amount is $1800.

Solution
$\dfrac{60}{100} = \dfrac{1800}{n}$
$60 \times n = 1800 \times 100$
$60 \times n = 180,000$
$n = 180,000 \div 60$
$n = 3000$
The cost of the computer 4 years ago was $3000.

28. **Strategy** To find the dollar value of all online transactions, write and solve the basic percent equation using n to represent the dollar value of all online transactions. The percent is 50.4% and the amount is $25.96 billion.

Solution
$0.504 \times n = \$25.96$ billion
$n = 25.96$ billion $\div\ 0.504$
$n \approx 51.5$ billion
The dollar value of all the online transactions made that year was $52 billion.

Chapter 5 Test

1. $97.3\% = 97.3 \times 0.01 = 0.973$

2. $83\dfrac{1}{3}\% = 83\dfrac{1}{3} \times \dfrac{1}{100} = \dfrac{250}{3} \times \dfrac{1}{100} = \dfrac{250}{300} = \dfrac{5}{6}$

3. $0.3 \times 100\% = 30\%$

4. $1.63 \times 100\% = 163\%$

5. $\dfrac{3}{2} \times 100\% = 1.5 \times 100\% = 150\%$

6. $\dfrac{2}{3} \times 100\% = \dfrac{200}{3}\% = 66\dfrac{2}{3}\%$

7. $77\% \times 65 = n$
$0.77 \times 65 = n$
$50.05 = n$

8. $47.2\% \times 130 = n$
$0.472 \times 130 = n$
$61.36 = n$

9. 7% of $120 = n$ or 76% of $13 = n$
$0.07 \times 120 = n$ $0.76 \times 13 = n$
$8.4 = n$ $9.88 = n$
$9.88 > 8.4$, so 76% of 13 is larger.

10. 13% of $200 = n$ or 212% of $12 = n$
$0.13 \times 200 = n$ $2.12 \times 12 = n$
$26 = n$ $25.44 = n$
$25.44 < 26$, so 212% of 12 is smaller.

11. **Strategy** To find the amount spent for advertising, write and solve the basic percent equation using n to represent the amount spent for advertising. The percent is 6% and the base is $75,000.

Solution
$6\% \times \$75,000 = n$
$0.06 \times 75,000 = n$
$4500 = n$
The amount spent for advertising is $4500.

12. Strategy   To find how many pounds of vegetables were not spoiled:
• Write and solve the basic percent equation using n to represent the number of pounds that were spoiled. The percent is 6.4% and the base is 1250.
• Find the number of pounds that were not spoiled by subtracting the number of pounds of spoiled vegetables from the total (1250 pounds).

Solution
$$6.4\% \times 1250 = n \qquad 1250$$
$$0.064 \times 1250 = n \qquad -\ \ 80$$
$$80 = n \qquad \overline{1170}$$

1170 pounds of vegetables were not spoiled.

13. $\dfrac{440}{3000} \approx 0.147 = 14.7\%$

14.7% of the daily recommended amount of potassium is provided.

14.   Total number of calories $= 180 + 20 = 200$.

The percent provided $= \dfrac{200}{2200} = 9.1\%$.

9.1% of the daily recommended number of calories is provided.

15. Strategy   To find what percent of the permanent employees is hired as temporary employees, write and solve the basic percent using n to represent the percent of the permanent employees. The base is 125 and the amount is 20.

Solution
$$n \times 125 = 20$$
$$n = 20 \div 125$$
$$n = 0.16$$
$$n = 16\%$$

16% of the permanent employees are hired.

16. Strategy   To find what percent of the questions the student answered correctly:
• Find how many questions the student answered correctly by subtracting the number missed (7) from the total number of questions (80).
• Write and solve the basic percent equation using n to represent the percent of questions answered correctly. The base is 80 and the amount is the number of questions answered correctly.

Solution
$$80 - 7 = 73$$
$$n \times 80 = 73$$
$$n = 73 \div 80$$
$$n = 0.9125$$
$$n \approx 91.3\%$$

The student answered approximately 91.3% of the questions correctly.

17.
$$15\% \times n = 12$$
$$0.15 \times n = 12$$
$$n = 12 \div 0.15$$
$$n = 80$$

18.
$$150\% \times n = 42.5$$
$$1.5 \times n = 42.5$$
$$n = 42.5 \div 1.5$$
$$n \approx 28.3$$

19. Strategy   To find the number of PDAs tested, write and solve the basic percent equation using n to represent the number of PDAs tested. The percent is 1.2% and the amount is 384.

Solution
$$1.2\% \times n = 384$$
$$0.012 \times n = 384$$
$$n = 384 \div 0.012$$
$$n = 32{,}000$$

32,000 PDAs were tested.

20. Strategy   To find what percent the increase is of the original price:
• Find the amount of the increase by subtracting the original value ($95,000) from the price 5 years later ($152,000).
• Write and solve the basic percent equation using n to represent the percent. The base is the original price ($95,000) and the amount is the amount of the increase.

Solution
$$\$152{,}000$$
$$-\ \ 95{,}000$$
$$\overline{\$57{,}000}$$
$$n \times \$95{,}000 = \$57{,}000$$
$$n = 57{,}000 \div 95{,}000$$
$$n = 0.60$$
$$n = 60\%$$

The increase is 60% of the original price.

21.
$$\dfrac{86}{100} = \dfrac{123}{n}$$
$$86 \times n = 123 \times 100$$
$$86 \times n = 12{,}300$$
$$n = 12{,}300 \div 86$$
$$n \approx 143.02$$
$$n \approx 143.0$$

22.

$$\frac{n}{100} = \frac{120}{12}$$

$$12 \times n = 100 \times 120$$
$$12 \times n = 12{,}000$$
$$n = 12{,}000 \div 12$$
$$n = 1000$$

1000% of 12 is 120.

23. Strategy To find the dollar increase in the hourly wage:
- Write and solve a proportion to find the hourly wage last year. Let n represent last year's wage. The amount is \$16.24 and the percent is 112%.
- Subtract last year's wage from this year's wage (\$16.24).

Solution

$$\frac{112}{100} = \frac{16.24}{n}$$
$$112 \times n = 16.24 \times 100$$
$$112 \times n = 1624$$
$$n = 1624 \div 112$$
$$n = 14.5$$

$$\begin{array}{r} \$16.24 \\ -\ 14.50 \\ \hline \$1.74 \end{array}$$

The dollar increase is \$1.74.

24. Strategy To find what percent the population now is of the population 10 years ago, write and solve a proportion using n to represent the percent. The base is 32,500 and the amount is 71,500.

Solution

$$\frac{n}{100} = \frac{71{,}500}{32{,}500}$$
$$32{,}500 \times n = 71{,}500 \times 100$$
$$32{,}500 \times n = 7{,}150{,}000$$
$$n = 7{,}150{,}000 \div 32{,}500$$
$$n = 220$$

The population now is 220% of what it was 10 years ago.

25. Strategy To find the value of the car, write and solve a proportion using n to represent the value of the car. The percent is 1.4% and the amount is \$175.

Solution

$$\frac{1.4}{100} = \frac{175}{n}$$
$$1.4 \times n = 175 \times 100$$
$$1.4 \times n = 17{,}500$$
$$n = 17{,}500 \div 1.4$$
$$n = 12{,}500$$

The value of the car is \$12,500.

Cumulative Review Exercises

1. $18 \div (7-4)^2 + 2 = 18 \div (3)^2 + 2$
$$= 18 \div 9 + 2$$
$$= 2 + 2 = 4$$

2.

		2	3	5
$16 =$	$2 \cdot 2 \cdot 2 \cdot 2$			
$24 =$	$2 \cdot 2 \cdot 2$	③		
$30 =$	2	3	⑤	

$$\text{GCF} = 2 \cdot 2 \cdot 2 \cdot 2 \cdot 3 \cdot 5 = 240$$

3.

$$2\frac{1}{3} = 2\frac{8}{24}$$
$$3\frac{1}{2} = 3\frac{12}{24}$$
$$+4\frac{5}{8} = 4\frac{15}{24}$$
$$9\frac{35}{24} = 10\frac{11}{24}$$

4.

$$25\frac{5}{12} = 27\frac{20}{48} = 26\frac{68}{48}$$
$$-14\frac{9}{16} = 14\frac{27}{48} = 14\frac{27}{48}$$
$$12\frac{41}{48}$$

5. $7\frac{1}{3} \times 1\frac{5}{7} = \frac{22}{3} \times \frac{12}{7}$

$$= \frac{22 \times 12}{3 \times 7}$$

$$= \frac{2 \cdot 11 \cdot 2 \cdot 2 \cdot \overset{1}{\cancel{3}}}{\underset{1}{\cancel{3}} \cdot 7}$$

$$= \frac{88}{7} = 12\frac{4}{7}$$

6. $\frac{14}{27} \div 1\frac{7}{9} = \frac{14}{27} \div \frac{16}{9}$

$$= \frac{14}{27} \times \frac{9}{16}$$

$$= \frac{14 \times 9}{27 \times 16}$$

$$= \frac{\overset{1}{\cancel{2}} \cdot 7 \cdot \overset{1}{\cancel{3}} \cdot \overset{1}{\cancel{3}}}{3 \cdot \underset{1}{\cancel{3}} \cdot \underset{1}{\cancel{3}} \cdot 2 \cdot 2 \cdot 2 \cdot 2}$$

$$= \frac{7}{24}$$

7. $\left(\frac{3}{4}\right)^3 \left(\frac{8}{9}\right)^2 = \left(\frac{3}{4} \cdot \frac{3}{4} \cdot \frac{3}{4}\right)\left(\frac{8}{9} \cdot \frac{8}{9}\right)$

$$= \frac{27}{64} \cdot \frac{64}{81}$$

$$= \frac{1}{3}$$

8. $\left(\dfrac{2}{3}\right)^2 - \left(\dfrac{3}{8} - \dfrac{1}{3}\right) \div \dfrac{1}{2} = \dfrac{4}{9} - \left(\dfrac{9}{24} - \dfrac{8}{24}\right) \div \dfrac{1}{2}$

$= \dfrac{4}{9} - \dfrac{1}{24} \div \dfrac{1}{2}$

$= \dfrac{4}{9} - \left(\dfrac{1}{24} \times \dfrac{2}{1}\right)$

$= \dfrac{4}{9} - \dfrac{1}{12}$

$= \dfrac{16}{36} - \dfrac{3}{36} = \dfrac{13}{36}$

9. ⌐——Given place value
3.07973
└——9 > 5
3.08

10.
$$\begin{array}{r} 2\ \ 10\ 8\ 10\ 12\\ \cancel{3}.\cancel{0}\cancel{9}\cancel{0}\ \cancel{2}\\ -1.9706\\ \hline 1.1196 \end{array}$$

11.
$$\begin{array}{r} 34.28125 \approx 34.2813\\ 0.032.\overline{)1.097.00000}\\ -96\\ \hline 137\\ -128\\ \hline 90\\ -64\\ \hline 260\\ -256\\ \hline 40\\ -32\\ \hline 80\\ -64\\ \hline 160\\ -160\\ \hline 0 \end{array}$$

12. $3\dfrac{5}{8} = \dfrac{29}{8}$

$$\begin{array}{r} 3.625\\ 8\overline{)29.000}\\ -24\\ \hline 50\\ -48\\ \hline 20\\ -16\\ \hline 40\\ -40\\ \hline 0 \end{array}$$

13. $1.75 = \dfrac{175}{100} = \dfrac{7}{4} = 1\dfrac{3}{4}$

14. $\dfrac{3}{8} = 0.375$

$\dfrac{3}{8} < 0.87$

15. $\dfrac{3}{8} = \dfrac{20}{n}$

$3 \times n = 8 \times 20$
$3 \times n = 160$
$n = 160 \div 3$
$n \approx 53.3$

16. $\dfrac{\$76.80}{8\ \text{hours}} = \$9.60/\text{hour}$

17. $18\dfrac{1}{3}\% = 18\dfrac{1}{3} \times \dfrac{1}{100} = \dfrac{55}{3} \times \dfrac{1}{100} = \dfrac{55}{300} = \dfrac{11}{60}$

18. $\dfrac{5}{6} \times 100\% = \dfrac{500}{6}\% = 83\dfrac{1}{3}\%$

19. $16.3\% \times 120 = n$
$0.163 \times 120 = n$
$19.56 = n$

20. $n \times 18 = 24$
$n = 24 \div 18$
$n = 1.33\ldots$
$n = 133\dfrac{1}{3}\%$

21. $125\% \times n = 12.4$
$1.25 \times n = 12.4$
$n = 12.4 \div 1.25$
$n = 9.92$

22. $n \times 35 = 120$
$n = 120 \div 35$
$n \approx 3.4285$
$n \approx 342.9\%$

23. **Strategy** To find Sergio's take-home pay:
• Find the amount deducted by multiplying the income (\$740) by $\dfrac{1}{5}$.
• Subtract the amount deducted from the income.

Solution $\dfrac{1}{5} \times \$740 = \$148$
$\$740 - \$148 = \$592$
Sergio's take-home pay is \$592.

24. **Strategy** To find the amount of the monthly payment:
• Find the amount that will be paid by payments by subtracting the down payment (\$1000) from the price of the car (\$8353).
• Divide the total amount remaining to be paid by the number of payments (36).

Solution
$$\begin{array}{r}\$8353\\ -1000\\ \hline \$7353\end{array} \qquad 36\overline{)\$7353.00}\ \ ^{\$204.25}$$
Each monthly payment is \$204.25.

25. Strategy To find the number of gallons of gasoline used during the month, divide the total paid in taxes ($79.80) by the tax paid per gallon ($.19).

Solution $79.80 ÷ $.19 = 420
420 gallons were used during the month.

26. Strategy To find the real estate tax on a house valued at $250,000, write and solve a proportion using n to represent the tax.

Solution
$$\frac{3440}{172,000} = \frac{n}{250,000}$$
$$3440 \times 250,000 = 172,000 \times n$$
$$860,000,000 = 172,000 \times n$$
$$860,000,000 \div 172,000 = n$$
$$5000 = n$$
The real estate tax is $5000.

27. Strategy To find what percent of the purchase price the sales tax is, write and solve the basic percent equation using n to represent the percent. The base is $490 and the amount is $29.40.

Solution $n \times \$490 = \$29.40$
$$n = 29.40 \div 490$$
$$n = 0.06$$
$$n = 6\%$$
The sales tax is 6% of the purchase price.

28. Strategy To find what percent of the people did not favor the candidate:
• Find the number of people who did not favor the candidate by subtracting the number of people who did favor the candidate (165) from the total surveyed (300).
• Write and solve the basic percent equation using n to represent the percent of people who did not favor the candidate. The base is 300 and the amount is the number of people who did not favor the candidate.

Solution
$\begin{array}{r} 300 \\ -165 \\ \hline 135 \end{array}$ $n \times 300 = 135$
$$n = 135 \div 300$$
$$n = 0.45$$
$$n = 45\%$$

45% of the people did not favor the candidate.

29. Strategy To find the average hours:
• Find the number of hours in a week by multiplying the number of hours in a day (24) by the number of days in a week (7).
• Write and solve the basic percent equation using n to represent the number of hours spent watching TV. The percent is 36.5% and the base is 168.

Solution
$\begin{array}{r} 24 \\ \times\ 7 \\ \hline 168 \end{array}$ $36.5\% \times 168 = n$
$$0.365 \times 168 = n$$
$$61.3 \approx n$$

The approximate average number of hours spent watching TV in a week is 61.3 hours.

30. Strategy To find what percent of the children tested had levels of lead that exceeded federal standards, write and solve a proportion using n to represent the percent who had levels of lead that exceeded federal standards. The base is 5500 and the amount is 990.

Solution
$$\frac{n}{100} = \frac{990}{5500}$$
$$n \times 5500 = 990 \times 100$$
$$n \times 5500 = 99,000$$
$$n = 99,000 \div 5500$$
$$n = 18$$
18% of the children tested had levels of lead that exceeded federal standards.

Chapter 6: Applications for Business and Consumers

Prep Test

1. 0.75

2. 52.05

3. 504.51

4. 9750

5. $1500 \times 0.06 \times 0.5 = 90 \times 0.5 = 45$

6. 1417.24

7.
```
   3.33
3)10.00
  -9
   10
   -9
   10
    -9
    1
```

8.
```
      0.605
570)345.000
   -3420
     300
    -  0
    3000
   -2850
     150
```

9. $0.379 < 0.397$

Go Figure

To find the price of the earrings including sales tax, note that multiplying 1.04 by the price before the sales tax is added (in dollars and cents) must yield a whole number. To solve for the number in dollars and cents, divide both sides by 1.04. So we want to find a whole number that when divided by 1.04 results in a terminating decimal with no more than 2 decimal places. To anticipate a solution, the factors of 104 are 2, 2, 2, and 13. Therefore, 2, 4, or 8 divided by 104 results in a terminating decimal; 13, 26, 52, or 104 divided by 104 results in a nonterminating decimal. So the whole number cannot be 2, 4, or 8 since the result would be a nonterminating decimal.
Using trial and error,

$\quad 1 \div 1.04 = 0.961538\ldots$ (nonterminating decimal)
$\quad 2 \div 1.04 = 1.923076\ldots$ (nonterminating decimal)
$\quad 3 \div 1.04 = 2.884615\ldots$ (nonterminating decimal)
$\quad 4 \div 1.04 = 3.846153\ldots$ (nonterminating decimal)
$\qquad \vdots$
$\quad 12 \div 1.04 = 11.53846\ldots$ (nonterminating decimal)
$\quad 13 \div 1.04 = 12.5$

The earrings sold for \$12.50 plus 4% tax (\$.50). So a customer paid \$13, including 4% tax for the pair of earrings.

Section 6.1

Objective A Exercises

1. **Strategy** To find the unit cost, divide the total cost (\$.99) by the number of units (18).

   Solution $.99 \div 18 = 0.055$
   The unit cost is \$.055 per ounce.

3. **Strategy** To find the unit cost, divide the total cost (\$2.99) by the number of units (8).

   Solution $2.99 \div 8 \approx 0.3737$
   The unit cost is \$.374 per ounce.

5. **Strategy** To find the unit cost, divide the total cost (\$3.99) by the number of units (50).

   Solution $3.99 \div 50 = 0.0798$
   The unit cost is \$.080 per tablet.

7. **Strategy** To find the unit cost, divide the total cost (\$13.95) by the number of units (2).

   Solution $13.95 \div 2 = 6.975$
   The unit price is \$6.975 per clamp.

9. **Strategy** To find the unit cost, divide the total cost (\$2.99) by the number of units (15).

   Solution $2.99 \div 15 \approx 0.1993$
   The unit cost is \$.199 per ounce.

11. **Strategy** To find the unit cost, divide the total cost (\$.95) by the number of units (8).

    Solution $0.95 \div 8 \approx 0.1187$
    The unit cost is \$.119 per screw.

Objective B Exercises

13. **Strategy** To find the more economical purchase, compare the unit costs.

    Solution Sutter Home: $3.29 \div 25.5 \approx 0.1290$
    Muir Glen: $3.79 \div 26 \approx 0.1458$
    $0.1290 < 0.1458$
    The Sutter Home pasta sauce is the more economical purchase.

15. **Strategy** To find the more economical purchase, compare the unit costs.

    Solution 20 ounces: $3.29 \div 20 = 0.1645$
    12 ounces: $1.99 \div 12 \approx 0.1658$
    $0.1645 < 0.1648$
    20 ounces is the more economical purchase.

17. **Strategy** To find the more economical purchase, compare the unit costs.

    Solution 200 tablets: $7.39 \div 200 \approx 0.0370$
    400 tablets: $12.99 \div 400 \approx 0.0325$
    $0.0325 < 0.0370$
    400 tablets is the more economical purchase.

19. **Strategy** To find the more economical purchase, compare the unit costs.

    Solution Kraft: $4.37 \div 16 \approx 0.2731$
    Land O' Lakes: $2.29 \div 9 \approx 0.2544$
    $0.2544 < 0.2731$
    Land O' Lakes cheddar cheese is the more economical purchase.

21. **Strategy** To find the more economical purchase, compare the unit costs.

    Solution Maxwell House: $3.99 \div 4 = 0.9975$
    Sanka: $2.39 \div 2 = 1.195$
    $0.9975 < 1.195$
    Maxwell House coffee is the more economical purchase.

23. **Strategy** To find the more economical purchase, compare the unit costs.

    Solution Purina: $4.19 \div 56 \approx 0.0748$
    Friskies: $3.37 \div 50.4 \approx 0.0669$
    $0.0669 < 0.0748$
    Friskies Chef's Blend is the more economical purchase.

Objective C Exercises

25. **Strategy** To find the total cost, multiply the unit cost ($4.59) by the number of units (3).

    Solution $4.59 \times 3 = 13.77$
    The total cost is $13.77.

27. **Strategy** To find the total cost, multiply the unit cost ($.23) by the number of units (8).

    Solution $0.23 \times 8 = 1.84$
    The total cost is $1.84.

29. **Strategy** To find the total cost, multiply the unit cost ($.98) by the number of units (6.5).

    Solution $0.98 \times 6.5 = 6.37$
    The total cost is $6.37.

31. **Strategy** To find the total cost, multiply the unit cost ($1.29) by the number of units (2.1).

    Solution $1.29 \times 2.1 = 2.709$
    The total cost is $2.71.

33. **Strategy** To find the total cost, multiply the unit cost ($7.95) by the number of units $\left(\frac{3}{4}\right)$.

    Solution $7.95 \times \frac{3}{4} = 7.95 \times 0.75 \approx 5.962$
    The total cost is $5.96.

Applying the Concepts

35. Students might explain that unit pricing is used in grocery stores. The unit price of a product is the total price divided by the number of units the product contains. Consumers can use this information to determine which size of a product is the more economical purchase.

Section 6.2

Objective A Exercises

1. **Strategy** To find the percent increase:
   ● Find the amount of the increase by subtracting the enrollment for 1988 (45.4 million) from the projected enrollment for 2008 (54.3 million).
   ● Write and solve the basic percent equation for percent. The base is 45.4 and the amount is the amount of the increase.

    Solution $54.3 - 45.4 = 8.9$
    Percent $\times$ base = amount
    $n \times 45.4 = 8.9$
    $n = 8.9 \div 45.4$
    $n \approx 0.196$
    The percent increase is 19.6%.

3. **Strategy** To find the percent increase:
• Subtract the number of stores before the increase (420) from the number of stores after the increase (914).
• Write and solve the basic percent equation for percent. The base is 420 and the amount is the amount of the increase.

Solution $914 - 420 = 494$
Percent × base = amount
$n \times 420 = 494$
$n = 494 \div 420$
$n \approx 1.176$
The percent increase is 117.6%.

5. **Strategy** To find the percent increase:
• Find the amount of the increase by subtracting the number of events in 1924 (14) from the number of events in 2002 (78).
• Write and solve the basic percent equation for percent. The base is 14 and the amount is the amount of the increase.

Solution $78 - 14 = 64$
Percent × base = amount
$n \times 14 = 64$
$n = 64 \div 14$
$n \approx 4.571$
The percent increase is 457.1%.

7. **Strategy** To find the population 8 years later:
• Write and solve the basic percent equation for the amount of increase. The percent is 24.3% and the base is 127,000.
• Add the increase to initial population (127,000).

Solution Percent × base = amount
$24.3\% \times 127,000 = n$
$0.243 \times 127,000 = n$
$30,861 = n$
$30,861 + 127,000 = 157,861$
The population 8 years later was 157,861 people.

9. **Strategy** To find the average age of American mothers giving birth to their first child:
• Find the increase by writing and solving the basic percent equation for amount of increase. The percent is 16.4% and the base is 21.4 years.
• Add the increase to the 1970 age (21.4)

Solution Percent × base = amount
$16.4\% \times 21.4 = n$
$0.164 \times 21.4 = n$
$3.5096 = n$
$21.4 + 3.5 = 24.9$ years
The average age of American mothers giving birth in 2000 was 24.9 years.

Objective B Exercises

11. **Strategy** To find the markup, solve the basic percent equation for amount.

Solution Percent × base = amount
$42\% \times 85 = n$
$0.42 \times 85 = n$
$35.70 = n$
The markup is $35.70.

13. **Strategy** To find the markup rate, solve the basic percent equation for percent. The base is $3250 and the amount is $975.

Solution Percent × base = amount
$n \times 3250 = 975$
$n = 975 \div 3250$
$n = 0.30 = 30\%$
The markup rate is 30%.

15a. **Strategy** To find the markup, solve the basic percent equation for amount.

Solution Percent × base = amount
$48\% \times 162 = n$
$0.48 \times 162 = n$
$77.76 = n$
The markup is $77.76.

b. **Strategy** To find the selling price, add the markup to the cost.

Solution $77.76 + 162 = 239.76$
The selling price is $239.76.

17a. **Strategy** To find the markup, solve the basic percent equation for amount.

Solution Percent × base = amount
$55\% \times 2 = n$
$0.55 \times 2 = n$
$1.1 = n$
The markup is $1.10.

b. **Strategy** To find the selling price, add the markup to the cost.

Solution $2.00 + 1.10 = 3.10$
The selling price is $3.10.

19. Strategy To find the selling price:
● Solve the basic percent equation for amount to find the amount of the markup.
● Add the amount of the markup to the cost ($50).

Solution Percent × base = amount
$48\% \times 50 = n$
$0.48 \times 50 = n$
$24 = n$
$50 + 24 = 74$
The selling price is $74.

Objective C Exercises

21. Strategy To find the percent decrease, solve the basic percent equation for percent. The base is $800 and the amount is $320.

Solution Percent × base = amount
$n \times 800 = 320$
$n = 320 \div 800$
$n = 0.40 = 40\%$
The amount represents a decrease of 40%.

23. Strategy To find the decrease in the number of employees, solve the percent equation for amount. The base is 1200 and the percent is 45%.

Solution Percent × base = amount
$45\% \times 1200 = n$
$0.45 \times 1200 = n$
$540 = n$
There is decrease of 540 employees.

25a. Strategy To find the amount of decrease, subtract the new value (39) from the original value (52).

Solution $52 - 39 = 13$
The amount of the decrease is 13 minutes.

b. Strategy To find the percent decrease, solve the basic percent equation for percent. The base is 52 and the amount is 13.

Solution Percent × base = amount
$n \times 52 = 13$
$n = 13 \div 52$
$n = 0.25 = 25\%$
The amount represents a decrease of 25%.

27a. Strategy To find the amount of decrease, solve the basic percent equation for amount. The base is $1.60 and the percent is 37.5%.

Solution Percent × base = amount
$37.5\% \times 1.60 = n$
$0.375 \times 1.60 = n$
$0.60 = n$
The amount of decrease is $.60.

b. Strategy To find the dividend this year, subtract the amount of the decrease ($0.60) from the dividend last year ($1.60).

Solution $1.60 - 0.60 = 1.00$
The dividend this year is $1.00.

29. Strategy To find the percent decrease, solve the basic percent equation for percent. The amount is the amount of the decrease (26) and the base is 394.

Solution Percent × base = amount
$n \times 394 = 26$
$n = 26 \div 394$
$n \approx 0.066 = 6.6\%$
The amount represents a decrease of 6.6%.

Objective D Exercises

31. Strategy To find the discount rate, solve the basic percent equation for percent. The base is $24 and the amount is $8.

Solution Percent × base = amount
$n \times 24 = 8$
$n = 8 \div 24$
$n = 0.333\ldots = 33\frac{1}{3}\%$
The discount rate is $33\frac{1}{3}\%$.

33. Strategy To find the discount, solve the basic percent equation for amount. The percent is 20% and the base is $340.

Solution Percent × base = amount
$20\% \times 340 = n$
$0.20 \times 340 = n$
$68 = n$
The discount is $68.

35. Strategy To find the discount rate, solve the basic percent equation for percent. The base is $140 and the amount is $42.

Solution Percent × base = amount
$n \times 140 = 42$
$n = 42 \div 140$
$n = 0.30 = 30\%$
The discount rate is 30%.

37a. **Strategy**   To find the discount, solve the basic percent equation for amount. The percent is 20% and the base is $1.25.

Solution   Percent × base = amount
$$20\% \times 1.25 = n$$
$$0.20 \times 1.25 = n$$
$$0.25 = n$$
The discount is $.25 per pound.

b. **Strategy**   To find the sale price, subtract the discount ($.25) from the original price ($1.25).

Solution $1.25 - 0.25 = 1.00$
The sale price is $1.00 per pound.

39a. **Strategy**   To find the discount, subtract the sale price ($16) from the original cost ($20).

Solution $20 - 16 = 4$
The amount of the discount is $4.

b. **Strategy**   To find the discount rate, solve the basic percent equation for percent. The base is $20 and the amount is $4.

Solution   Percent × base = amount
$$n \times 20 = 4$$
$$n = 4 \div 20$$
$$n = 0.20 = 20\%$$
The discount rate is 20%.

Applying the Concepts

41. $12 × 0.10 + $12 = $1.20 + $12 = $13.20
$12(1.10) = $13.20
Yes

43.   No. Suppose the regular price is $100. Then the sale price is $75 because $100 - 0.25(100) = 100 - 25 = 75$. The promotional sale offers 25% off the sale price of $75, so the price is lowered to $56.25 because $75 - 0.25(75) = 75 - 18.75 = 56.25$. A sale that offers 50% off the regular price of $100 offers the product for $50. Because $50 < $56.25, the better price is the one that is 50% off the regular price.

Section 6.3

Objective A Exercises

1a. $10,000

b. $850

c.   4.25%

d.   2 years

3. **Strategy**   To find the simple interest, multiply the principal by the annual interest rate by the time (in years).

Solution $8000 \times 0.06 \times 2 = 960$
The simple interest owed is $960.

5. **Strategy**   To find the simple interest, multiply the principal by the annual interest rate by the time (in years).

Solution $100,000 \times 0.045 \times \frac{9}{12} = 3375$
The simple interest due is $3375.

7. **Strategy**   To find the simple interest, multiply the principal by the annual interest rate by the time (in years).

Solution $20,000 \times 0.088 \times \frac{9}{12} = 1320$
The simple interest due is $1320.

9. **Strategy**   To find the simple interest, multiply the principal by the annual interest rate by the time (in years).

Solution $5000 \times 0.075 \times \frac{90}{365} \approx 92.47$
The simple interest due is $92.47.

11. **Strategy**   To find the simple interest, multiply the principal by the annual interest rate by the time (in years).

Solution $7500 \times 0.055 \times \frac{75}{365} \approx 84.76$
The simple interest due is $84.76.

13. **Strategy**   To find the maturity value of the loan, add the principal and the simple interest.

Solution $4800 + 320 = 5120$
The maturity value of the loan is $5120.

15. **Strategy**   To find the maturity value:
• Find the simple interest due by multiplying the principal by the annual interest rate by the time (in years).
• Find the maturity value by adding the principal and the simple interest.

Solution $150,000 \times 0.095 \times 1 = 14,250$
$$150,000 + 14,250 = 164,250$$
The maturity value is $164,250.

17. Strategy   To find the total amount due:
• Find the simple interest due by multiplying the principal by the annual interest rate by the time (in years).
• Find the total amount due by adding the principal and the simple interest.

Solution $12,500 \times 0.045 \times \dfrac{8}{12} \approx 375$

$12,500 + 375 = 12,875$

The total amount due on the loan is $12,875.

19. Strategy   To find the maturity value:
• Find the simple interest due by multiplying the principal by the annual interest rate by the time (in years).
• Find the maturity value by adding the principal and the simple interest.

Solution $14,000 \times 0.0525 \times \dfrac{270}{365} \approx 543.70$

$14,000 + 543.70 = 14,543.70$

The maturity value is $14,543.70.

21. Strategy   To find the monthly payment, divide the sum of the loan amount ($225,000) and the interest ($72,000) by the number of payments (48).

Solution $\dfrac{225,000 + 72,000}{48} = 6187.50$

The monthly payment is $6187.50.

23a. Strategy   To find the simple interest charged, multiply the principal ($12,000) by the annual interest rate by the time (in years).

Solution $12,000 \times 0.045 \times 2 = 1080$
The interest charged is $1080.

b. Strategy   To find the monthly payment, divide the sum of the loan amount ($12,000) and the interest ($1080) by the number of payments (24).

Solution $\dfrac{12,000 + 1080}{24} = 545$

The monthly payment is $545.

25. Strategy   To find the monthly payment:
• Find the simple interest due by multiplying the principal by the annual interest rate by the time (in years).
• Find the monthly payment by adding the interest due to the loan amount ($42,000) and dividing that sum by the number of payments (42).

Solution $42,000 \times 0.095 \times 3.5 = 13,965$
The monthly payment is
$\dfrac{42,000 + 13,965}{42} = \$1332.50.$

Objective B Exercises

27. Strategy   To find the finance charge, multiply the unpaid balance by the monthly interest rate by the number of months.

Solution $118.72 \times 0.0125 \times 1 = 1.48$
The finance charge is $1.48.

29. Strategy   To find the finance charge, multiply the unpaid balance by the monthly interest rate by the number of months.

Solution $12,368.92 \times 0.015 \times 1 = 185.53$
The finance charge is $185.53.

31. Strategy   To find the difference in finance charges:
• Find the difference in monthly interest rates by subtracting the smaller rate (1.15%) from the larger rate (1.85%).
• To find the difference in finance charges, multiply the unpaid balance by the difference in monthly interest rates by the number of months.

Solution $0.0185 - 0.0115 = 0.007 = 0.7\%$
$1438.20 \times 0.007 \times 1 \approx 10.07$
The difference in finance charges is $10.07.

Objective C Exercises

33. Strategy   To find the value of the investment in 1 year, multiply the original investment by the compound interest factor.

Solution $750 \times 1.04080 = 780.60$
The value of the investment after 1 year is $780.60.

35. Strategy   To find the value of the investment after 15 years, multiply the original investment by the compound interest factor.

Solution $3000 \times 2.42726 = 7281.78$
The value of the investment after 15 years is $7281.78.

37a. Strategy To find the value of the investment in 5 years, multiply the original investment by the compound interest factor.

Solution $75,000 \times 1.48595 = 111,446.25$
The value of the investment in 5 years will be $111,446.25.

b. Strategy To find the amount of interest that will be earned, subtract the original investment from the new value of the investment.

Solution $111,446.25 - 75,000 = 36,446.25$
The amount of interest earned will be $36,446.25.

39. Strategy To find the amount of interest earned over a 20-year period:
• Find the value of the investment after 20 years by multiplying the original investment by the compound interest factor.
• Subtract the original value of the investment ($2500) from the new value of the investment.

Solution $2500 \times 3.31979 = 8299.48$
$8299.48 - 2500 = 5799.48$
The amount of interest earned is $5799.48.

Applying the Concepts

41. Strategy To find the simple interest owed to the credit union, multiply the principal by the monthly interest rate by the time (in months).

Solution $800 \times 0.02 \times 1 = 16$
The interest owed is $16.

43. You received less interest during the second month because there are fewer days in the month of September (30 days) than in the month of August (31 days). Using the simple interest formula:
$500 \times 0.05 \times \frac{31}{365} \approx 2.12$
$502.12 \times 0.05 \times \frac{30}{365} \approx 2.06$
Even though the principal is greater during the second month, the interest earned is less because there are fewer days in the month.

Section 6.4

Objective A Exercises

1. Strategy To find the mortgage, subtract the down payment from the purchase price.

Solution $97,000 - 14,550 = 82,450$
The mortgage is $82,450.

3. Strategy To find the down payment, solve the basic percent equation for amount. The base is $25,000 and the percent is 30%.

Solution Percent $\times$ base = amount
$0.30 \times 25,000 = 7500$
The down payment is $7500.

5. Strategy To find the down payment, solve the basic percent equation for amount. The base is $850,000 and the percent is 25%.

Solution Percent $\times$ base = amount
$0.25 \times 850,000 = 212,500$
The down payment is $212,500.

7. Strategy To find the loan origination fee, solve the basic percent equation for amount. The base is $150,000 and the percent is $2\frac{1}{2}$%.

Solution Percent $\times$ base = amount
$0.025 \times 150,000 = 3750$
The loan origination fee is $3750.

9a. Strategy To find the down payment, solve the basic percent equation for amount. The base is $150,000 and the percent is 5%.

Solution Percent $\times$ base = amount
$0.05 \times 150,000 = 7,500$
The down payment is $7,500.

b. Strategy To find the mortgage, subtract the down payment from the purchase price.

Solution $150,000 - 7,500 = 142,500$
The mortgage is $142,500.

11. Strategy To find the mortgage:
• Find the down payment by solving the basic percent equation for amount. The percent is 10% and the base is $210,000.
• Subtract the down payment from the purchase price.

Solution Percent $\times$ base = amount
$0.10 \times 210,000 = 21,000$
$210,000 - 21,000 = 189,000$
The mortgage is $189,000.

Objective B Exercises

13. Strategy To find the monthly mortgage payment, multiply the mortgage by the monthly mortgage factor.

Solution $150,000 \times 0.0077182 = 1157.73$
The monthly mortgage payment is $1157.73.

15. Strategy To determine whether the couple can afford to buy the house:
• Find the monthly mortgage payment by multiplying the mortgage amount by the monthly mortgage factor.
• Compare the monthly mortgage payment with $800.

Solution $110,000 \times 0.0073376 \approx 807.14$
$800 < \$807.14$
No, the couple cannot afford to buy the house.

17. Strategy To find the monthly property tax payment, divide the annual property tax by 12.

Solution $1348.20 \div 12 = 112.35$
The monthly property tax is $112.35.

19a. Strategy To find the monthly mortgage payment, multiply the mortgage by the monthly mortgage factor.

Solution $200,000 \times 0.0083920 = 1678.40$
The monthly mortgage payment is $1678.40.

b. Strategy To find how much of the monthly mortgage payment is interest, subtract the amount that is principal ($941.72) from the monthly mortgage payment.

Solution $1678.40 - 941.72 = 736.68$
The interest payment is $736.68.

21. Strategy To find the monthly payment for property tax and home mortgage:
• Find the monthly tax payment by dividing the annual property tax by 12.
• Find the monthly mortgage payment by dividing the annual mortgage payment by 12.
• Add the monthly tax payment to the monthly mortgage payment.

Solution $948 \div 12 = 79 =$ monthly tax payment
$10,844.40 \div 12 = 903.70$
$903.70 + 79 = 982.70$
The total monthly payment for the property tax and home mortgage is $982.70.

23. Strategy To find the monthly mortgage payment:
• Find the mortgage amount by subtracting the down payment from the purchase price.
• Multiply the mortgage amount by the monthly mortgage factor.

Solution $210,000 - 15,000 = 195,000$
$195,000 \times 0.0084386 \approx 1645.53$
The monthly mortgage payment is $1645.53.

Applying the Concepts

25. Choice 1: 8% for 20 years
$100,000 \times 0.0083644 = 836.44$/month
Payback $= 836.44 \times 240$ months $= \$200,745.60$, or 100,745.60 in interest
Choice 2: 8% for 30 years
$100,000 \times 0.0073376 = 733.76$
Payback $= 733.76 \times 360$ months $= \$264,153.60$ or 164,153.60 in interest
$164,153.60 - 100,745.60 = \$63,408$
By using the 20-year loan, the couple will save $63,408.

Section 6.5

Objective A Exercises

1. Strategy To determine whether Amanda has enough money for the down payment:
• Find the down payment by solving the basic percent equation for amount. The base is $7100 and the percent is 12%.
• Compare the required down payment with $780.

Solution Percent × base = amount
$0.12 \times 7100 = 852$ down payment
$852 > \$780$
No, Amanda does not have enough for the down payment.

3. Strategy To find how much sales tax is paid, solve the basic percent equation for amount. The base is $26,500 and the percent is 4.5%.

Solution $0.045 \times 26,500 = 1192.5$
The sales tax is $1192.50.

5. Strategy To find the state license fee, solve the basic percent equation for amount. The base is $22,500 and the percent is 2%.

Solution Percent × base = amount
$0.02 \times 22,500 = 450$
The license fee is $450.

7a. **Strategy** To find the sales tax, solve the basic percent equation for amount. The base is $32,000 and the percent is 3.5%.

Solution Percent × base = amount
$0.035 × 32,000 = 1120$
The sales tax is $1120.

b. **Strategy** To find the total cost of the sales tax and license fee, add the sales tax ($1120) and the license fee ($275).

Solution $1120 + 275 = 1395$
The total cost of the sales tax and license fee is $1395.

9a. **Strategy** To find the down payment, solve the basic percent equation for amount. The base is $16,200 and the percent is 25%.

Solution Percent × base = amount
$0.25 × 16,200 = 4050$
The down payment is $4050.

b. **Strategy** To find the amount financed, subtract the down payment from the purchase price.

Solution $16,200 − 4050 = 12,150$
The amount financed is $12,150.

11. **Strategy** To find the amount financed:
• Find the amount of the down payment by solving the basic percent equation for amount. The base is $35,000 and the percent is 20%.
• Subtract the down payment from the purchase price ($35,000).

Solution Percent × base = amount
$0.20 × 35,000 = 7000$
$35,000 − 7000 = 28,000$
The amount financed is $28,000.

Objective B Exercises

13. **Strategy** To find the monthly truck payment, multiply the amount financed by the monthly payment factor.

Solution $24,000 × 0.0230293 ≈ 552.70$
The monthly truck payment is $552.70.

15. **Strategy** To find how much it costs to operate a car, multiply the number of miles (16,000) by the cost per mile ($.32).

Solution $0.32 × 16,000 = 5120$
The cost is $5120.

17. **Strategy** To find the cost per mile, divide the total cost by the number of miles.

Solution $1600 ÷ 14,000 ≈ 0.11$
The cost is about $.11 per mile.

19. **Strategy** To find the amount of interest, subtract the amount of principal from the monthly car payment.

Solution $143.50 − 68.75 = 74.75$
The amount of interest is $74.75.

21a. **Strategy** To find the amount financed, subtract the down payment ($10,800) from the purchase price ($164,000).

Solution $164,000 − 10,800 = 153,200$
The amount financed is $153,200.

b. **Strategy** To find the monthly truck payment, multiply the amount financed by the monthly payment factor.

Solution $153,200 × 0.0193328 = 2961.78$
The monthly payment is $2961.78.

23. **Strategy** To find the monthly car payment:
• Find the amount financed by subtracting the down payment from the purchase price.
• Multiply the amount financed by the monthly payment factor.

Solution $27,500 − 5500 = 22,000$
$22,000 × 0.0295240 ≈ 649.53$
The monthly payment is $649.53.

Applying the Concepts

25. The total loan cost for the 7% loan plus application fee = $5800 × 0.0239462 × 48 + $45 − $5800 = $911.62
The total loan cost for the 8% loan with no fee = $5800 × 0.0244129 × 48 − $5800 = $996.55
The 7% loan has the lesser loan cost.

Section 6.6

Objective A Exercises

1. **Strategy** To find the earnings, multiply the hourly wage by the number of hours.

Solution $9.50 × 40 = 380$
Lewis earns $380.

3. **Strategy** To find the commission, solve the basic percent equation for amount. The base is $131,000 and the percent is 3%.

Solution Percent × base = amount
$0.03 × 131,000 = 3930$
The real estate agent's commission is $3930.

5. **Strategy** To find the commission, solve the basic percent equation for amount. The base is $5600 and the percent is 1.5%.

   Solution Percent × base = amount
   0.015 × 5600 = 84
   The stockbroker's commission is $84.

7. **Strategy** To find the monthly salary, divide the annual salary by 12.

   Solution 38,928 ÷ 12 = 3244
   Keisha receives $3244 a month.

9. **Strategy** To find the electrician's hourly wage for overtime, multiply the regular hourly wage by 2 (double time).

   Solution 25.80 × 2 = 51.60
   The overtime wage is $51.60/hour.

11. **Strategy** To find the commission, solve the basic percent equation for amount. The base is $450 and the percent is 25%.

    Solution Percent × base = amount
    0.25 × 450 = 112.5
    The golf pro's commission was $112.50.

13. **Strategy** To find the earnings, multiply the earnings per page by the number of pages.

    Solution 2.75 × 225 = 618.75
    The typist earns $618.75.

15. **Strategy** To find the hourly wage, divide the total wage by the number of hours.

    Solution 3400 ÷ 40 = 85
    Maxine's hourly wage is $85.

17a. **Strategy** To find the hourly wage on Saturday, multiply the regular hourly wage by 1.5 (time and a half).

     Solution 15.90 × 1.5 = 23.85
     Mark's hourly wage on Saturday is $23.85.

b. **Strategy** To find the earnings for Saturday, multiply the hourly wage by the number of hours.

   Solution 23.85 × 8 = 190.8
   Mark earns $190.80.

19a. **Strategy** To find the increase in pay, solve the basic percent equation for amount. The base is $21.50 and the percent is 10%.

     Solution Percent × base = amount
     0.10 × 21.50 = $2.15.
     The nurse's increase in pay is $2.15.

b. **Strategy** To find the hourly wage for the night shift, add the increase in pay to the regular hourly wage.

   Solution 21.50 + 2.15 = 23.65
   The nurse's hourly pay is $23.65.

21. **Strategy** To find the earnings for the week:
    • Find the amount of sales over $1500 by subtracting $1500 from the total sales ($3000).
    • Find the commission by solving the basic percent equation for amount. The base is the sales over $1500 and the percent is 15%.
    • Add the commission to the weekly salary ($250).

    Solution 3000 − 1500 = 1500 sales over 1500
    Percent × base = amount
    0.15 × 1500 = 225 commission
    250 + 225 = 475
    Nicole's earnings were $475.

Applying the Concepts

23. Let n represent the previous year's salary.
    Percent × base = amount
    1.018 × n = 52,169
    n = 52,169 ÷ 1.018
    $n ≈$ 51,247
    Increase in salary: 52,169 − 51,247 = $922
    The amount of increase in the starting salary for a chemical engineer from the previous year was $922.

25. Let n represent the previous year's salary.
    Percent × base = amount
    0.874 × n = 28,546
    n = 28,546 ÷ 0.874
    $n ≈$ 32,661
    Decrease in salary: 32,661 − 28,546 = 4115
    The amount of decrease in the starting salary for a political science major from the previous year was $4115.

Section 6.7

Objective A Exercises

1. **Strategy** To find your current checking balance, add the deposit to the old balance.

   Solution 342.51 + 143.81 = 486.32
   Your current checking account balance is $486.32.

3. Strategy To find the current checking account balance, subtract the amount of the check from the old balance.

Solution $2431.76 - 1209.29 = 1222.47$
The real estate firm's current checking account balance is $1222.47.

5. Strategy To find the current checking account balance, subtract the amount of each check from the original balance.

Solution
```
  1204.63
-  119.27
  1085.36
-  260.09
   825.27
```
The nutritionist's current balance is $825.27.

7. Strategy To find the current checking account balance, add the amount of the deposit to the old balance. Then subtract the amount of each check.

Solution
```
  3476.85
+ 1048.53
  4525.38
-  848.37
  3677.01
-  676.19
  3000.82
```
The current checking account balance is $3000.82.

9. Strategy To determine whether there is enough money in the account, compare $675 with the current balance after finding the current checking account balance.

Solution
```
  404.96
+ 350.00
  754.96
-  71.29
  683.67
```
$683.67 > $675
Yes, there is enough money in the carpenter's account to purchase the refrigerator.

11. Strategy To determine whether there is enough money in the account to make the two purchases, add the amounts of the two purchases and compare the total with the current checking account balance.

Solution $3500 + 2050 = 5550$ total of purchases
$5550 < $5625.42
Yes, there is enough money in the account to make the two purchases.

13. Solution
Current checkbook balance:	989.86
Checks: 228	419.32
233	166.40
235	+288.39
	1863.97
Interest:	+13.22
	1877.19
Service charge:	−0.00
	1877.19
Deposits:	−0.00
Checkbook balance:	1877.19

Current bank balance from bank statement: $1877.19.
Checkbook balance: $1877.19.
The bank statement and checkbook balance.

15. Solution
Current checkbook balance:	1051.92
Checks: 223	414.83
224	113.37
Interest:	+5.15
	1585.27
Service charge:	−0.00
	1585.27
Deposits:	−0.00
Checkbook balance:	1585.27

Current bank balance from bank statement: $1585.27.
Checkbook balance: $1585.27.
The bank statement and checkbook balance.

Applying the Concepts

17. added

19. subtract

Chapter 6 Review Exercises

1. Strategy To find the unit cost, divide the total cost ($3.90) by the number of units (20).

Solution $3.90 \div 20 = 0.195$
The unit cost is $.195 per ounce or 19.5¢ per ounce.

2. Strategy To find the cost per mile:
• Find the total cost by adding the amounts spent ($1025.58, $605.82, $37.92, and $188.27).
• Divide the total cost by the number of miles (11,320).

Solution $1025.58 + 605.82 + 37.92 + 188.27 = 1857.59$
$1857.59 \div 11,320 \approx 0.164$
The cost is $.164 or 16.4¢ per mile.

3. Strategy To find the percent increase:
- Find the amount of the increase by subtracting the original price ($42.375) from the increased price ($55.25).
- Solve the basic percent equation for percent. The base is $42.375 and the amount is the amount of the increase.

Solution
$55.25 - 42.375 = 12.875$
$n \times 42.375 = 12.875$
$n = 12.875 \div 42.375 \approx 0.304$
$= 30.4\%$
The percent increase is 30.4%.

4. Strategy To find the markup, solve the basic percent equation for amount. The base is $180 and the percent is 40%.

Solution
$0.40 \times 180 = n$
$72 = n$
The markup is $72.

5. Strategy To find the simple interest, multiply the principal by the annual interest rate by the time (in years).

Solution
$100,000 \times 0.04 \times \frac{9}{12} = 3000$
The simple interest due is $3000.

6. Strategy To find the value of the investment in 10 years, multiply the original investment by the compound interest factor.

Solution
$25,000 \times 1.82203 = 45,550.75$
The value of the investment after 10 years is $45,550.75.

7. Strategy To find the percent increase:
- Find the amount of the increase by subtracting the original amount ($4.12) from the increased amount ($4.73).
- Solve the basic percent equation for percent. The base is $4.12 and the amount is the increased amount.

Solution
$4.73 - 4.12 = 0.61$
$n \times 4.12 = 0.61$
$n = 0.61 \div 4.12 \approx 0.15 = 15\%$
The percent increase is 15%.

8. Strategy To find the total monthly payment:
- Find the monthly property tax by dividing the annual tax ($658.32) by 12.
- Add the monthly property tax payment to the monthly mortgage payment ($523.67).

Solution
$658.32 \div 12 = 54.86$
$54.86 + 523.67 = 578.53$
The total monthly payment for the mortgage and property tax is $578.53.

9. Strategy To find the monthly payment:
- Find the down payment by solving the basic percent equation for amount. The percent is 8% and the base is $24,450.
- Find the amount financed by subtracting the down payment from the purchase price ($24,450).
- Multiply the amount financed by the monthly payment factor.

Solution
$0.08 \times 24,450 = 1956$
$24,450 - 1956 = 22,494$
$22,494 \times 0.0230293 \approx 518.02$
The monthly payment is $518.02.

10. Strategy To find the value of the investment in 1 year, multiply the original investment by the compound interest factor.

Solution
$50,000 \times 1.07186 = 53,593$
The value of the investment will be $53,593.

11. Strategy To find the down payment, solve the basic percent equation for amount. The base is $125,000 and the percent is 15%.

Solution
$0.15 \times 125,000 = 18,750$
The down payment is $18,750.

12. Strategy To find the total cost of the sales tax and license fee:
- Find the sales tax by solving the basic percent equation for amount. The base is $18,500 and the percent is 6.25%.
- Add the sales tax and the license fee ($315).

Solution
$0.0625 \times 18,500 = 1156.25$
$1156.25 + 315 = 1471.25$
The total cost of the sales tax and license fee is $1471.25.

13. Strategy To find the selling price:
- Find the markup by solving the basic percent equation for amount. The percent is 35% and the base is $1540.
- Find the selling price by adding the markup to the cost.

Solution
$0.35 \times 1540 = 539$
$539 + 1540 = 2079$
The selling price is $2079.

14. **Strategy** To find how much of the payment is interest, subtract the principal ($25.45) from the total payment ($122.78).

Solution $122.78 - 25.45 = 97.33$
The interest paid is $97.33.

15. **Strategy** To find the commission, solve the basic percent equation for amount. The base is $108,000 and the percent is 3%.

Solution $0.03 \times 108,000 = n$
$3240 = n$
The commission was $3240.

16. **Strategy** To find the sale price:
• Find the amount of the discount by solving the basic percent equation for amount. The base is $235 and the percent is 40%.
• Subtract the discount from the original price.

Solution $0.40 \times 235 = n$
$94 = n$
$235 - 94 = 141$
The discount price is $141.

17. **Strategy** To find the current checkbook balance, subtract the amount of each check and add the amount of the deposit.

Solution
```
 1568.45
-123.76
 1444.69
-756.45
 688.24
-88.77
 599.47
+344.21
 943.68
```
The current checkbook balance is $943.68.

18. **Strategy** To find the maturity value:
• Find the simple interest due by multiplying the principal by the annual interest rate by the time (in years).
• Find the maturity value by adding the principal and the simple interest.

Solution $30,000 \times 0.08 \times \frac{6}{12} = 1200$
$30,000 + 1200 = 31,200$
The maturity value is $31,200.

19. **Strategy** To find the origination fee, solve the basic percent equation for amount. The base is $75,000 and the percent is $2\frac{1}{2}\%$.

Solution $0.025 \times 75,000 = 1875$
The origination fee is $1875.

20. **Strategy** To find the more economical purchase, compare the unit costs.

Solution $3.49 \div 16 \approx 0.218$
$6.99 \div 33 \approx 0.212$
The more economical purchase is 33 ounces for $6.99.

21. **Strategy** To find the monthly mortgage payment:
• Find the down payment by solving the basic percent equation for amount. The base is $156,000 and the percent is 10%.
• Find the amount financed by subtracting the down payment from the purchase price.
• Find the monthly mortgage payment by multiplying the amount financed by the monthly mortgage factor.

Solution $0.10 \times 156,000 = 15,600$
$156,000 - 15,600 = 140,400$
$140,400 \times 0.006653 = 934.08$
The monthly mortgage payment is $934.08.

22. **Strategy** To find the total income:
• Find the overtime wage by multiplying the regular wage by 1.5 (time and half).
• Find the number of overtime hours worked by subtracting the regular weekly schedule (40) from the total hours worked (48).
• Find the wages earned for overtime by multiplying the overtime wage by the number of overtime hours worked.
• Find the wages for the 40-hour week by multiplying the hourly rate ($12.60) by 40.
• Add the pay from the overtime hours to the pay from the regular week.

Solution $1.5 \times 12.60 = 18.90$
$48 - 40 = 8; 8 \times 18.90 = 151.20$
$40 \times 12.60 = 504$
$504 + 151.20 = 655.20$
The total income was $655.20.

23. Strategy To find the donut shop's current checkbook balance, subtract the amount of each check and add the amount of each deposit.

Solution

$$
\begin{array}{r}
9567.44 \\
- 1023.55 \\
\hline
8543.89 \\
- 345.44 \\
\hline
8198.45 \\
- 23.67 \\
\hline
8174.78 \\
+ 555.89 \\
\hline
8730.67 \\
+ 135.91 \\
\hline
8866.58
\end{array}
$$

The donut shop's checkbook balance is $8866.58.

24. Strategy To find the monthly payment, divide the sum of the loan amount ($55,000) and the interest ($1375) by the number of payments (4).

Solution $\dfrac{55,000 + 1375}{4} = 14,093.75$

The monthly payment is $14,093.75.

25. Strategy To find the finance charge, multiply the unpaid balance by the monthly interest rate by the number of months.

Solution $576 \times 0.0125 \times 1 = 7.2$
The finance charge is $7.20.

Chapter 6 Test

1. Strategy To find the cost per foot, divide the total cost ($138.40) by the number of feet (20).

Solution $138.40 \div 20 = 6.92$
The cost per foot is $6.92.

2. Strategy To find the more economical purchase, compare the unit prices of the items.

Solution $7.49 \div 3$ or $12.59 \div 5$
$7.49 \div 3 \approx 2.50$; $12.59 \div 5 \approx 2.52$
The more economical purchase is 3 pounds for $7.49.

3. Strategy To find the total cost, multiply the cost per pound ($4.15) by the number of pounds (3.5).

Solution $4.15 \times 3.5 \approx 14.53$
The total cost is $14.53.

4. Strategy To find the percent increase:
- Find the amount of the increase by subtracting the original price ($415) from the increased price ($498).
- Solve the basic percent equation for percent. The base is $415 and the amount is the amount of the increase.

Solution $498 - 415 = 83$
$n \times 415 = 83$
$n = 83 \div 415$
$n = 0.20 = 20\%$
The percent increase in the cost of the exercise bicycle is 20%.

5. Strategy To find the selling price:
- Find the amount of the markup by solving the basic percent equation for amount. The percent is 40% and the base is $215.
- Add the markup to the cost ($215).

Solution $0.40 \times 215 = 86$
$215 + 86 = 301$
The selling price of a compact disc player is $301.

6. Strategy To find the percent decrease:
- Find the amount of the decrease by subtracting the decreased value ($360) from the original value ($390).
- Solve the basic percent equation for percent. The base is ($390) and the amount is the amount of the decrease.

Solution $390 - 360 = 30$
$n \times 390 = 30$
$n = 30 \div 390$
$n \approx 0.077 = 7.7\%$
The percent decrease is 7.7%.

7. Strategy To find the percent decrease:
- Find the amount of the decrease by subtracting the decreased value ($896) from the original value ($1120).
- Solve the basic percent equation for percent. The base is $1120 and the amount is the amount of the decrease.

Solution $1120 - 896 = 224$
$n \times 1120 = 224$
$n = 224 \div 1120 = 0.20 = 20\%$
The percent decrease is 20%.

8. **Strategy** To find the sale price:
   • Find the amount of the discount by solving the basic percent equation for amount. The base is $299 and the percent is 30%.
   • Subtract the amount of the discount from the regular price ($299).

   Solution
   $0.30 \times 299 = n$
   $89.7 = n$
   $299 - 89.70 = 209.30$
   The sale price of the corner hutch is $209.30.

9. **Strategy** To find the discount rate:
   • Find the amount of the discount by subtracting the sale price ($5.70) from the regular price ($9.50).
   • Solve the basic percent equation for percent. The base is $9.50 and the amount is the amount of the discount.

   Solution
   $9.50 - 5.70 = 3.80$
   $n \times 9.50 = 3.80$
   $n = 3.80 \div 9.50 = 0.40 = 40\%$
   The discount rate is 40%.

10. **Strategy** To find the simple interest due, multiply the principal by the annual interest rate by the time in years.

    Solution $75,000 \times 0.08 \times \dfrac{4}{12} = 2000$
    The simple interest due is $2000.

11. **Strategy** To find the maturity value:
    • Find the simple interest due by multiplying the principal by the annual interest rate by the time (in years).
    • Find the maturity value by adding the principal and the simple interest.

    Solution $25,000 \times 0.092 \times \dfrac{9}{12} = 1725$
    $25,000 + 1725 = 26,725$
    The maturity value is $26,725.

12. **Strategy** To find the finance charge, multiply the unpaid balance by the monthly interest rate by the number of months.

    Solution $374.95 \times 0.012 \times 1 = 4.50$
    The finance charge is $4.50.

13. **Strategy** To find the interest earned:
    • Find the value of the investment in 10 years by multiplying the original investment by the compound interest factor.
    • Find the interest earned by subtracting the original investment from the new value of the investment.

Solution $30,000 \times 1.81402 = 54,420.6$
$54,420.60 - 30,000 = 24,420.60$
The amount of interest earned in 10 years will be $24,420.60.

14. **Strategy** To find the loan origination fee, solve the basic percent equation for amount. The base is $134,000 and the percent is $2\dfrac{1}{2}\%$.

    Solution $0.025 \times 134,000 = 3350$
    The origination fee is $3350.

15. **Strategy** To find the monthly mortgage payment, multiply the mortgage amount by the monthly mortgage factor.

    Solution $222,000 \times 0.0077182 \approx 1713.44$
    The monthly mortgage payment is $1713.44.

16. **Strategy** To find the amount financed:
    • Find the amount of the down payment by solving the basic percent equation for amount. The base is $23,750 and the percent is 20%.
    • Subtract the down payment from the purchase price.

    Solution $0.20 \times 23,750 = 4,750$
    $23,750 - 4,750 = 19,000$
    The amount financed is $19,000.

17. **Strategy** To find the monthly car payment:
    • Find the amount of the down payment by solving the basic percent equation for amount. The base is $23,714 and the percent is 15%.
    • Find the amount financed by subtracting the down payment from the purchase price ($23,714).
    • Multiply the amount financed by the monthly mortgage factor.

    Solution
    $0.15 \times 23,714 = 3,557.10$
    $23,714 - 3,557.10 = 20,156.90$
    $20,156.90 \times 0.0239462 \approx 482.681$
    The monthly car payment is $482.68.

18. **Strategy** To find Shaney's total weekly earnings:
● Find the hourly overtime wage for multiplying the hourly wage ($20.40) by 1.5 (time and a half).
● Find the earnings for overtime by multiplying the number of overtime hours (15) by the hourly overtime wage.
● Find the earnings for the normal hours worked by multiplying the number of hours worked (30) by the hourly rate ($20.40).
● Add the earnings from the night hours to the salary from the normal hours.

Solution
$20.40 \times 1.5 = 30.60$
$15 \times 30.60 = 459$
$30 \times 20.40 = 612$
$459 + 612 = 1071$
Shaney earns $1071.

19. **Strategy** Find the current checkbook balance by subtracting the checks written and adding the deposit to the original balance.

Solution
$$
\begin{array}{r}
7349.44 \\
-\ 1349.67 \\
\hline
5999.77 \\
-\ 344.12 \\
\hline
5655.65 \\
+\ 956.60 \\
\hline
6612.25
\end{array}
$$

The current checkbook balance is $6612.25.

20. **Solution**

Current checkbook balance:	1106.31
Checks:	322.37
	413.45
	+78.20
	1920.33
Service charge:	−0.00
	1920.33
Deposits:	−0.00
Checkbook balance:	1920.33

Current bank balance from bank statement: $1920.33.
Checkbook balance: $1920.33.
The bank statement and checkbook balance.

Cumulative Review Exercises

1.
$12 - (10 - 8)^2 \div 2 + 3$
$12 - 2^2 \div 2 + 3$
$12 - 4 \div 2 + 3$
$12 - 2 + 3$
$10 + 3 = 13$

2.
$$
\begin{array}{r}
3\frac{1}{3} = 3\frac{8}{24} \\
4\frac{1}{8} = 4\frac{3}{24} \\
+1\frac{1}{12} = 1\frac{2}{24} \\
\hline
8\frac{13}{24}
\end{array}
$$

3.
$$
\begin{array}{r}
12\frac{3}{16} = 12\frac{9}{48} = 11\frac{57}{48} \\
-9\frac{5}{12} = 9\frac{20}{48} = 9\frac{20}{48} \\
\hline
2\frac{37}{48}
\end{array}
$$

4.
$5\frac{5}{8} \times 1\frac{9}{15} = \frac{45}{8} \times \frac{24}{15}$

$= \frac{45 \times 24}{8 \times 15}$

$= \frac{\overset{1}{\cancel{3}} \cdot 3 \cdot \overset{1}{\cancel{5}} \cdot \overset{1}{\cancel{2}} \cdot \overset{1}{\cancel{2}} \cdot \overset{1}{\cancel{2}} \cdot 3}{\underset{1}{\cancel{2}} \cdot \underset{1}{\cancel{2}} \cdot \underset{1}{\cancel{2}} \cdot \underset{1}{\cancel{3}} \cdot \underset{1}{\cancel{5}}} = 9$

5. $3\frac{1}{2} \div 1\frac{3}{4} = \frac{7}{2} \div \frac{7}{4} = \frac{7}{2} \times \frac{4}{7} = \frac{\overset{1}{\cancel{7}} \cdot \overset{1}{\cancel{2}} \cdot 2}{\underset{1}{\cancel{2}} \cdot \underset{1}{\cancel{7}}} = 2$

6. $\left(\frac{3}{4}\right)^2 \div \left(\frac{3}{8} - \frac{1}{4}\right) + \frac{1}{2}$

$\left(\frac{3}{4} \cdot \frac{3}{4}\right) \div \left(\frac{3}{8} - \frac{2}{8}\right) + \frac{1}{2}$

$\frac{9}{16} \div \frac{1}{8} + \frac{1}{2}$

$\frac{9}{16} \times \frac{8}{1} + \frac{1}{2}$

$\frac{9}{2} + \frac{1}{2} = \frac{10}{2} = 5$

7.
$$
\begin{array}{r}
52.18 \approx 52.2 \\
0.059.\overline{)3.079.20} \\
-295 \\
\hline
129 \\
-118 \\
\hline
112 \\
-59 \\
\hline
530 \\
-472 \\
\hline
58
\end{array}
$$

8. $\dfrac{17}{12} = 17 \div 12$

$$1.4166 \approx 1.417$$
$$12\overline{)17.0000}$$
$$\underline{-12}$$
$$50$$
$$\underline{-48}$$
$$20$$
$$\underline{-12}$$
$$80$$
$$\underline{-72}$$
$$80$$
$$\underline{-72}$$
$$8$$

9. $\dfrac{\$410}{8 \text{ hours}} = \$51.25/\text{hour}$

10.
$$\frac{5}{n} = \frac{16}{35}$$
$$5 \times 35 = n \times 16$$
$$175 = n \times 16$$
$$175 \div 16 = n$$
$$10.9375 = n$$
$$10.94 \approx n$$

11. $\dfrac{5}{8} \times 100\% = \dfrac{500}{8}\% = 62.5\%$

12. 6.5% of $420 = 0.065 \times 420 = 27.3$

13. $18.2 \times 0.01 = 0.182$

14. $n \times 20 = 8.4$
$n = 8.4 \div 20 = 0.42 = 42\%$

15. $0.12 \times n = 30$
$n = 30 \div 0.12 = 250$

16. $0.42 \times n = 65$
$n = 65 \div 0.42 \approx 154.76$

17. **Strategy** To find the total rainfall for the 3 weeks, add the 3 weekly amounts $\left(3\dfrac{3}{4}, 8\dfrac{1}{2}, \text{ and } 1\dfrac{2}{3} \text{ inches}\right)$.

Solution
$$3\frac{3}{4} = 3\frac{9}{12}$$
$$8\frac{1}{2} = 8\frac{6}{12}$$
$$+1\frac{2}{3} = 1\frac{8}{12}$$
$$12\frac{23}{12} = 13\frac{11}{12}$$

The total rainfall is $13\dfrac{11}{12}$ inches.

18. **Strategy** Find the amount paid in taxes by multiplying the total monthly income ($4850) by the portion paid in taxes $\left(\dfrac{1}{5}\right)$.

Solution $4850 \times \dfrac{1}{5} = 970$
The amount paid in taxes is $970.

19. **Strategy** To find the ratio:
• Find the amount of the decrease by subtracting the decreased price ($30) from the original price ($75).
• Write in simplest form the ratio of the decrease to the original price.

Solution $75 - 30 = 45$
$$\frac{45}{75} = \frac{3}{5}$$
The ratio is $\dfrac{3}{5}$.

20. **Strategy** To find the number of miles driven per gallon of gasoline, divide the number of miles driven (417.5) by the number of gallons used (12.5).

Solution $417.5 \div 12.5 = 33.4$
The mileage was 33.4 miles per gallon.

21. **Strategy** To find the unit cost, divide the total cost ($12.96) by the number of pounds (14).

Solution $12.96 \div 14 \approx 0.93$
The cost is $.93 per pound.

22. **Strategy** To find the dividend on 200 shares, write and solve a proportion.

Solution
$$\frac{80}{112} = \frac{200}{n}$$
$$80 \times n = 112 \times 200$$
$$80 \times n = 22,400$$
$$n = 22,400 \div 80$$
$$n = 280$$
The dividend is $280.

23. **Strategy** To find the sale price:
• Solve the basic percent equation for amount to find the amount of the discount. The base is $900 and the percent is 20%.
• Subtract the discount from the regular price.

Solution $0.20 \times 900 = 180$
$900 - 180 = 720$
The sale price is $720.

24. Strategy   To find the selling price:
● Find the amount of markup by solving the basic percent equation for amount. The base is $85 and the percent is 40%.
● Add the markup to the cost.

Solution $0.40 \times 85 = 34$
$85 + 34 = 119$
The selling price of the disc player is $119.

25. Strategy   To find the percent increase:
● Find the amount of the increase by subtracting the original value from the value after the increase.
● Solve the basic percent equation for the percent. The base is $2800 and the amount is the amount of the increase.

Solution $3024 - 2800 = 224$
$n \times 2800 = 224$
$n = 224 \div 2800 = 0.08 = 8\%$
The percent increase in Sook Kim's salary is 8%.

26. Strategy   To find the simple interest due, multiply the principal by the annual rate by the time (in years).

Solution $120{,}000 \times 0.045 \times \dfrac{6}{12} = 2700$
The simple interest due is $2700.

27. Strategy   To find the monthly payment:
● Find the amount financed by subtracting the down payment from the purchase price.
● Multiply the amount financed by the monthly mortgage factor.

Solution $26{,}900 - 2{,}000 = 24{,}900$
$24{,}900 \times 0.0317997 \approx 791.812$
The monthly car payment is $791.81.

28. Strategy   To find the new checking account balance, add the deposit to the original balance and subtract the check amounts.

Solution
1846.78
$+568.30$
2415.08
-123.98
2291.10
-47.33
2243.77
The family's new checking account balance is $2243.77.

29. Strategy   To find the cost per mile:
● Find the total cost by adding the expenses ($840, $520, $185, and $432).
● Divide the total cost by the number of miles driven (10,000).

Solution
840
520
185
$+432$
1977
$1977 \div 10{,}000 = 0.1977$
The cost per mile is about $.20.

30. Strategy   To find the monthly mortgage payment, multiply the mortgage amount by the monthly mortgage factor.

Solution $172{,}000 \times 0.0071643 \approx 1232.26$
The monthly mortgage payment is $1232.26.

Chapter 7: Statistics and Probability

Prep Test

1. $\dfrac{49 \text{ billion}}{102 \text{ billion}} \approx 0.480 = 48.0\%$ was bill-related mail.

2. Between 2005 and 2006
$\$74{,}418 - 70{,}206 = \$4212$
Between 2006 and 2007
$\$78{,}883 - 74{,}418 = \$4465$
Between 2007 and 2008
$\$83{,}616 - 78{,}883 = \$4733$
Between 2008 and 2009
$\$88{,}633 - \$83{,}616 = \$5017$
Between 2009 and 2010
$\$93{,}951 - 88{,}633 = \$5318$
a. The greatest cost increase is between 2009 and 2010.
b. Between those years, there was an increase of $5318.

3a. $\dfrac{45 \text{ gold}}{27 \text{ silver}} = \dfrac{45}{27} = \dfrac{5}{3}$

b. 27 silver : 27 bronze $= 27 : 27 = 1 : 1$.

4a. 3.9, 3.9, 4.2, 4.5, 5.2, 5.5, 7.1

b. $\dfrac{3.9 + 4.5 + 4.2 + 3.9 + 5.2 + 7.1 + 5.5}{7} = 4.9$ million

5a. $5\% \times 90{,}000 = 0.05 \times 90{,}000$
$= 4500$ women are in the Marine Corps.

b. $5\% = 0.05 = \dfrac{5}{100} = \dfrac{1}{20}$

or $\dfrac{4500}{90{,}000} = \dfrac{1}{20}$ of the women in the military are in the Marine Corps.

Go Figure

If my father's parents have 10 grandchildren, and I have 2 brothers and 1 sister, then my siblings and I account for 4 out of the 10 grandchildren. We have 6 first cousins on my father's side.
If my mother's parents have 11 grandchildren, and I have 2 brothers and 1 sister, then my siblings and I account for 4 out of the 11 grandchildren. We have 7 first cousins on my mother's side.
Adding the 6 first cousins on my father's side to the 7 first cousins on my mother's side, results in my having 13 first cousins.

Section 7.1

Objective A Exercises

1. **Strategy** To find the gross revenue:
   • Read the pictograph to determine the gross revenue of the four movies.
   • Add the four numbers.

   Solution
   150 million
   300 million
   200 million
   $\underline{+\ 100 \text{ million}}$
   750 million
   The gross revenue is $750 million.

3. **Strategy** To find the percent, solve the basic percent equation for percent. The base is 750 million (from Exercise 1) and the amount is the revenue from *The Lion King* (300 million).

   Solution Percent $\times$ base $=$ amount
   $n \times 750 \text{ million} = 300 \text{ million}$
   $n = 300 \div 750$
   $n = 0.40$
   The percent is 40%.

5. **Strategy** To find how many more people agreed that humanity should explore planets than agreed that space exploration impacts daily life, subtract the number that agreed that space exploration impacts daily life (600) from the number that agreed that humanity should explore planets (650).

   Solution $650 - 600 = 50$
   50 more people agreed that humanity should explore space than agreed that space exploration impacts daily life.

7. **Strategy** To find the number of children who said they hid vegetables under a napkin, write and solve the basic percent equation for amount. The percent is 30% and the base is 500.

   Solution Percent $\times$ base $=$ amount
   $0.30 \times 500 = 150$
   150 children said they hid their vegetables under a napkin.

9. No, the sum of the percents given in the graph is only 80%, not 100%.

Objective B Exercises

11. Strategy To find the ratio:
● Read the circle graph to determine the units in Finance and Accounting.
● Write in the simplest form the ratio of the number of units in Finance to the number of units in Accounting.

Solution Number of units in Finance: 15
Number of units in Accounting: 45
$\frac{15}{45} = \frac{1}{3}$
The ratio is $\frac{1}{3}$.

13. Strategy To find the percent, solve the basic percent equation for percent. The base is the total number of units needed (128 units) and the amount is the number of units in math (12 units).

Solution Percent × base = amount
$n \times 128 = 12$
$n = 12 \div 128 = 0.094 = 9.4\%$
The percent is 9.4%.

15. Strategy To find the number of people surveyed:
● Read the circle graph to determine the number of responses.
● Add the five numbers.

Solution
High ticket prices: 33
People talking: 42
Uncomfortable seats: 17
Dirty floors: 27
High food prices : + 31
150
The number of people surveyed is 150 people.

17. Strategy To find the percent, solve the basic percent equation for percent. The base is the total number of responses (150) and the amount is the number of "people talking" responses (42).

Solution Percent × base = amount
$n \times 150 = 42$
$n = 42 \div 150 = 0.28 = 28\%$
The percent is 28%.

19. Strategy To find the amount of money spent:
● Read the circle graph to find the percent of money spent on portable game machines.
● Use the basic percent equation to find the amount.

Solution 9% is spent on portable game machines.
Percent × base = amount
$0.09 \times 3,100,000,000 = n$
$279,000,000 = n$
Americans spend $279,000,000 on portable game machines.

21. Strategy To determine whether the amount spent for TV game machines is more than three times the amount spent for portable game machines:
● Multiply by 3 the amount spent for portable game machines (Use amount from Exercise 19).
● Compare the result with the amount spent for TV game machines. (Use amount from Exercise 18).

Solution $3 \times 279,000,000 = 837,000,000$
$837,000,000 < 1,085,000,000$
Yes, the amount spent for TV game machines is more than three times the amount spent for portable game machines.

23. Strategy To find whether the number of homeless who are aged 25 to 34 is more or less than twice the number of homeless under the age of 25, read the pictograph and determine whether the number of homeless aged 25 to 34 (25%) is more or less than twice the number of homeless under 25 (12%).

Solution $2 \times 12\% = 24\%$
$25\% > 24\%$
The number of homeless aged 25 to 34 is more than twice the number of homeless under the age of 25.

25. Strategy To find how many of every 100,000 homeless people are over age 54:
● Locate the percent of homeless over age 54.
● Solve the basic percent equation for amount. The base is 100,000.

Solution Percent homeless over age 54: 8%
Percent × base = amount
$0.08 \times 100,000 = n$
$8,000 = n$
Out of every 100,000 homeless people, there are 8000 people over age 54.

27. Strategy To find the difference, subtract the area of South America (6,870,000 square miles) from the area of North America (9,420,000 square miles).

Solution
$$\begin{array}{rl} 9,420,000 & \text{North America} \\ -\ 8,870,000 & \text{South America} \\ \hline 2,550,000 & \end{array}$$
North America is 2,550,000 square miles larger than South America.

29. Strategy To find the percent:
• Read the circle graph to determine the land area of Australia.
• Write and solve the basic percent equation for percent. The amount is the land area of Australia and the base is the total land area of the seven continents. (57,240,000 square miles).

Solution The area of Australia is 2,970,000 square miles.
Percent × base = amount
$n \times 57,240,000 = 2,970,000$
$n = 2,970,000 \div 57,240,000$
≈ 0.052
Australia is 5.2% of the total land area.

31. Strategy To find the amount spent on health care:
• Locate the percent of the after-tax income that is spent on health care.
• Solve the basic percent equation for amount.

Solution Spent on health care: 5%
Percent × base = amount
$0.05 \times 40,550 = 2027.5$
$2027.50 is spent on health care.

33. Strategy To find whether the amount for housing is more than twice the amount spent on transportation, since the base is the same ($40,550) in each case, compare the percents.

Solution Spent on housing: 32%
Spent on transportation: 17%
$32\% < 2 \times 17\% = 34\%$
No, the amount spent on housing is not more than twice the amount spent on transportation.

Applying the Concepts

35. Answers will vary. For example: The couple's largest single expense was rent.
Food represents approximately one-quarter of the month's expenditures.
Rent represents approximately one-third of the month's expenditures.
The expenditure for food is approximately the same as the expenditures for entertainment and transportation.
The couple spent more for transportation than for entertainment.

Section 7.2

Objective A Exercises

1. Strategy To find the total passenger cars produced worldwide:
• Read the bar graph to determine the number of cars produced (in millions) in each region.
• Add the five numbers

Solution
$$\begin{array}{rl} 15 & \text{Western Europe} \\ 11 & \text{Asia} \\ 8 & \text{Eastern Europe/Russia} \\ 3 & \text{North America} \\ +\ 2 & \text{Latin America} \\ \hline 39 & \end{array}$$
39 million passenger cars were produced worldwide.

3. Strategy To find the percent, solve the basic percent equation for percent. The base is the total number of cars produced (39 million, from Exercise 1) and the amount is the number of cars produced in Asia (11 million)

Solution Percent × base = amount
$n \times 39 = 11$
$n = 11 \div 39 \approx 0.28 = 28\%$
The percent is 28%.

5. Strategy To find the difference:
• Read the double bar graph for the Mini Cooper to find the fuel efficiency in the city (28 MPG) and the fuel efficiency on the highway (38 MPG).
• Subtract the two numbers.

Solution
$$\begin{array}{rl} 38 & \text{Highway} \\ -\ 28 & \text{City} \\ \hline 10 & \end{array}$$
The difference is approximately 10 miles per gallon.

7. Strategy To estimate the difference between the maximum salaries in New York:
• Read the double-bar graph for the maximum salaries for city and suburb police officers.
• Subtract to find the difference between the two salaries.

Solution Suburb salary: 60,000
City salary: − 44,000
16,000
The maximum salary of police officers in the suburbs is $16,000 higher than the maximum salary of police officers in the city.

9. Strategy To find which city has the greatest difference between the maximum salary in the city and in the suburb:
• Read the double-bar graph to find maximum salaries for in the city and the suburb.
• Subtract the maximum salary in the city from the maximum salary in the suburb.

Solution Washington, D.C.:
51,000 − 41,000 = 10,000
Detroit:
46,000 − 38,000 = 8,000
New York:
60,000 − 44,000 = 16,000
Philadelphia:
56,000 − 38,000 = 18,000
Los Angeles:
52,000 − 49,000 = 3,000
The greatest difference in salaries is in Philadelphia.

Objective B Exercises

11. Strategy To find the amount of snowfall during January, read the broken-line graph for January.

Solution The amount of snowfall during January was 20 inches.

13. Strategy To find the total snowfall during March and April:
• Read the broken-line graph to find the snowfall amounts for March and April.
• Add the two amounts.

Solution March 17
April + 8
25
The snowfall during March and April was 25 inches.

15. Strategy To find the difference:
• Read the broken-line graph to find the number of Calories recommended for men and the number recommended for women 19–22 years of age.
• Subtract the number recommended for women from the number recommended for men.

Solution For men: 2900
For women: − 2100
800

The difference is 800 Calories.

17. Strategy To find the ratio:
• Read the double broken-line graph to find the number of Calories recommended for women 15–18 years old and the number recommended for women 51–74 years old.
• Write in simplest form the ratio of the number of Calories recommended for women 15–18 years old to the number recommended for women 51–74 years old.

Solution Women 15–18 years old: 2100
Women 51–74 years old: 1800
$\frac{2100}{1800} = \frac{7}{6}$
The ratio is $\frac{7}{6}$.

Applying the Concepts

19. **Strategy** To create each entry of the table, read the values (in billions of dollars) from the graph of the foreign aid and domestic aid, and find the difference.

    Solution For 1991, the total aid:
    $2.1 - 0.8 = $1.3 billion
    For 1992, the total aid:
    $1.9 - 0.8 = $1.1 billion
    For 1993, the total aid:
    $1.7 - 0.7 = $1.0 billion
    For 1994, the total aid:
    $1.5 - 0.6 = $0.9 billion
    For 1995, the total aid:
    $1.3 - 0.5 = $0.8 billion
    For 1996, the total aid:
    $1.3 - 0.5 = $0.8 billion
    For 1997, the total aid:
    $1.8 - 0.7 = $1.1 billion
    For 1998, the total aid:
    $1.7 - 0.8 = $0.9 billion
    For 1999, the total aid:
    $2.3 - 1.3 = $1.0 billion

Year	Difference
1991	$1.3 billion
1992	$1.1 billion
1993	$1.0 billion
1994	$0.9 billion
1995	$0.8 billion
1996	$0.8 billion
1997	$1.1 billion
1998	$0.9 billion
1999	$1.0 billion

Section 7.3

Objective A Exercises

1. **Strategy** Read the histogram to find the number of students who have a tuition between $3000 and $6000.

    Solution 44 students have a tuition that is between $3000 and $6000.

3. **Strategy** To find the number of students who pay more than $12,000 for tuition:
   • Read the histogram to find the number of students whose tuition is between $12,000 and $15,000 and the number whose tuition is between $15,000 and $18,000.
   • Add the two numbers.

    Solution $12,000 – $15,000   10 students
    $15,000 – $18,000   + 8 students
    18 students
    18 students paid more than $12,000.

5. **Strategy** To find the number of cars between 6 and 12 years old:
   • Read the histogram to find the number of cars between 6 and 9 years old and the number between 9 and 12 years old.
   • Add the two numbers.

    Solution 6 to 9 years: 220 cars
    9 to 12 years: + 190 cars
    410 cars

    There are 410 cars between 6 and 12 years old.

7. **Strategy** To find the number of cars more than 12 years old:
   • Read the histogram to find the number of cars 12 to 15 years old and the number 15 to 18 years old.
   • Add the two numbers.

    Solution 12 to 15 years: 90 cars
    15 to 18 years: + 140 cars
    230 cars

    230 cars are more than 12 years old.

9. **Strategy** To find the number of adults who spend between 1 and 2 hours at the mall, read the histogram.

    Solution 54 adults spend between 1 and 2 hours at the mall.

11. **Strategy** To find the percent:
    • Read the histogram to find the number of adults who spend less than 1 hour at the mall.
    • Solve the basic percent equation for percent. The base is 100 and the amount is the number of adults who spend less than one hour at the mall.

    Solution Number of adults who spend less then one hour at the mall: 22

    Percent × base = amount
    $$n \times 100 = 22$$
    $$n = 22 \div 100$$
    $$n = 0.22$$
    The percent is 22%.

Objective B Exercises

13. Strategy To find how many entrants were in the discus finals:
● Read the number of entrants for each of the points on the frequency polygon.
● Add the five numbers.

Solution

8	between 150 and 160 feet
9	between 160 and 170 feet
4	between 170 and 180 feet
2	between 180 and 190 feet
+ 1	between 190 and 200 feet
24	

There were 24 entrants in the discus finals.

15. Strategy To find the percent:
● Read the frequency polygon to find the number of entrants that had distances between 160 feet and 170 feet (9).
● Solve the basic percent equation for percent. The base is the total number of entrants (24) and the amount is the number of entrants that had distances between 160 and 170 feet (9).

Solution
Percent × base = amount
$n \times 24 = 9$
$n = 9 \div 24$
$n = 0.375$
37.5% of the entrants had distances between 160 feet and 170 feet.

17. Strategy To find the percent:
● Read the frequency polygon to find how many people purchased between 20 and 30 tickets.
● Solve the basic percent equation for percent. The base is 74 and the amount is the number of people who purchased between 20 and 30 tickets.

Solution
Between 20 and 30 tickets: 8
Percent × base = amount
$n \times 74 = 8$
$n = 8 \div 74$
$n \approx 0.108$
The percent is 10.8%.

19. No, a frequency polygon shows only the number of occurrences in a class. It does not show the number of occurrences for any particular value.

21. Strategy To find the percent:
● Read the frequency polygon to find the number of students that scored between 800 and 1000.
● Solve the basic percent equation for percent. The base is 1,080,000 and the amount is the number of students scoring between 800 and 1000.

Solution
Number of student scoring between 800 and 1000: 350,000
Percent × base = amount
$n \times 1,080,000 = 350,000$
$n = 350,000 \div 1,080,000$
$n \approx 0.324$
The percent is 32.4%.

23. Strategy To find the number of students:
● Read the frequency polygon to find the number of students who scored between 800 and 1000, between 1000 and 1200, between 1200 and 1400, and between 1400 and 1600.
● Add the four numbers.

Solution

Between 800 and 1000:	350,000
Between 1000 and 1200:	350,000
Between 1200 and 1400:	170,000
Between 1400 and 1600:	+ 30,000
	900,000

900,000 students scored above 800.

Applying the Concepts

25. A frequency table is another method of organizing data. In a frequency table, or a frequency distribution, data are combined into categories called classes, and the frequency, which is the number of pieces of data in each class, is shown.

Section 7.4

Objective A Exercises

1a. Median

b. Mean

c. Mode

d. Median

e. Mode

f. Mean

3. **Strategy**   To find the mean value of the number of seats occupied:
- Find the sum of the number of seats occupied.
- Divide the sum by the number of flights (16).

Solution

309
422
389
412
401
352
367
319
410
391
330
408
399
387
411
+398
6,105

$$16\overline{)6,105} = 381.5625$$

The mean of the number of seats filled is 381.5625 seats.

Strategy   To find the median value of the number of seats occupied, arrange the numbers in order from smallest to largest. The median is the mean of the two middle numbers.

Solution

309 ⎫
319 ⎪
330 ⎪
352 ⎬ 7 numbers
367 ⎪
387 ⎪
389 ⎭

391 ⎫
398 ⎭ Middle numbers

399 ⎫
401 ⎪
408 ⎪
410 ⎬ 7 numbers
411 ⎪
412 ⎪
422 ⎭

$$\frac{391 + 398}{2} = 394.5$$

The median of the number of seats filled is 394.5 seats.

Strategy   To find the mode, look at the number of seats occupied and locate the number that occurs most frequently.

Solution   Since each number occurs only once, there is no mode.

5. **Strategy**   To find the mean cost:
- Find the sum of the costs.
- Divide the total costs by the number of purchases (8).

Solution

$45.89
$52.12
$41.43
$40.67
$48.73
$42.45
$47.81
+ $45.82
$364.92

$$8\overline{)364.920} = 45.615$$

The mean cost is $45.615.

Strategy   To find the median cost, arrange the costs in order from smallest to largest. The median is the mean of the two middle numbers.

Solution

$40.67 ⎫
$41.43 ⎬ 3 numbers
$42.45 ⎭

$45.82 ⎫
$45.89 ⎬ middle numbers

$47.81 ⎫
$48.73 ⎬ 3 numbers
$52.12 ⎭

$$\frac{\$45.82 + 45.88}{2} = \$45.855$$

The median cost is $45.855.

7. **Strategy**   To find the mean monthly rate:
- Find the sum of the monthly rates.
- Divide the sum by the number of plans (8).

Solution

$423
$390
$405
$396
$426
$355
$404
+ $430
$3,229

$$8\overline{)\$3,229.000} = \$403.625$$

The mean monthly rate is $403.625.

Strategy   To find the median monthly rate, write the rates in order from smallest to largest. The median is the mean of the two middle terms.

Solution

$$
\left.
\begin{array}{l}
\$355 \\
\$390 \\
\$396
\end{array}
\right\} \text{3 numbers}
$$

$$
\left.
\begin{array}{l}
\$404 \\
\$405
\end{array}
\right\} \text{middle numbers}
$$

$$
\left.
\begin{array}{l}
\$423 \\
\$426 \\
\$430
\end{array}
\right\} \text{3 numbers}
$$

$$\frac{\$404 + 405}{2} = \$404.50$$

The median monthly rate is $404.50.

9. Strategy   To find the mean life expectancy:
- Find the sum of the years.
- Divide the sum by the number of countries (10).

Solution

$$
\begin{array}{r}
62 \\
75 \\
78 \\
70 \\
64 \\
75 \\
66 \\
71 \\
74 \\
+\ 73 \\
\hline
708
\end{array}
$$

$$
\begin{array}{r}
70.8 \\
10\overline{)708.0}
\end{array}
$$

The mean life expectancy is 70.8 years.

Strategy   To find the median life expectancy, write the years in order from lowest to highest. The median is the mean of the two middle numbers.

Solution

$$
\left.
\begin{array}{l}
62 \\
64 \\
66 \\
70
\end{array}
\right\} \text{4 numbers}
$$

$$
\left.
\begin{array}{l}
71 \\
73
\end{array}
\right\} \text{middle numbers}
$$

$$
\left.
\begin{array}{l}
74 \\
75 \\
75 \\
78
\end{array}
\right\} \text{4 numbers}
$$

$$\frac{71 + 73}{2} = 72$$

The median life expectancy is 72 years.

11. No, the mean scores of the two students are not the same. The mean score of the second student is 5 points higher than the mean score of the first student.

Objective B Exercises

13a. 25%

b. 75%

c. 75%

d. 25%

15. Strategy
- Read the lowest value, the highest value, the first quartile, the third quartile, the median directly from the box-and-whiskers plot.
- Find the range by subtracting the lowest from the highest.
- Interquartile range $= Q_3 - Q_1$.

Solution   Lowest is $46,596.
Highest is $82,879.
$Q_1 = \$56,067$
$Q_3 = \$66,507$
Median = $61,036
Range: $82{,}879 - 46{,}596 = \$36{,}283$
Interquartile range:
$66{,}507 - 56{,}067 = \$10{,}440$

17a. Strategy   To find the number of adults who had a cholesterol level above 217, the median, solve the basic percent equation for the amount, where the base is 80 and the percent is 50%.

Solution   Percent × base = amount
$0.50 \times 80 = 40$
There were 40 adults who had cholesterol levels above 217.

b. Strategy   To find the number of adults who had a cholesterol level below 254, the third quartile, solve the basic percent equation for the amount, where the base is 80 and the percent is 75%.

Solution   Percent × base = amount
$0.75 \times 80 = 60$
There were 60 adults who had cholesterol levels below 254.

c. Strategy   To find the number of cholesterol levels represented in each quartile, solve the basic percent equation for the amount, where the base is 80 and the percent is 25%.

Solution   Percent × base = amount
$0.25 \times 80 = 20$
There are 20 cholesterol levels in each quartile.

d. The first quartile is at 198. So 25% of the adults had cholesterol levels not more than 198.

19a.   Strategy • Arrange the data from smallest to largest.
• Find the range.
• Find Q_1, the median of the lower half of the data.
• Find Q_3, the median of the upper half of the data.
• Interquartile range = $Q_3 - Q_1$.

Solution

0.41	0.41	0.56	0.61	0.76
0.87	1.06	2.10	2.60	4.80

Range: $4.80 - 0.41 = 4.39$ million metric tons
$Q_1 = 0.56$ million metric tons
$Q_3 = 2.10$ million metric tons
Interquartile range = $Q_3 - Q_1$
$$= 2.10 - 0.56$$
$$= 1.54 \text{ million metric tons}$$

b.

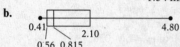

0.41   2.10   4.80
0.56 0.815

c. 4.80

21a.   Strategy To determine whether the difference in means is greater than 1 inch:
• Find the sum of the rainfall in Orlando.
• Divide the sum by the number of months (12) to find the mean.
• Find the sum of the rainfall in Portland.
• Divide the sum by the number of months (12) to find the mean.
• Find the difference in the means.

Solution

Orlando			Portland		
2.1			6.2		
2.8			3.9		
3.2			3.6		
2.2			2.3		
4.0			2.1		
7.4	4.0		1.5	3.1	
7.8	12)47.8		0.5	12)37.5	
6.3			1.1		
5.6			1.6		
2.8			3.1		
1.8			5.2		
+1.8			+6.4		
47.8			37.5		

$4.0 - 3.1 = 0.9$
No, the difference in the means is not greater than 1 inch.

b.   Strategy To find the difference between the medians, write the rainfall in order from lowest to highest. The median is the mean of the two middle numbers. Find the difference between the Orlando median and Portland median.

Solution

Orlando		Portland	
1.8		0.5	
1.8		1.1	
2.1	5 numbers	1.5	5 numbers
2.2		1.6	
2.8		2.1	
2.8	middle	2.3	middle
3.2		3.1	
4.0		3.6	
5.6		3.9	
6.3	5 numbers	5.2	5 numbers
7.4		6.2	
7.8		6.4	

For Orlando, $\dfrac{2.8 + 3.2}{2} = 3.0$

For Portland, $\dfrac{2.3 + 3.1}{2} = 2.7$
$3.0 - 2.7 = 0.3$
The difference in medians is 0.3 inch.

c.   Strategy To draw box-and-whiskers:
• Find Q_1 and Q_3 in Orlando.
• Find Q_1 and Q_3 in Portland.

Solution For Orlando, $Q_1 = \dfrac{2.1 + 2.2}{2} = 2.15$,

$Q_3 = \dfrac{5.6 + 6.3}{2} = 5.95$

For Portland, $Q_1 = \dfrac{1.5 + 1.6}{2} = 1.55$,

$Q_3 = \dfrac{3.9 + 5.2}{2} = 4.55$

1.8   3.0   5.95   7.8
2.15

0.5   2.7   6.4
1.55   4.55

d. Answers will vary. For example, the distribution of the data is relatively similar for the two cities. However, the value of each of the 5 points on the boxplot for the Portland data is less than the corresponding value on the boxplot for the Orlando data. The average monthly rainfall in Portland is less than the average monthly rainfall in Orlando.

Applying the Concepts

23. Answers will vary. For example, 55, 55, 55, 55, 55, or 50, 55, 55, 55, 60

25a. Q_1 is the number that one-quarter of the data lie below.

b. Q_3 is the number that one-quarter of the data lie above.

c. $\bar{x}$ is the symbol for the mean of a set of data.

27. The box does represent 50% of the data, but it provides a picture of the spread of the data. The box is not one-half the entire length of the box-and-whiskers plot because there is a greater spread of data in the interquartile range than in the first quarter or fourth quarter of the data.

Section 7.5

Objective A Exercises

1. The possible outcomes of tossing a coin four times: {(HHHH), (HHHT), (HHTT), (HHTH), (HTTT), (HTHH), (HTTH), (HTHT), (TTTT), (TTTH), (TTHH), (THHH), (TTHT), (THHT), (THTT), (THTH)}

3. The possible outcomes of tossing two tetrahedral dice: {(1, 1), (1, 2), (1, 3), (1, 4), (2, 1), (2, 2), (2, 3), (2, 4), (3, 1), (3, 2), (3, 3), (3, 4), (4, 1), (4, 2), (4, 3), (4, 4)}

5a. The sample space is {1, 2, 3, 4, 5, 6, 7, 8}.

b. The outcomes in the event the number is less than 4 is {1, 2, 3}.

7a. **Strategy** To calculate the probability:
• Count the number of possible outcomes. See the table on p. 323.
• Count the number of favorable outcomes.
• Use the probability formula.

Solution There are 36 possible outcomes.
There are 4 favorable outcomes: (1, 4), (4, 1), (2, 3), (3, 2).
Probability $= \dfrac{4}{36} = \dfrac{1}{9}$
The probability that the sum is 5 is $\dfrac{1}{9}$.

b. **Strategy** To calculate the probability:
• Count the number of possible outcomes. See the table on p. 323.
• Count the number of favorable outcomes.
• Use the probability formula.

Solution There are 36 possible outcomes.
There are 0 favorable outcomes.
Probability $= \dfrac{0}{36} = 0$
The probability that the sum is 15 is 0.

c. **Strategy** To calculate the probability:
• Count the number of possible outcomes. See the table on p. 323.
• Count the number of favorable outcomes.
• Use the probability formula.

Solution There are 36 possible outcomes.
There are 36 favorable outcomes.
Probability $= \dfrac{36}{36} = 1$
The probability that the sum is less than 15 is 1.

d. **Strategy** To calculate the probability:
• Count the number of possible outcomes. See the table on p. 323.
• Count the number of favorable outcomes.
• Use the probability formula.

Solution There are 36 possible outcomes.
There is 1 favorable outcome: (1, 1).
Probability $= \dfrac{1}{36}$
The probability that the sum is 2 is $\dfrac{1}{36}$.

9a. **Strategy** To calculate the probability:
• Count the number of possible outcomes.
• Count the number of favorable outcomes.
• Use the probability formula.

Solution A dodecahedral die has 12 sides. There are 3 favorable outcomes: 4, 8, 12.
Probability $= \dfrac{3}{12} = \dfrac{1}{4}$
The probability that the number is divisible by 4 is $\dfrac{1}{4}$.

b. **Strategy** To calculate the probability:
• Count the number of possible outcomes.
• Count the number of favorable outcomes.
• Use the probability formula.

Solution A dodecahedral die has 12 sides. There are 4 favorable outcomes: 3, 6, 9, 12.
Probability $= \dfrac{4}{12} = \dfrac{1}{3}$
The probability that the number is a multiple of 3 is $\dfrac{1}{3}$.

11. Strategy To calculate the probability:
- Count the number of possible outcomes. See the table on p. 323.
- Count the number of favorable outcomes.
- Use the probability formula.
- Compare the probabilities.

Solution There are 36 possible outcomes.
For a sum of 10, there are 3 favorable outcomes: $(5, 5)$, $(4, 6)$, $(6, 4)$.

Probability $= \dfrac{3}{36}$

For a sum of 5, there are 4 favorable outcomes:
$(1, 4)$, $(4, 1)$, $(2, 3)$, $(3, 2)$.

Probability $= \dfrac{4}{36}$

$\dfrac{4}{36} > \dfrac{3}{36}$

The probability of throwing a sum of 5 is greater.

13a. Strategy To calculate the probability:
- Count the number of possible outcomes.
- Count the number of favorable outcomes.
- Use the probability formula.

Solution There are 11 possible outcomes.
There are 4 favorable outcomes.

Probability $= \dfrac{4}{11}$

The probability is $\dfrac{4}{11}$ that the letter I is drawn.

b. Strategy To calculate the probability:
- Count the number of possible outcomes.
- Count the number of favorable outcomes.
- Use the probability formula.
- Compare the probabilities.

Solution There are 11 possible outcomes.
There are 4 favorable outcomes of choosing an S.

Probability $= \dfrac{4}{11}$

There are 2 favorable outcomes of choosing a P.

Probability $= \dfrac{2}{11}$

$\dfrac{4}{11} > \dfrac{2}{11}$

The probability of choosing an S is greater.

15a. Strategy To calculate the probability:
- Count the number of possible outcomes.
- Count the number of favorable outcomes.
- Use the probability formula.

Solution There are 12 possible outcomes (3 blue + 4 green + 5 red).
There are 4 favorable outcomes.

Probability $= \dfrac{4}{12} = \dfrac{1}{3}$

The probability is $\dfrac{1}{3}$ that the marble chosen is green.

b. Strategy To calculate the probability:
- Count the number of possible outcomes.
- Count the number of favorable outcomes.
- Use the probability formula.
- Compare the probabilities.

Solution There are 12 possible outcomes.
There are 3 favorable outcomes of choosing a blue marble.

Probability $= \dfrac{3}{12} = \dfrac{1}{4}$

There are 5 favorable outcomes of choosing a red marble.

Probability $= \dfrac{5}{12}$

$\dfrac{5}{12} > \dfrac{3}{12}$

The probability of choosing a red marble is greater.

17. Strategy To calculate the probability:
- Count the number of possible outcomes.
- Count the number of favorable outcomes.
- Use the probability formula.

Solution There are 47 possible outcomes $(4 + 8 + 22 + 10 + 3)$.
There are 8 favorable outcomes.

Probability $= \dfrac{8}{47}$

The probability is $\dfrac{8}{47}$ that the paper has a B grade.

19. **Strategy**   To calculate the empirical probability, use the probability formula and divide the number of observations (587) by the total number of observations (725).

   Solution   Probability $= \dfrac{587}{725} \approx 0.81$
   The probability is 0.81 that an employee has a group health insurance plan.

Applying the Concepts

21. No, the numbers 1 through 5 are not equally likely because the sizes of the sectors are different.

23. The mathematical probability of an event is a number from 0 to 1. Because $\dfrac{5}{3} = 1\dfrac{2}{3}, \dfrac{5}{3} > 1$. The probability of an even cannot be $\dfrac{5}{3}$ because the probability cannot be greater than 1.

Chapter 7 Review Exercises

1. **Strategy**   To find the amount of money:
   • Read the circle graph to determine the amounts of money spent.
   • Add the amounts.

   Solution

Defense:	$148 million
Agriculture:	$15 million
EPA:	$24 million
Commerce:	$27 million
NASA:	$31 million
Other:	+ 104 million
	$349 million

   The agencies spent $349 million in maintaining websites.

2. **Strategy**   To find the ratio:
   • Read the circle graph to find the amount spent by the Department of Commerce and by the EPA.
   • Write, in simplest form, the ratio of the amount spent by the Department of Commerce to the amount spent by the EPA.

   Solution

Commerce:	$27 million
EPA:	$24 million

   $\dfrac{\$27 \text{ million}}{\$24 \text{ million}} = \dfrac{9}{8}$

   The ratio is $\dfrac{9}{8}$.

3. **Strategy**   To find the percent, solve the basic percent equation for percent. The base is the total amount spent ($349 million) and the amount is the amount spent by NASA ($31 million)

   Solution   Percent × base = amount
   $n \times \$349 \text{ million} = \$31 \text{ million}$
   $n = \$31 \text{ million} \div \$349 \text{ million}$
   $n \approx 0.0888$
   8.9% of the total mount of money was spent by NASA.

4. Texas had the larger population.

5. **Strategy**   To find the difference in populations:
   • Read the double broken-line graph to find the Texas population and the California population in 2000.
   • Subtract the population of Texas from the population of California.

   Solution

California:	32.5
Texas:	− 20.0
	12.5

   The population of California is 12.5 million people more than the population of Texas.

6. **Strategy**   To find which 25-year period Texas had the smallest increase in population.
   • Read the double-line graph to find the population for each 25-year period.
   • Subtract the two numbers.

   Solution   1900 to 1925:
   $6 − 2.5 = 3.5$ million
   1925 to 1950:
   $8 − 6 = 2$ million
   1950 to 1975:
   $12 − 8 = 4$ million
   1975 to 2000:
   $21 − 12 = 9$ million
   The Texas population increased the least from 1925 to 1950.

7. Strategy   To find the number of games in which the Knicks scored fewer than 100 points:
● Read the frequency polygon to find the number of games in which the Knicks scored 60–70 points, 70–80 points, 80–90 points, and 90–100 points.
● Add the four numbers.

Solution   60–70 points: 1 game
70–80 points: 7 games
80–90 points: 15 games
90–100 points: + 31 games
                          54

There were 54 games in which the Knicks scored fewer than 100 points.

8. Strategy   To find the ratio:
● Read the frequency polygon to find the number of games in which the Knicks scored between 90 and 100 points and between 110 and 120 points.
● Write in simplest form the ratio of the number of games in which the Knicks scored between 90 and 100 points to the number of games in which they scored between 110 and 120 points.

Solution   90 to 100 points: 31 games
110 to 120 points: 8 games
$$\frac{31 \text{ games}}{8 \text{ games}} = \frac{31}{8}$$
The ratio is $\frac{31}{8}$.

9. Strategy   To find the percent:
● Read the frequency polygon to find the number of games in which the Knicks scored 110 to 120 points and 120 to 130 points.
● Add the two numbers.
● Solve the basic percent equation for percent. The base is 80 and the amount is the number of games in which more than 110 points were scored.

Solution   110–120 points: 8 games
120–130 points: + 1 game
                          9 games
Percent × base = amount
$n \times 80 = 9$
$n = 9 \div 80$
$n = 0.1125$
The percent is 11.3%.

10.   From the pictograph, O'Hare airport has 10 million more passengers than Los Angeles airport.

11. Strategy   To find the ratio:
● Read the pictograph to find the number of passengers going through San Francisco airport and the number of passengers going through Dallas/Ft. Worth airport.
● Write in simplest form the ratio of the number of passengers going through San Francisco airport to the number of passengers going through Dallas/Ft. Worth airport.

Solution   San Francisco: 40
Dallas/Ft. Worth: 60
$$\frac{40}{60} = \frac{2}{3} = 2:3$$
The ratio is 2 : 3.

12. Strategy   To find the difference between the total days of operation and days of full operation of the Midwest ski areas:
● Read the double-bar graph for the number of days that the resorts were open and the days of full operation.
● Subtract the two numbers.

Solution   Days open: 90
Days of full operation: − 40
                                       50
The difference was 50 days.

13. Strategy   To find the percent:
● Read the double-bar graph to find the number of days that the Rocky Mountain ski areas were open and the number of days of full operation.
● Solve the basic percent equation for percent. The base is the number of days open and the amount is the number of days of full operation.

Solution   Days open: 140
Days of full operation: 70
Percent × base = amount
$n \times 140 = 70$
$n = 70 \div 140$
$n = 0.5$
The percent is 50%.

14a. Strategy To determine which region has the lowest number of days of full operation, read the bar graph and select the lowest graph that shows days of full operation.

Solution The Southeast had the lowest number of days of full operation.

b. Strategy Read the number of days from the lowest graph.

Solution This region had 30 days of full operation.

15. Strategy To calculate the probability:
• Count the number of possible outcomes.
• Count the number of favorable outcomes.
• Use the probability formula.

Solution There are 16 possible outcomes.
There are 4 favorable outcomes:
THHH, HHHT, HHTH, HTHH.

Probability $= \dfrac{4}{16} = \dfrac{1}{4}$

The probability of one tail and three heads is $\dfrac{1}{4}$.

16. Strategy To find the number of people who slept 8 hours or more:
• Read the histogram to find the number of people who slept 8 hours, 9 hours, or more than 9 hours.
• Add the three numbers.

Solution
Slept 8 hours: 12
Slept 9 hours: 2
Slept more than 9 hours: $+1$
15

There were 15 people who slept 8 or more hours.

17. Strategy To find the percent:
• Read the histogram to find the number of people who slept 7 hours.
• Solve the basic percent equation for percent. The base is 46 and the amount is the number of people who slept 7 hours.

Solution
Slept 7 hours: 13
Percent × base = amount
$n \times 46 = 13$
$n = 13 \div 46$
$n \approx 0.2826$
The percent is 28.3%.

18a. Strategy To find the mean heart rates:
• Find the sum of the heart rates.
• Divide the sum by the number of women.

Solution
80
82
99
91
93
87
103
94
73
96
86
80
97
94
108
81
100
109
91
84
78
96
96
$+ 100$
2198

The mean heart rate is 91.6 heartbeats per minute.

Strategy To find the median heart rate: write the heart rates in order from smallest to largest. The median is the mean of the two middle numbers.

Solution

73
78
80
80
81
82 ⎱ 11 numbers
84
86
87
91
91

93 ⎱ middle numbers
94

94
96
96
96
97
99 ⎱ 11 numbers
100
100
103
108
109

$$\frac{93 + 94}{2} = 93.5$$

The median heart rate is 93.5 heartbeats per minute.

Strategy To find the mode, look at the heart rates and identify the number that occurs most frequently.

Solution The mode is 96 heartbeats per minute, the number that occurs most frequently.

b. Strategy
- Arrange the data from smallest to largest. Then find the range.
- Find Q_1, the median of the lower half of the data.
- Find Q_3, the median of the upper half of the data.
- Interquartile range = $Q_3 - Q_1$.

Solution Use the list in part a.
Range = 109 − 73 = 36
The range is 36 heartbeats per minute.

$$Q_1 = \frac{82 + 84}{2} = 83$$

$$Q_3 = \frac{97 + 99}{2} = 98$$

$$Q_3 - Q_1 = 98 - 83 = 15$$

The interquartile range is 15 heartbeats per minute.

Chapter 7 Test

1. Strategy To find the number of students who spent between $15 and $25 each week:
- Read the frequency polygon to find the number of students who spent between $15 and $20 and the number who spent between $20 and $25.
- Add the two numbers.

Solution
Number between $15–$20: 12
Number between $20–$25: + 7
                         19

19 students spent between $15 and $25 each week.

2. Strategy To find the ratio:
- Read the frequency polygon to find the number of students who spent between $10 and $15 and the number who spent between $15 and $20.
- Write in simplest form the ratio of the number of students who spent between $10 and $15 to the number of students who spent between $15 and $20.

Solution
Between $10 and $15: 8 students
Between $15 and $20: 12 students

$$\frac{8 \text{ students}}{12 \text{ students}} = \frac{2}{3}$$

The ratio is $\frac{2}{3}$.

3. Strategy To find the percent:
- Read the frequency polygon to find the number of students who spent between $0 to $5, between $5 and $10, and between $10 and $15 each week.
- Add the three numbers.
- Solve the basic percent equation for percent. The base is 40 and the amount is the number of students who spent less than $15 per week.

Solution
Between $0 and $5:   4 students
Between $5 and $10: 6 students
Between $10 and $15: + 8 students
                      18 students

Percent × base = amount
$n \times 40 = 18$
$n = 18 \div 40$
$n = 0.45$
The percent is 45%.

4. Strategy To find the number of people surveyed:
- Read the pictograph to determine the number of people for each letter grade.
- Add the four numbers.

Solution Number of A grades: 21
Number of B grades: 10
Number of C grades: 4
Number of D grades: $+1$
36

There were 36 people that were surveyed for the Gallup poll.

5. Strategy To find the ratio:
- Read the pictograph to find the number of people who gave their marriage a B grade and the number who gave their marriage a C grade.
- Write in simplest form the ratio of the number of people who gave their marriage a B grade to the number of people who gave their marriage a C grade.

Solution Number of B grades: 10 people
Number of C grades: 4 people
$$\frac{10 \text{ people}}{4 \text{ people}} = \frac{5}{2}$$
The ratio is $\frac{5}{2}$.

6. Strategy To find the percent:
- Read the pictograph to find the number of people who gave their marriage an A grade.
- Solve the basic percent equation for percent. The base is 36 (from Exercise 4) and the amount is the number of people who gave their marriage an A grade.

Solution Number of A grades: 21 people
Percent × base = amount
$n \times 36 = 21$
$n = 21 \div 36$
$n \approx 0.583$
The percent is 58.3%.

7. Strategy Read the bar graph to find the two consecutive years that the number of fatalities were the same.

Solution During 1995 and 1996, the number of fatalities was the same.

8. Strategy To find the total fatalities on amusement rides during 1991 to 1999:
- Read the bar graph to determine the number of fatalities for each year.
- Add the nine numbers.

Solution
3 1991
2 1992
4 1993
2 1994
3 1995
3 1996
4 1997
5 1998
$+6$ 1999
32

There were 32 fatal accidents from 1991 to 1999.

9. Strategy To find how many more fatalities in 1995 to 1998 than 1991 to 1994:
- Add the number of fatalities for 1995 to 1998.
- Add the number of fatalities for 1991 to 1994.
- Subtract the two numbers.

Solution
3 1995   3 1991
3 1996   2 1992
4 1997   4 1993
$+5$ 1998 $+2$ 1994
15   11

$15 - 11 = 4$

There were 4 more fatalities from 1995 to 1998.

10. Strategy To find how many more R-rated films than PG:
- Read the circle graph to find the number of films rated R and PG.
- Subtract the two numbers.

Solution R: 427
PG: -72
355
There were 355 more films rated R.

11. Strategy To find how many times more PG-13 films were released than NC-17:
- Read the circle graph to find the number of films rated PG-13 and NC-17.
- Divide the two numbers.

Solution PG-13: 112
NC-17: 7
$$7\overline{)112} = 16$$
There were 16 times more films rated PG-13.

12.   Strategy   To find the percent of films rated G:
● Read the circle graph to find the number of G-rated films.
● Write and solve the basic percent equation for the percent. The base is the total number of films (655) and the amount is the number of G-rated films.

Solution   G: 37
Percent × base = amount
$$n \times 655 = 37$$
$$n = 37 \div 655$$
$$n \approx 0.056$$
The percent of films rated G was 5.6%.

13.   Strategy   To find the number of states with median income between $40,000 and $60,000.
● Read the histogram to find the number of states with per capita income between $40,000 and $50,000 and between $50,000 and $60,000.
● Add the two numbers.

Solution $40,000 to $50,000:   5 states
$50,000 to $60,000:   + 19 states
                              24 states
There are 24 states that have a median income between $40,000 and $60,000.

14.   Strategy   To find the percent of the states with a median income between $50,000 and $70,000:
● Read the histogram to find the number of states with median incomes between $50,000 and $60,000 and between $60,000 and $70,000.
● Add the two numbers.
● Solve the basic percent equation for percent. The base is 50 and the amount is the number of states with a median income between $50,000 and $70,000.

Solution $50,000 to $60,000:   19 states
$60,000 to $70,000:   + 17 states
                              36 states
Percent × base = amount
$$n \times 50 = 36$$
$$n = 36 \div 50$$
$$n = 0.72$$
The percent is 72%.

15.   Strategy   To find the percent:
● Read the histogram to find the number of states that have a median income between $70,000 and $80,000 and between $80,000 and $90,000.
● Add the two numbers.
● Solve the basic percent equation for percent. The base is 50 and the amount is the number of states with a median income above $70,000.

Solution $70,000–$80,000:   5 states
$80,000–$90,000:   + 4 states
                              9 states

Percent × base = amount
$$n \times 50 = 9$$
$$n = 9 \div 50$$
$$n = 0.18$$
The percent is 18%.

16.   Strategy   To calculate the probability:
● Count the number of possible outcomes.
● Count the number of favorable outcomes.
● Use the probability formula.

Solution   There are 50 possible outcomes.
There are 15 favorable outcomes.

$$\text{Probability} = \frac{15}{50} = \frac{3}{10}$$

The probability is $\frac{3}{10}$ that the ball chosen is red.

17.   Strategy   To find which decade had the smallest increase in enrollment.
● Read the line graph to find the enrollment for each decade.
● Subtract the two numbers.

Solution   1960 to 1970:
$8 - 4 = 4$ million
1970 to 1980:
$12 - 8 = 4$ million
1980 to 1990:
$14 - 12 = 2$ million
1990 to 2000:
$15 - 14 = 1$ million
The student enrollment increased the least during the 1990s.

18. **Strategy** To approximate the increase in enrollment:
● Read the enrollment for 1960 and 2000.
● Subtract the two numbers.

Solution
2000:   15 million
1960:   − 4 million
         11 million
The increase in enrollment was 11 million students.

19a. **Strategy** To find the mean lifetime of the batteries:
● Find the sum of the times.
● Divide the sum by the number of batteries tested (20).

Solution
2.9
2.4
3.1
2.5
2.6
2.0
3.0
2.3
2.4
2.7
2.0   2.53
2.4   20)50.60
2.6
2.7
2.1
2.9
2.8
2.4
2.0
+ 2.8
50.6

The mean time is 2.53 days.

b. **Strategy** To find the median lifetime of the batteries, write times in order from lowest to highest. The median is the mean of the two middle numbers.

Solution

$$\left.\begin{array}{l}2.0\\2.0\\2.0\\2.1\\2.3\\2.4\\2.4\\2.4\\2.4\end{array}\right\}\text{9 numbers}$$

$$\left.\begin{array}{l}2.5\\2.6\end{array}\right\}\text{middle numbers}$$

$$\left.\begin{array}{l}2.6\\2.7\\2.7\\2.8\\2.8\\2.9\\2.9\\3.0\\3.1\end{array}\right\}\text{9 numbers}$$

$$\frac{2.5 + 2.6}{2} = 2.55$$

The median time is 2.55 days.

c. **Strategy** The data is arranged from smallest to largest in part b.
● Find Q_1, the median of the lower half of the data.
● Find Q_3, the median of the upper half of the data.
● Draw the box-and-whiskers plot.

Solution
$$Q_1 = \frac{2.3 + 2.4}{2} = 2.35$$

$$Q_3 = \frac{2.8 + 2.8}{2} = 2.8$$

2.0   2.35 2.55   2.8   3.1

Cumulative Review Exercises

1. $2^2 \cdot 3^3 \cdot 5 = (2 \cdot 2) \cdot (3 \cdot 3 \cdot 3) \cdot (5)$
              $= 4 \cdot 27 \cdot 5 = 540$

2. $3^2 \cdot (5 - 2) \div 3 + 5$
    $9 \cdot (3) \div 3 + 5$
    $27 \div 3 + 5$
    $9 + 5$
    14

3.
              2   3   5
24 = | (2 · 2 · 2) | (3) | |
40 = | 2 · 2 · 2 | | (5) |

LCM $= 2 \cdot 2 \cdot 2 \cdot 3 \cdot 5 = 120$

4. $\dfrac{60}{144} = \dfrac{\overset{1}{2} \cdot \overset{1}{2} \cdot \overset{1}{3} \cdot 5}{\underset{1}{2} \cdot \underset{1}{2} \cdot 2 \cdot 2 \cdot \underset{1}{3} \cdot 3} = \dfrac{5}{12}$

5.
$$4\frac{1}{2} = 4\frac{20}{40}$$
$$2\frac{3}{8} = 2\frac{15}{40}$$
$$+5\frac{1}{5} = 5\frac{8}{40}$$
$$11\frac{43}{40} = 12\frac{3}{40}$$

6.
$$12\frac{5}{8} = 12\frac{15}{24} = 11\frac{39}{24}$$
$$-7\frac{11}{12} = 7\frac{22}{24} = 7\frac{22}{24}$$
$$4\frac{17}{24}$$

7.
$$\frac{5}{8} \times 3\frac{1}{5} = \frac{5}{8} \times \frac{16}{5}$$
$$= \frac{5 \cdot 16}{8 \cdot 5}$$
$$= \frac{\overset{1}{\cancel{5}}\cdot\overset{1}{\cancel{2}}\cdot\overset{1}{\cancel{2}}\cdot\overset{1}{\cancel{2}}\cdot 2}{\underset{1}{\cancel{2}}\cdot\underset{1}{\cancel{2}}\cdot\underset{1}{\cancel{2}}\cdot\underset{1}{\cancel{5}}} = 2$$

8. $3\frac{1}{5} \div 4\frac{1}{4} = \frac{16}{5} \div \frac{17}{4} = \frac{16}{5} \times \frac{4}{17} = \frac{16 \cdot 4}{5 \cdot 17} = \frac{64}{85}$

9. $\frac{5}{8} \div \left(\frac{3}{4} - \frac{2}{3}\right) + \frac{3}{4} = \frac{5}{8} \div \left(\frac{9}{12} - \frac{8}{12}\right) + \frac{3}{4}$
$$= \frac{5}{8} \div \frac{1}{12} + \frac{3}{4}$$
$$= \frac{5}{8} \times \frac{12}{1} + \frac{3}{4}$$
$$= \frac{5 \cdot \overset{1}{\cancel{2}}\cdot\overset{1}{\cancel{2}}\cdot 3}{\underset{1}{\cancel{2}}\cdot\underset{1}{\cancel{2}}\cdot 2} + \frac{3}{4}$$
$$= \frac{15}{2} + \frac{3}{4} = \frac{30}{4} + \frac{3}{4} = \frac{33}{4} = 8\frac{1}{4}$$

10. 209.305

11.
$$\begin{array}{r} 4.092 \\ \times\ 0.69 \\ \hline 36828 \\ 24552 \\ \hline 2.82348 \end{array}$$

12. $16\frac{2}{3} = \frac{50}{3}$
$$\begin{array}{r} 16.666 \approx 16.67 \\ 3)\overline{50.000} \\ \underline{-3} \\ 20 \\ \underline{-18} \\ 20 \\ \underline{-18} \\ 20 \\ \underline{-18} \\ 20 \\ \underline{-18} \\ 20 \end{array}$$

13. $\dfrac{330 \text{ miles}}{12.5 \text{ gal}} = 26.4$ mpg

14.
$$\frac{n}{5} = \frac{16}{25}$$
$$n \times 25 = 5 \times 16$$
$$n \times 25 = 80$$
$$n = 80 \div 25 = 3.2$$

15. $\dfrac{4}{5} \times 100\% = 80\%$

16. $10\% \times n = 8$
$$0.10 \times n = 8$$
$$n = 8 \div 0.10 = 80$$

17. $38\% \times 43 = n$
$$0.38 \times 43 = n$$
$$16.34 = n$$

18. $n \times 75 = 30$
$$n = 30 \div 75 = 0.40 = 40\%$$

19. **Strategy** To find the income for the week:
● Find the commission earned on sales by solving the basic percent equation for amount. The base is $27,500 and the percent is 2%.
● Find the total income by adding the base salary ($100) to the commission.

Solution
$$2\% \times 27,500 = n$$
$$0.02 \times 27,500 = n$$
$$550 = n$$

$100 + 550 = 650$
The salesperson's income for the week was $650.

20. **Strategy** To find the cost, write and solve a proportion.

Solution
$$\frac{8.15}{1000} = \frac{n}{50,000}$$
$$8.15 \times 50,000 = n \times 1000$$
$$407,500 = n \times 1000$$
$$407,500 \div 1000 = n$$
$$407.50 = n$$
The cost is $407.50.

21. Strategy To find the interest due, multiply the principal by the annual interest rate and the time (in years).

Solution $125{,}000 \times 0.06 \times \dfrac{6}{12} = 3750$
The interest due is $3750.

22. Strategy To find the markup rate of the compact disc player:
• Find the markup amount by subtracting the cost ($180) from the selling price ($279).
• Solve the basic percent equation for percent. The base is $180 and the amount is the amount of the markup.

Solution
$279 - 180 = 99$
Percent × base = amount
$n \times 180 = 99$
$n = 99 \div 180 = 0.55 = 55\%$
The markup rate is 55%.

23. Strategy To find how much is budgeted for food:
• Read the circle graph to find what percent of the budget is spent on food.
• Solve the basic percent equation for amount. The base is $3000 and the rate is the percent of the budget that is spent on food.

Solution Amount spent on food: 19%

Percent × base = amount
$19\% \times 3000 =$ amount
$0.19 \times 3000 = 570$
The amount budgeted for food is $570.

24. Strategy To find the difference:
• Read the double-broken-line graph to find the number of problems student 1 answered correctly on test 1 and the number of problems student 2 answered correctly on test 1.
• Subtract the student 1 total from the student 2 total to find the difference.

Solution student 2: 27 answered correctly
student 1: −15 answered correctly
12 answered correctly
The difference in the number answered correctly is 12 problems.

25. Strategy To find the mean high temperature:
• Find the sum of the high temperatures.
• Divide the sum of the high temperatures by the number of temperatures (7).

Solution
56°
72°
80°
75°
68°
62°
+ 74°
487° sum of high temperatures

69.57
7)487.00

The mean high temperatures is 69.6°F.

26. Strategy To calculate the probability:
• Count the number of possible outcomes.
• Count the number of favorable outcomes.
• Use the probability formula.

Solution There are 36 possible outcomes. There are 5 favorable outcomes: (2, 6), (6, 2), (3, 5), (5, 3), (4, 4).

Probability $= \dfrac{5}{36}$

The probability is $\dfrac{5}{36}$ that the sum of the dots on the two dice is 8.

Chapter 8: U.S. Customary Units of Measurement

Prep Test

1. 702

2. 58

3. 4

4. $\dfrac{5}{3} \times 6 = \dfrac{5 \cdot 2 \cdot \overset{1}{\cancel{3}}}{\underset{1}{\cancel{3}}} = 10$

5. $400 \times \dfrac{1}{8} \times \dfrac{1}{2} = \dfrac{\overset{1}{\cancel{2}} \cdot \overset{1}{\cancel{2}} \cdot \overset{1}{\cancel{2}} \cdot \overset{1}{\cancel{2}} \cdot 5 \cdot 5}{\underset{1}{\cancel{2}} \cdot \underset{1}{\cancel{2}} \cdot \underset{1}{\cancel{2}} \cdot \underset{1}{\cancel{2}}} = 25$

6. $5\dfrac{3}{4} \times 8 = \dfrac{23}{4} \times 8 = \dfrac{23 \cdot \overset{1}{\cancel{2}} \cdot \overset{1}{\cancel{2}} \cdot 2}{\underset{1}{\cancel{2}} \cdot \underset{1}{\cancel{2}}} = 46$

7. $3\overline{)714}$ 　→　238

8. $12\overline{)18.0}$ 　→　1.5

Go Figure

To go from the bank to the bookstore takes 10 minutes. She goes a distance of
$\dfrac{3}{4} - \dfrac{1}{3} = \dfrac{9}{12} - \dfrac{4}{12} = \dfrac{5}{12}$.
We want to find how long it takes her to reach the halfway mark on her way to work. The distance between the bank and the halfway point is
$\dfrac{1}{2} - \dfrac{1}{3} = \dfrac{3}{6} - \dfrac{2}{6} = \dfrac{1}{6} = \dfrac{2}{12}$.

We can say that $\dfrac{1}{12} = 1$ unit. This is similar to the inches and foot relationship. Using a proportion: it takes 10 minutes to go 5 units, how many minutes does it take to go 2 units?
$\dfrac{10 \text{ min}}{5 \text{ units}} = \dfrac{n \text{ min}}{2 \text{ units}}$
$10 \times 2 = 5 \times n$
$20 = 5 \times n$
$20 \div 5 = n$
$4 = n$
It takes Mandy 4 minutes to go 2 units. It takes 4 minutes to go from the bank to the halfway point. 7:52 A.M. + 4 minute = 7:56 A.M. She reaches the halfway point at 7:56 A.M.

Section 8.1

Objective A Exercises

1. $6 \text{ ft} = 6 \,\cancel{\text{ft}} \times \dfrac{12 \text{ in.}}{1 \,\cancel{\text{ft}}} = 72 \text{ in.}$

3. $30 \text{ in.} = 30 \,\cancel{\text{in.}} \times \dfrac{1 \text{ ft}}{12 \,\cancel{\text{in.}}} = 2\dfrac{1}{2} \text{ ft}$

5. $13 \text{ yd} = 13 \,\cancel{\text{yd}} \times \dfrac{3 \text{ ft}}{1 \,\cancel{\text{yd}}} = 39 \text{ ft}$

7. $16 \text{ ft} = 16 \,\cancel{\text{ft}} \times \dfrac{1 \text{ yd}}{3 \,\cancel{\text{ft}}} = 5\dfrac{1}{3} \text{ yd}$

9. $2\dfrac{1}{3} \text{ yd} = 2\dfrac{1}{3} \,\cancel{\text{yd}} \times \dfrac{36 \text{ in.}}{1 \,\cancel{\text{yd}}} = 84 \text{ in.}$

11. $120 \text{ in.} = 120 \,\cancel{\text{in.}} \times \dfrac{1 \text{ yd}}{36 \,\cancel{\text{in.}}} = 3\dfrac{1}{3} \text{ yd}$

13. $2 \text{ mi} = 2 \,\cancel{\text{mi}} \times \dfrac{5280 \text{ ft}}{1 \,\cancel{\text{mi}}} = 10{,}560 \text{ ft}$

15. $7\dfrac{1}{2} \text{ in.} = 7\dfrac{1}{2} \,\cancel{\text{in.}} \times \dfrac{1 \text{ ft}}{12 \,\cancel{\text{in.}}} = \dfrac{5}{8} \text{ ft}$

Objective B Exercises

17.
```
        1 mi 1120 ft
  5280)6400
      −5280
        1120
```
6400 ft = 1 mi 1120 ft

19.
```
  6 ft  7 in.
+ 3 ft  4 in.
  9 ft 11 in.
```

21.
```
   4 ft 15 in.
   5 ft  3 in.
 − 2 ft  6 in.
   2 ft  9 in.
```

23.
```
   2 ft  5 in.
 ×         6
  12 ft 30 in. = 14 ft 6 in.
```

25.
```
          2 ft 8 in.
  2)5 ft 4 in.
   −4 ft
     1 ft = 12 in.
           16 in.
          −16 in.
             0
```

27. $4\dfrac{2}{3} \text{ ft} + 6\dfrac{1}{2} \text{ ft} = 4\dfrac{4}{6} \text{ ft} + 6\dfrac{3}{6} \text{ ft}$
$= 10\dfrac{7}{6} \text{ ft}$
$= 11\dfrac{1}{6} \text{ ft}$

29.
$$\begin{array}{r} 1 \text{ mi} \quad 4200 \text{ ft} \\ + 2 \text{ mi} \quad 3600 \text{ ft} \\ \hline 3 \text{ mi} \quad 7800 \text{ ft} = 4 \text{ mi } 2520 \text{ ft} \end{array}$$

Objective C Exercises

31. **Strategy**   To find how many 4-inch tiles can be placed along the counter:
• Convert the length (4 ft 8 in.) to inches.
• Divide the total length by the length of one tile (4 in.).

     Solution $4 \text{ ft} = 4 \text{ ft} \times \dfrac{12 \text{ in.}}{1 \text{ ft}} = 48 \text{ in.}$

4 ft 8 in. = 48 in. + 8 in. = 56 in.
$56 \div 4 = 14$
14 tiles can be placed along one row.

33. **Strategy**   To find the missing dimension:
• Find the total of the two given dimensions by adding the two lengths $\left(1\dfrac{1}{3} \text{ ft and } 1\dfrac{1}{3} \text{ ft}\right)$.
• Subtract the total of the two given dimensions from the entire length $\left(4\dfrac{1}{2} \text{ ft}\right)$.

     Solution $1\dfrac{1}{3} \text{ ft} + 1\dfrac{1}{3} \text{ ft} = 2\dfrac{2}{3} \text{ ft}$

$$\begin{array}{r} 4\dfrac{1}{2} \text{ ft} = 4\dfrac{3}{6} = 3\dfrac{9}{6} \\ -2\dfrac{2}{3} \text{ ft} = 2\dfrac{4}{6} = 2\dfrac{4}{6} \\ \hline 1\dfrac{5}{6} \end{array}$$

The missing dimension is $1\dfrac{5}{6}$ ft.

35. **Strategy**   To find the length of material needed, add the diameters of the two holes (3 in. each) and the lengths of the three spaces left in between.

     Solution
$$\begin{array}{r} 3 \text{ in.} \\ 3 \text{ in.} \\ \dfrac{1}{2} \text{ in.} \\ \dfrac{1}{2} \text{ in.} \\ + \dfrac{1}{2} \text{ in.} \\ \hline 6\dfrac{3}{2} \text{ in.} = 7\dfrac{1}{2} \text{ in.} \end{array}$$

The length of material needed is $7\dfrac{1}{2}$ in.

37. **Strategy**   To find the length of each piece, divide the total length $\left(6\dfrac{2}{3} \text{ ft}\right)$ by the number of equal pieces (4).

     Solution $6\dfrac{2}{3} \text{ ft} \div 4 = \dfrac{20}{3} \div 4$

$$= \dfrac{20}{3} \times \dfrac{1}{4} = \dfrac{2 \cdot 2 \cdot 5}{3 \cdot 2 \cdot 2}$$

$$= \dfrac{5}{3} \text{ ft} = 1\dfrac{2}{3} \text{ ft}$$

The length of each piece is $1\dfrac{2}{3}$ ft

39. **Strategy**   To find the length of framing needed, add the lengths of the four sides of the frame (1 ft 9 in., 1 ft 6 in., 1 ft 9 in., and 1 ft 6 in.).

     Solution
$$\begin{array}{r} 1 \text{ ft} \quad 9 \text{ in.} \\ 1 \text{ ft} \quad 6 \text{ in.} \\ 1 \text{ ft} \quad 9 \text{ in.} \\ +1 \text{ ft} \quad 6 \text{ in.} \\ \hline 4 \text{ ft } 30 \text{ in.} = 6 \text{ ft } 6 \text{ in.} \end{array}$$
The length of framing needed is 6 ft 6 in.

41. **Strategy**   To find the total length of the wall in feet:
• Multiply the length of each brick (9 in.) by the number of bricks (45) to find the total length in inches.
• Convert the total length in inches to feet.

     Solution
$$\begin{array}{r} 9 \text{ in.} \\ \times \quad 45 \\ \hline 405 \text{ in.} \end{array} \quad 405 \text{ in.} \times \dfrac{1 \text{ ft}}{12 \text{ in.}} = 33\dfrac{3}{4} \text{ ft}$$

The total length of the wall is $33\dfrac{3}{4}$ ft.

Applying the Concepts

43. $\dfrac{19 \text{ in.}}{1} \times \dfrac{1 \text{ ft}}{12 \text{ in.}} \times \dfrac{1 \text{ mi}}{5280 \text{ ft}} \times 200{,}000{,}000$
$\approx 59{,}975 \text{ mi}$
Yes, since $59{,}975 > 25{,}000$ the line would reach around the Earth at the equator.

Section 8.2

Objective A Exercises

1. $64 \text{ oz} = 64 \text{ oz} \times \dfrac{1 \text{ lb}}{16 \text{ oz}} = 4 \text{ lb}$

3. $8 \text{ lb} = 8 \text{ lb} \times \dfrac{16 \text{ oz}}{1 \text{ lb}} = 128 \text{ oz}$

5. $3200 \text{ lb} = 3200 \text{ lb} \times \dfrac{1 \text{ ton}}{2000 \text{ lb}} = 1\dfrac{3}{5} \text{ tons}$

7. $6 \text{ tons} = 6 \text{ tons} \times \dfrac{2000 \text{ lb}}{1 \text{ ton}} = 12{,}000 \text{ lb}$

9. $66 \text{ oz} = 66 \cancel{\text{ oz}} \times \dfrac{1 \text{ lb}}{16 \cancel{\text{ oz}}} = 4\dfrac{1}{8} \text{ lb}$

11. $1\dfrac{1}{2} \text{ lb} = 1\dfrac{1}{2} \cancel{\text{ lb}} \times \dfrac{16 \text{ oz}}{1 \cancel{\text{ lb}}} = 24 \text{ oz}$

13. $1\dfrac{3}{10} \text{ tons} = 1\dfrac{3}{10} \cancel{\text{ tons}} \times \dfrac{2000 \text{ lb}}{1 \cancel{\text{ ton}}} = 2600 \text{ lb}$

15. $500 \text{ lb} = 500 \cancel{\text{ lb}} \times \dfrac{1 \text{ ton}}{2000 \cancel{\text{ lb}}} = \dfrac{1}{4} \text{ ton}$

17. $180 \text{ oz} = 180 \cancel{\text{ oz}} \times \dfrac{1 \text{ lb}}{16 \cancel{\text{ oz}}} = 11\dfrac{1}{4} \text{ lb}$

Objective B Exercises

19.
$$
\begin{array}{r}
4 \text{ tons } 1000 \text{ lb} \\
2000\overline{)9000} \\
-8000 \\
\hline
1000
\end{array}
$$
$9000 \text{ lb} = 4 \text{ tons } 1000 \text{ lb}$

21.
$$
\begin{array}{r}
2 \text{ lb } 8 \text{ oz} \\
16\overline{)40} \\
-32 \\
\hline
8
\end{array}
$$
$40 \text{ oz} = 2 \text{ lb } 8 \text{ oz}$

23.
$$
\begin{array}{r}
1 \text{ ton } 800 \text{ lb} \\
+3 \text{ tons } 1600 \text{ lb} \\
\hline
4 \text{ tons } 2400 \text{ lb} = 5 \text{ tons } 400 \text{ lb}
\end{array}
$$

25.
$$
\begin{array}{r}
\overset{2}{} \quad \overset{2500}{} \\
\cancel{3} \text{ tons } \cancel{500} \text{ lb} \\
-1 \text{ ton } 800 \text{ lb} \\
\hline
1 \text{ ton } 1700 \text{ lb}
\end{array}
$$

27. $5\dfrac{1}{2} \text{ lb} \times 6 = \dfrac{11}{2} \text{ lb} \times 6$

$\qquad = \dfrac{66}{2} \text{ lb}$

$\qquad = 33 \text{ lb}$

29. $4\dfrac{2}{3} \text{ lb} \times 3 = \dfrac{14}{3} \text{ lb} \times 3$

$\qquad = \dfrac{42}{3} \text{ lb}$

$\qquad = 14 \text{ lb}$

31.
$$
\begin{array}{r}
6\dfrac{1}{2} \text{ oz} \\
+2\dfrac{1}{2} \text{ oz} \\
\hline
8\dfrac{2}{2} \text{ oz} = 9 \text{ oz}
\end{array}
$$

33.
$$
\begin{array}{r}
1 \text{ lb } 7 \text{ oz} \\
4\overline{)5 \text{ lb } 12 \text{ oz}} \\
-4 \text{ lb} \\
\hline
1 \text{ lb} = 16 \text{ oz} \\
28 \text{ oz} \\
-28 \text{ oz} \\
\hline
0
\end{array}
$$

Objective C Exercises

35. **Strategy** To find the weight of the load, multiply the number of bricks (800) by the weight of one brick $\left(2\dfrac{1}{2} \text{ lb}\right)$.

Solution $800 \times 2\dfrac{1}{2} \text{ lb} = 800 \times \dfrac{5}{2} \text{ lb}$

$\qquad = \dfrac{4000}{2} \text{ lb}$

$\qquad = 2000 \text{ lb}$

The load of bricks weighs 2000 lb.

37. **Strategy** To find the weight of the package in pounds:
• Multiply the number of tiles (144) by the weight of one tile (7 oz).
• Convert the number of ounces to pounds.

Solution
$$
\begin{array}{r}
144 \\
\times 7 \text{ oz} \\
\hline
1008 \text{ oz}
\end{array}
$$
$1008 \cancel{\text{ oz}} \times \dfrac{1 \text{ lb}}{16 \cancel{\text{ oz}}} = 63 \text{ lb}$

The package of tiles weighs 63 lb.

39. **Strategy** To find the weight of the case in pounds:
• Find the weight in ounces by multiplying the weight of each can (6 oz) by the number of cans (24).
• Convert the weight in ounces to pounds.

Solution
$$
\begin{array}{r}
6 \text{ oz} \\
\times 24 \\
\hline
144 \text{ oz}
\end{array}
$$
$144 \cancel{\text{ oz}} \times \dfrac{1 \text{ lb}}{16 \cancel{\text{ oz}}} = 9 \text{ lb}$

The weight of the case of soft drinks is 9 lb.

41. **Strategy** To find how much shampoo is in each container, divide the total weight of the shampoo (5 lb 4 oz) by the number of containers (4).

Solution
$$
\begin{array}{r}
1 \text{ lb } 5 \text{ oz} \\
4\overline{)5 \text{ lb } 4 \text{ oz}} \\
-4 \text{ lb} \\
\hline
1 \text{ lb} = 16 \text{ oz} \\
20 \text{ oz} \\
-20 \text{ oz} \\
\hline
0
\end{array}
$$

Each container holds 1 lb 5 oz of shampoo.

43. Strategy To find the cost of the ham roast:
- Convert 5 lb 10 oz to pounds.
- Multiply the number of pounds by the price per pound ($4.80).

Solution 5 lb 10 oz $= 5\dfrac{5}{8}$ lb $= 5.625$ lb

$$\begin{array}{r} \$\,4.80 \\ \times\quad 5.625 \\ \hline \$\,27.00000 \end{array}$$

The ham roast costs $27.

45. Strategy To find the cost of mailing the manuscript:
- Convert 2 lb 3 oz to ounces.
- Multiply the number of ounces by the postage rate per ounce ($.25)

Solution 2 lb 3 oz $= 35$ oz

$$\begin{array}{r} \$0.25 \\ \times\quad 35 \\ \hline \$8.75 \end{array}$$

The cost of mailing the manuscript is $8.75.

Applying the Concepts

47. Answers will vary.

Section 8.3

Objective A Exercises

1. 60 fl oz $= 60 \text{ fl oz} \times \dfrac{1 \text{ c}}{8 \text{ fl oz}} = 7\dfrac{1}{2}$ c

3. 3 c $= 3 \text{ c} \times \dfrac{8 \text{ fl oz}}{1 \text{ c}} = 24$ fl oz

5. 8 c $= 8 \text{ c} \times \dfrac{1 \text{ pt}}{2 \text{ c}} = 4$ pt

7. $3\dfrac{1}{2}$ pt $= 3\dfrac{1}{2} \text{ pt} \times \dfrac{2 \text{ c}}{1 \text{ pt}} = 7$ c

9. 22 qt $= 22 \text{ qt} \times \dfrac{1 \text{ gal}}{4 \text{ qt}} = 5\dfrac{1}{2}$ gal

11. $2\dfrac{1}{4}$ gal $= 2\dfrac{1}{4} \text{ gal} \times \dfrac{4 \text{ qt}}{1 \text{ gal}} = 9$ qt

13. $7\dfrac{1}{2}$ pt $= 7\dfrac{1}{2} \text{ pt} \times \dfrac{1 \text{ qt}}{2 \text{ pt}} = 3\dfrac{3}{4}$ qt

15. 20 fl oz $= 20 \text{ fl oz} \times \dfrac{1 \text{ c}}{8 \text{ fl oz}} \times \dfrac{1 \text{ pt}}{2 \text{ c}} = \dfrac{5}{4}$ pt $= 1\dfrac{1}{4}$ pt

17. 17 c $= 17 \text{ c} \times \dfrac{1 \text{ pt}}{2 \text{ c}} \times \dfrac{1 \text{ qt}}{2 \text{ pt}} = 4\dfrac{1}{4}$ qt

Objective B Exercises

19.
$$\begin{array}{r} 3 \text{ gal } 2 \text{ qt} \\ 4\overline{)14} \\ \underline{-12} \\ 2 \end{array}$$
14 qt $= 3$ gal 2 qt

21.
$$\begin{array}{r} 2 \text{ qt } 1 \text{ pt} \\ 2\overline{)5} \\ \underline{-4} \\ 1 \end{array}$$
5 pt $= 2$ qt 1 pt

23.
$$\begin{array}{r} 4 \text{ qt } 1 \text{ pt} \\ +2 \text{ qt } 1 \text{ pt} \\ \hline 6 \text{ qt } 2 \text{ pt} = 7 \text{ qt} \end{array}$$

25.
$$\begin{array}{r} 2 \text{ c } 11 \text{ fl oz} \\ \cancel{3 \text{ c } 3 \text{ fl oz}} \\ -2 \text{ c } 5 \text{ fl oz} \\ \hline 6 \text{ fl oz} \end{array}$$

27. $3\dfrac{1}{2}$ pt $\times 5 = \dfrac{7}{2}$ pt $\times 5 = \dfrac{35}{2}$ pt $= 17\dfrac{1}{2}$ pt

29. $3\dfrac{1}{2}$ gal $\div 4 = 3\dfrac{1}{2}$ gal $\times \dfrac{1}{4}$
$= \dfrac{7}{2}$ gal $\times \dfrac{1}{4}$
$= \dfrac{7}{8}$ gal

31.
$$\begin{array}{r} 3 \text{ gal } 3 \text{ qt} \\ +1 \text{ gal } 2 \text{ qt} \\ \hline 4 \text{ gal } 5 \text{ qt} = 5 \text{ gal } 1 \text{ qt} \end{array}$$

33.
$$\begin{array}{r} 2 \text{ gal } 4 \text{ qt} \\ \cancel{3 \text{ gal}} \\ -1 \text{ gal } 2 \text{ qt} \\ \hline 1 \text{ gal } 2 \text{ qt} \end{array}$$

35. $4\dfrac{1}{2}$ gal $- 1\dfrac{3}{4}$ gal $= 4\dfrac{2}{4}$ gal $- 1\dfrac{3}{4}$ gal
$= 3\dfrac{6}{4}$ gal $- 1\dfrac{3}{4}$ gal
$= 2\dfrac{3}{4}$ gal

Objective C Exercises

37. Strategy To find how many gallons of coffee should be prepared:
● Find how many cups of coffee should be prepared by multiplying the number of adults attending (60) by the number of cups each adult will drink (2).
● Convert the number of cups to gallons.

Solution $2 \text{ c} \times 60 = 120 \text{ c}$

$$120 \text{ c} \times \frac{1 \text{ pt}}{2 \text{ c}} \times \frac{1 \text{ qt}}{2 \text{ pt}} \times \frac{1 \text{ gal}}{4 \text{ qt}} = \frac{120 \text{ gal}}{16}$$
$$= 7\frac{1}{2} \text{ gal}$$

$7\frac{1}{2}$ gal of coffee should be prepared.

39. Strategy To find the number of quarts of final solution:
● Find the total number of ounces in the solution by adding the number of ounces in the three components (72, 16, and 48 oz).
● Convert the number of ounces to quarts.

Solution
$$\begin{array}{r} 72 \text{ fl oz} \\ 16 \text{ fl oz} \\ + \ 48 \text{ fl oz} \\ \hline 136 \text{ fl oz} \end{array}$$

$$136 \text{ fl oz} \times \frac{1 \text{ c}}{8 \text{ fl oz}} \times \frac{1 \text{ pt}}{2 \text{ c}} \times \frac{1 \text{ qt}}{2 \text{ pt}} = \frac{136}{32} \text{ qt}$$
$$= 4\frac{1}{4} \text{ qt}$$

There are $4\frac{1}{4}$ qt of final solution.

41. Strategy To find the number of gallons of oil the farmer used:
● Multiply 5 qt by 7 to find the number of quarts used.
● Convert the number of quarts to gallons.

Solution
$$\begin{array}{r} 5 \text{ qt} \\ \times \ 7 \\ \hline 35 \text{ qt of oil} \end{array}$$

$$35 \text{ qt} = 35 \text{ qt} \times \frac{1 \text{ gal}}{4 \text{ qt}} = 8\frac{3}{4} \text{ gal}$$

The farmer used $8\frac{3}{4}$ gal of oil.

43. Strategy To find the more economical purchase:
● Convert 1 qt to ounces.
● Compare the price per ounce of each brand of tomato juice.

Solution $1 \text{ qt} = 1 \text{ qt} \times \frac{2 \text{ pt}}{1 \text{ qt}} \times \frac{2 \text{ c}}{1 \text{ pt}} \times \frac{8 \text{ oz}}{1 \text{ c}} = 32 \text{ oz}$

First brand: $32 \overline{)\ \dfrac{\$0.0497}{\$1.5900}}$

Second brand: $24 \overline{)\ \dfrac{\$0.052}{\$1.250}}$

$\$0.0497 < \$0.052$

The more economical purchase is 1 qt for \$1.59 (the first brand).

45. Strategy To find the profit:
● Convert 5 qt to fluid ounces.
● Divide the number of fluid ounces by 8 to find the number of bottles.
● Multiply the number of bottles by \$8.25 to find the total income.
● Subtract the original cost (\$81.50) from the total income to find the profit.

Solution $5 \text{ qt} = 5 \text{ qt} \times \frac{2 \text{ pt}}{1 \text{ qt}} \times \frac{2 \text{ c}}{1 \text{ pt}} \times \frac{8 \text{ oz}}{1 \text{ c}}$
$$= 160 \text{ fl oz}$$

$8 \overline{)\ \dfrac{20}{160 \text{ fl oz}}}$ number of bottles

$$\begin{array}{r} \$8.25 \\ \times \quad 20 \\ \hline \$165.00 \end{array} \quad \text{total income}$$

$$\begin{array}{r} \$165.00 \\ - \ 81.50 \\ \hline \$ \ 83.50 \end{array}$$

The profit made was \$83.50.

Applying the Concepts

47. Grain: A unit of weight in the U.S. Customary System; an avoirdupois unit equal to 0.002286 ounce, or 0.036 dram
Dram: A unit of weight in the U.S. Customary System; an avoirdupois unit equal to 0.0625 ounce, or 27.344 grains
Furlong: A unit of measuring distance, equal to 0.125 mile, or 220 yards
Rod: A linear measure equal to 5.5 yards, to 16.5 feet, and to 5.03 meters
Examples will vary.

Section 8.4

Objective A Exercises

1. $98 \text{ days} = 98 \text{ days} \times \frac{1 \text{ week}}{7 \text{ days}} = 14 \text{ weeks}$

3. $6\frac{1}{4} \text{ days} = \frac{25}{4} \text{ days} \times \frac{24 \text{ h}}{1 \text{ day}} = 150 \text{ h}$

5. $555 \text{ min} = 555 \text{ min} \times \frac{1 \text{ h}}{60 \text{ min}} = 9\frac{1}{4} \text{ h}$

7. $18\frac{1}{2} \text{ min} = 18\frac{1}{2} \text{ min} \times \frac{60 \text{ s}}{1 \text{ min}} = 1110 \text{ s}$

9. $12,600 \text{ s} = 12,600 \text{ s} \times \dfrac{1 \text{ min}}{60 \text{ s}} \times \dfrac{1 \text{ h}}{60 \text{ min}} = 3\dfrac{1}{2} \text{ h}$

11. $6\dfrac{1}{2} \text{ h} = 6\dfrac{1}{2} \text{ h} \times \dfrac{60 \text{ min}}{1 \text{ h}} \times \dfrac{60 \text{ s}}{1 \text{ min}} = 23,400 \text{ s}$

13. $5040 \text{ min} = 5040 \text{ min} \times \dfrac{1 \text{ h}}{60 \text{ min}} \times \dfrac{1 \text{ day}}{24 \text{ h}} = 3\dfrac{1}{2} \text{ days}$

15. $2\dfrac{1}{2} \text{ days} = 2\dfrac{1}{2} \text{ days} \times \dfrac{24 \text{ h}}{1 \text{ day}} \times \dfrac{60 \text{ min}}{1 \text{ h}} = 3600 \text{ min}$

17. $672 \text{ h} = 672 \text{ h} \times \dfrac{1 \text{ day}}{24 \text{ h}} \times \dfrac{1 \text{ week}}{7 \text{ days}} = 4 \text{ weeks}$

19. $3 \text{ weeks} = 3 \text{ weeks} \times \dfrac{7 \text{ days}}{1 \text{ week}} \times \dfrac{24 \text{ h}}{1 \text{ day}} = 504 \text{ h}$

21. $172,800 \text{ s} = 172,800 \text{ s} \times \dfrac{1 \text{ min}}{60 \text{ s}} \times \dfrac{1 \text{ h}}{60 \text{ min}} \times \dfrac{1 \text{ day}}{24 \text{ h}}$
$= 2 \text{ days}$

23. $3 \text{ days} = 3 \text{ days} \times \dfrac{24 \text{ h}}{1 \text{ day}} \times \dfrac{60 \text{ min}}{1 \text{ h}} \times \dfrac{60 \text{ s}}{1 \text{ min}}$
$= 259,200 \text{ s}$

Applying the Concepts

25. Yes, the year 1984 is divisible by 4.

27. Yes, the year 2144 is divisible by 4.

29. $2555 \text{ days} = 2555 \text{ days} \times \dfrac{1 \text{ year}}{365 \text{ days}} = 7 \text{ years}$

31. $6 \text{ years} = 6 \text{ years} \times \dfrac{365 \text{ days}}{1 \text{ year}} = 2190 \text{ days}$

33. $3\dfrac{1}{2} \text{ years} = 3\dfrac{1}{2} \text{ years} \times \dfrac{365 \text{ days}}{1 \text{ year}} \times \dfrac{24 \text{ h}}{1 \text{ day}} = 30,660 \text{ h}$

Section 8.5

Objective A Exercises

1. $25 \text{ Btu} = 25 \text{ Btu} \times \dfrac{778 \text{ ft} \cdot \text{lb}}{1 \text{ Btu}}$
$= 19,450 \text{ ft} \cdot \text{lb}$

3. $25,000 \text{ Btu} = 25,000 \text{ Btu} \times \dfrac{778 \text{ ft} \cdot \text{lb}}{1 \text{ Btu}}$
$= 19,450,000 \text{ ft} \cdot \text{lb}$

5. $\text{Energy} = 150 \text{ lb} \times 10 \text{ ft}$
$= 1500 \text{ ft} \cdot \text{lb}$

7. $\text{Energy} = 3300 \text{ lb} \times 9 \text{ ft}$
$= 29,700 \text{ ft} \cdot \text{lb}$

9. $3 \text{ tons} = 6000 \text{ lb}$
$\text{Energy} = 6000 \text{ lb} \times 5 \text{ ft}$
$= 30,000 \text{ ft} \cdot \text{lb}$

11. $850 \times 3 \text{ lb} = 2550 \text{ lb}$
$\text{Energy} = 2550 \text{ lb} \times 10 \text{ ft} = 25,500 \text{ ft} \cdot \text{lb}$

13. $45,000 \text{ Btu} = 45,000 \text{ Btu} \times \dfrac{778 \text{ ft} \cdot \text{lb}}{1 \text{ Btu}}$
$= 35,010,000 \text{ ft} \cdot \text{lb}$

15. $12,000 \text{ Btu} = 12,000 \text{ Btu} \times \dfrac{778 \text{ ft} \cdot \text{lb}}{1 \text{ Btu}}$
$= 9,336,000 \text{ ft} \cdot \text{lb}$

Objective B Exercises

17. $\dfrac{1100}{550} = 2 \text{ hp}$

19. $\dfrac{4400}{550} = 8 \text{ hp}$

21. $9 \times 550 \dfrac{\text{ft} \cdot \text{lb}}{\text{s}} = 4950 \dfrac{\text{ft} \cdot \text{lb}}{\text{s}}$

23. $7 \times 550 \dfrac{\text{ft} \cdot \text{lb}}{\text{s}} = 3850 \dfrac{\text{ft} \cdot \text{lb}}{\text{s}}$

25. $\text{Power} = \dfrac{125 \text{ lb} \times 12 \text{ ft}}{3 \text{ s}}$
$= 500 \dfrac{\text{ft} \cdot \text{lb}}{\text{s}}$

27. $\text{Power} = \dfrac{3000 \text{ lb} \times 40 \text{ ft}}{25 \text{ s}} = 4800 \dfrac{\text{ft} \cdot \text{lb}}{\text{s}}$

29. $\text{Power} = \dfrac{180 \text{ lb} \times 40 \text{ ft}}{5 \text{ s}} = 1440 \dfrac{\text{ft} \cdot \text{lb}}{\text{s}}$

31. $\dfrac{4950}{550} = 9 \text{ hp}$

33. $\dfrac{6600}{550} = 12 \text{ hp}$

Chapter 8 Review Exercises

1. $4 \text{ ft} = 4 \text{ ft} \times \dfrac{12 \text{ in.}}{1 \text{ ft}} = 48 \text{ in.}$

2.
```
      2 ft 6 in.
   3)7 ft   6 in.
    -6 ft
      1 ft = 12 in.
             18 in.
            -18 in.
                 0
```

3. $\text{Energy} = 200 \text{ lb} \times 8 \text{ ft} = 1600 \text{ ft} \cdot \text{lb}$

4. $2\dfrac{1}{2} \text{ pt} = 2\dfrac{1}{2} \text{ pt} \times \dfrac{2 \text{ c}}{1 \text{ pt}} \times \dfrac{8 \text{ fl oz}}{1 \text{ c}}$
$= 2\dfrac{1}{2} \times 16 \text{ fl oz}$
$= \dfrac{5}{2} \times 16 \text{ fl oz}$
$= 40 \text{ fl oz}$

5. $14 \text{ ft} = 14 \text{ ft} \times \dfrac{1 \text{ yd}}{3 \text{ ft}}$

   $= \dfrac{14}{3} \text{ yd}$

   $= 4\dfrac{2}{3} \text{ yd}$

6. $2400 \text{ lb} = 2400 \text{ lb} \times \dfrac{1 \text{ ton}}{2000 \text{ lb}}$

   $= \dfrac{2400}{2000} \text{ tons} = 1\dfrac{1}{5} \text{ tons}$

7.
   ```
         2 lb   7 oz
     3)7 lb   5 oz
      −6 lb
       1 lb = 16 oz
               21 oz
              −21 oz
                 0
   ```

8. $3\dfrac{3}{8} \text{ lb} = 3\dfrac{3}{8} \text{ lb} \times \dfrac{16 \text{ oz}}{1 \text{ lb}}$

   $= 3\dfrac{3}{8} \times 16 \text{ oz}$

   $= \dfrac{27}{8} \times 16 \text{ oz} = 54 \text{ oz}$

9.
   ```
      3 ft   9 in.
    + 5 ft   6 in.
      8 ft  15 in. = 9 ft 3 in.
   ```

10.
   ```
      2      2500
    3 tons  500 lb
   −1 ton  1500 lb
    1 ton  1000 lb
   ```

11.
   ```
      4 c   7 fl oz
    +2 c   3 fl oz
      6 c  10 fl oz = 7 c 2 fl oz
   ```

12.
   ```
      4    4
    5 yd 1 ft
   −3 yd 2 ft
    1 yd 2 ft
   ```

13. $12 \text{ c} \times \dfrac{1 \text{ pt}}{2 \text{ c}} \times \dfrac{1 \text{ qt}}{2 \text{ pt}} = \dfrac{12}{4} \text{ qt} = 3 \text{ qt}$

14. $375 \text{ min} \times \dfrac{1 \text{ h}}{60 \text{ min}} = 6\dfrac{1}{4} \text{ h}$

15. $2.5 \text{ hp} \times 550 \dfrac{\text{ft} \cdot \text{lb}}{\text{s}} = 1375 \dfrac{\text{ft} \cdot \text{lb}}{\text{s}}$

16.
   ```
        5 lb 8 oz
    ×         8
     40 lb 64 oz = 44 lb
   ```

17. $50 \text{ Btu} = 50 \text{ Btu} \times \dfrac{778 \text{ ft} \cdot \text{lb}}{1 \text{ Btu}} = 38,900 \text{ ft} \cdot \text{lb}$

18. $\dfrac{3850}{550} = 7 \text{ hp}$

19. **Strategy** To find the length of the remaining piece of board, subtract the length of the piece cut (6 ft 11 in.) from the total length (10 ft 5 in.).

    Solution
    ```
        9    17
     10 ft   5 in.
    −  6 ft  11 in.
       3 ft   6 in.
    ```
    The length of the remaining piece is 3 ft 6 in.

20. **Strategy** To find the cost of mailing the book:
    • Find the weight of the book in ounces.
    • Multiply the weight of the book in ounces by the price per ounce for postage ($0.24).

    Solution
    2 lb 3 oz $= 35 \text{ oz}$
    ```
         $0.24
    ×       35
         $8.40
    ```
    The cost of mailing the book is $8.40.

21. **Strategy** To find the number of quarts in a case:
    • Find the number of ounces in a case by multiplying the number of ounces in a can (18 fl oz) by the number of cans in a case (24).
    • Convert the number of ounces to quarts.

    Solution
    ```
        18 fl oz
    ×      24
       432 fl oz
    ```
    $432 \text{ fl oz} \times \dfrac{1 \text{ c}}{8 \text{ fl oz}} \times \dfrac{1 \text{ pt}}{2 \text{ c}} \times \dfrac{1 \text{ qt}}{2 \text{ pt}}$

    $= \dfrac{432}{32} \text{ qt} = 13\dfrac{1}{2} \text{ qt}$

    There are $13\dfrac{1}{2}$ qt in a case.

22. **Strategy** To find how many gallons of milk were sold:
● Find the number of cups sold by multiplying the number of cartons (256) by the number of cups per carton (1).
● Convert the number of cups to gallons.

Solution 256 cartons × 1 c = 256 c

$256 \text{ c} = 256 \text{ c} \times \dfrac{1 \text{ pt}}{2 \text{ c}} \times \dfrac{1 \text{ qt}}{2 \text{ pt}} \times \dfrac{1 \text{ gal}}{4 \text{ qt}}$

$= \dfrac{256}{16} \text{ gal} = 16 \text{ gal}$

16 gal of milk were sold that day.

23. $35,0000 \text{ Btu} = 35,000 \text{ Btu} \times \dfrac{778 \text{ ft} \cdot \text{lb}}{1 \text{ Btu}}$

$= 27,230,000 \text{ ft} \cdot \text{lb}$

24. $\text{Power} = \dfrac{800 \text{ lb} \times 15 \text{ ft}}{25 \text{ s}} = 480 \dfrac{\text{ft} \cdot \text{lb}}{\text{s}}$

Chapter 8 Test

1. $2\dfrac{1}{2} \text{ ft} = 2\dfrac{1}{2} \text{ ft} \times \dfrac{12 \text{ in.}}{1 \text{ ft}} = 2\dfrac{1}{2} \times 12 \text{ in.}$

$= \dfrac{5}{2} \times 12 \text{ in.}$

$= 30 \text{ in.}$

2.
$\overset{3 \quad 14}{4 \text{ ft } 2 \text{ in.}}$
$\underline{-1 \text{ ft } 9 \text{ in.}}$
$2 \text{ ft } 5 \text{ in.}$

3. **Strategy** To find the length of each equal piece, divide the total length $\left(6\dfrac{2}{3} \text{ ft}\right)$ by the number of pieces (5).

Solution $6\dfrac{2}{3} \text{ ft} \div 5 = \dfrac{20}{3} \text{ ft} \div 5$

$= \dfrac{20}{3} \text{ ft} \times \dfrac{1}{5}$

$= \dfrac{4}{3} \text{ ft} = 1\dfrac{1}{3} \text{ ft}$

4. **Strategy** To find the length of the wall in feet:
● Find the length of the wall in inches by multiplying the length of one brick (8 in.) by the number of bricks (72).
● Convert the length in inches to feet.

Solution
$\begin{array}{r} 8 \text{ in.} \\ \times \quad 72 \\ \hline 576 \text{ in.} \end{array}$
$\quad 576 \text{ in.} = 576 \text{ in.} \times \dfrac{1 \text{ ft}}{12 \text{ in.}}$
$= 48 \text{ ft}$
The wall is 48 ft long.

5. $2\dfrac{7}{8} \text{ lb} = 2\dfrac{7}{8} \text{ lb} \times \dfrac{16 \text{ oz}}{1 \text{ lb}}$

$= 2\dfrac{7}{8} \times 16 \text{ oz} = \dfrac{23}{8} \times 16 \text{ oz}$

$= 46 \text{ oz}$

6.
$\begin{array}{r} 2 \text{ lb } 8 \text{ oz} \\ 16\overline{)40} \\ \underline{-32} \\ 8 \end{array}$
$\qquad 40 \text{ oz} = 2 \text{ lb } 8 \text{ oz}$

7.
$\begin{array}{r} 9 \text{ lb } \ 6 \text{ oz} \\ +\ 7 \text{ lb } 11 \text{ oz} \\ \hline 16 \text{ lb } 17 \text{ oz} = 17 \text{ lb } 1 \text{ oz} \end{array}$

8.
$\begin{array}{r} 1 \text{ lb} \quad 11 \text{ oz} \\ 4\overline{)6 \text{ lb} \quad 12 \text{ oz}} \\ \underline{-4 \text{ lb}} \\ 2 \text{ lb} = \underline{32 \text{ oz}} \\ 44 \text{ oz} \\ \underline{-44 \text{ oz}} \\ 0 \end{array}$

9. **Strategy** To find the total weight of the workbooks in pounds:
● Find the total weight of the workbooks in ounces by multiplying the number of workbooks (1000) by the weight per workbook (12 oz).
● Convert the weight in ounces to pounds.

Solution $1000 \times 12 \text{ oz} = 12,000 \text{ oz}$

$12,000 \text{ oz} = 12,000 \text{ oz} \times \dfrac{1 \text{ lb}}{16 \text{ oz}} = 750 \text{ lb}$

The total weight of the workbooks is 750 lb.

10. **Strategy** To find the amount received for recycling the cans:
● Find the weight in ounces of the cans by solving a proportion.
● Convert the weight in ounces to pounds.
● Multiply the weight in pounds by the price paid per pound.

Solution $\dfrac{4 \text{ cans}}{3 \text{ oz}} = \dfrac{800 \text{ cans}}{n}$

$4 \times n = 3 \times 800$
$4 \times n = 2400$
$n = 2400 \div 4 = 600$
The cans weigh 600 oz.

$600 \text{ oz} = 600 \text{ oz} \times \dfrac{1 \text{ lb}}{16 \text{ oz}} = 37.5 \text{ lb}$

$37.5 \text{ lb} \times \$0.75 = \$28.13$
The amount the class received for recycling was \$28.13.

11. $13 \text{ qt} = 13 \text{ qt} \times \dfrac{1 \text{ gal}}{4 \text{ qt}} = \dfrac{13}{4} \text{ gal} = 3\dfrac{1}{4} \text{ gal}$

12. $3\dfrac{1}{2} \text{ gal} = 3\dfrac{1}{2} \text{ gal} \times \dfrac{4 \text{ qt}}{1 \text{ gal}} \times \dfrac{2 \text{ pt}}{1 \text{ qt}}$

$\qquad = 3\dfrac{1}{2} \times 8 \text{ pt} = 28 \text{ pt}$

13. $1\dfrac{3}{4} \text{ gal} \times 7 = \dfrac{7}{4} \text{ gal} \times 7 = \dfrac{49}{4} \text{ gal}$

$\qquad = 12\dfrac{1}{4} \text{ gal}$

14. $\quad$ 5 gal 2 qt
$\quad \underline{+ \ 2 \text{ gal } 3 \text{ qt}}$
$\quad$ 7 gal 5 qt $= 8$ gal 1 qt

15. $756 \text{ h} = 756 \text{ h} \times \dfrac{1 \text{ day}}{24 \text{ h}} \times \dfrac{1 \text{ week}}{7 \text{ days}} = 4\dfrac{1}{2} \text{ weeks}$

16. $3\dfrac{1}{4} \text{ days} = 3\dfrac{1}{4} \text{ days} \times \dfrac{24 \text{ h}}{1 \text{ day}} \times \dfrac{60 \text{ min}}{1 \text{ h}} = 4680 \text{ min}$

17. **Strategy** To find the number of cups of grapefruit juice in a case:
● Find the number of ounces of juice in a case by multiplying the number of cans in a case (24) by the number of ounces in a can (20).
● Convert the number of ounces to cups.

Solution $24 \times 20 \text{ oz} = 480 \text{ oz}$

$480 \text{ oz} = 480 \text{ oz} \times \dfrac{1 \text{ c}}{8 \text{ oz}} = 60 \text{ c}$

There are 60 c in a case.

18. **Strategy** To find the profit:
● Convert 40 gal to quarts.
● To find the total income for the sale of the oil, multiply the number of quarts by the sale price per quart ($2.15).
● To find the profit, subtract the price the mechanic pays for the oil ($200) from the total income.

Solution $40 \text{ gal} = 40 \text{ gal} \times \dfrac{4 \text{ qt}}{1 \text{ gal}} = 160 \text{ qt}$

$160 \text{ qt} \times \$2.15 = \$344 \text{ total income}$
$\$344 - \$200 = \$144$
Nick's profit is $144.

19. Energy $= 250 \text{ lb} \times 15 \text{ ft} = 3750 \text{ ft} \cdot \text{lb}$

20. $40,000 \text{ Btu} = 40,000 \text{ Btu} \times \dfrac{778 \text{ ft} \cdot \text{lb}}{1 \text{ Btu}}$

$\qquad = 31,120,000 \text{ ft} \cdot \text{lb}$

21. $\text{Power} = \dfrac{200 \text{ lb} \times 20 \text{ ft}}{25 \text{ s}} = 160 \dfrac{\text{ft} \cdot \text{lb}}{\text{s}}$

22. $\dfrac{2200}{550} = 4 \text{ hp}$

Cumulative Review Exercises

1.

	2	3	5
$9 =$		$3 \cdot 3$	
$12 =$	$2 \cdot 2$	3	
$15 =$		3	5

$\text{LCM} = 2 \cdot 2 \cdot 3 \cdot 3 \cdot 5 = 180$

2. $\dfrac{43}{8} = \begin{array}{r} 5\dfrac{3}{8} \\ 8)\overline{43} \\ \underline{-40} \\ 3 \end{array}$

3. $\quad 5\dfrac{7}{8} = 5\dfrac{21}{24}$
$\quad \underline{-2\dfrac{7}{12} = 2\dfrac{14}{24}}$
$\qquad\qquad 3\dfrac{7}{24}$

4. $5\dfrac{1}{3} \div 2\dfrac{2}{3} = \dfrac{16}{3} \div \dfrac{8}{3} = \dfrac{16}{3} \times \dfrac{3}{8} = 2$

5. $\dfrac{5}{8} \div \left(\dfrac{3}{8} - \dfrac{1}{4} \right) - \dfrac{5}{8}$

$\dfrac{5}{8} \div \left(\dfrac{3}{8} - \dfrac{2}{8} \right) - \dfrac{5}{8}$

$\dfrac{5}{8} \div \dfrac{1}{8} - \dfrac{5}{8}$

$\dfrac{5}{8} \times \dfrac{8}{1} - \dfrac{5}{8}$

$5 - \dfrac{5}{8}$

$4\dfrac{8}{8} - \dfrac{5}{8} = 4\dfrac{3}{8}$

6. $\quad$ ┌─Given place value
$2.0\overset{}{9}72$
$\quad$ └─$7 > 5$
2.10

7. $\quad$ 0.0792
$\quad \underline{\times \qquad 0.49}$
$\qquad$ 7128
$\quad \underline{\ 3168\quad}$
$\quad$ 0.038808

8. $\dfrac{n}{12} = \dfrac{44}{60}$
$n \times 60 = 12 \times 44$
$n \times 60 = 528$
$\quad n = 528 \div 60 = 8.8$

9. $2\dfrac{1}{2}\% \times 50 = n$
$0.025 \times 50 = n$
$\qquad 1.25 = n$

10. $42\% \times n = 18$
$0.42 \times n = 18$
$n = 18 \div 0.42$
$n \approx 42.86$

11. $\$37.08 \div 7.2 \text{ lb} = \$5.15/\text{lb}$

12.
$$3\frac{2}{5} \text{ in.} = 3\frac{6}{15} \text{ in.}$$
$$+5\frac{1}{3} \text{ in.} = 5\frac{5}{15} \text{ in.}$$
$$8\frac{11}{15} \text{ in.}$$

13.
$$\begin{array}{r} 1 \text{ lb 8 oz} \\ 16\overline{)24} \\ \underline{-16} \\ 8 \end{array} \qquad 24 \text{ oz} = 1 \text{ lb 8 oz}$$

14.
$$\begin{array}{r} 3 \text{ lb} \ \ 8 \text{ oz} \\ \times \qquad 9 \\ \hline 27 \text{ lb 72 oz} = 31 \text{ lb 8 oz} \end{array}$$

15.
$$4\frac{1}{3} \text{ qt} = 4\frac{2}{6} \text{ qt} = 3\frac{8}{6} \text{ qt}$$
$$-1\frac{5}{6} \text{ qt} = 1\frac{5}{6} \text{ qt} = 1\frac{5}{6} \text{ qt}$$
$$2\frac{3}{6} \text{ qt} = 2\frac{1}{2} \text{ qt}$$

16.
$$\begin{array}{r} \overset{3}{4} \text{ lb } \overset{22}{6} \text{ oz} \\ -2 \text{ lb 10 oz} \\ \hline 1 \text{ lb 12 oz} \end{array}$$

17. **Strategy** To find the dividend, solve a proportion.

Solution
$$\frac{\$56}{40 \text{ shares}} = \frac{n}{200 \text{ shares}}$$
$$56 \times 200 = 40 \times n$$
$$11{,}200 = 40 \times n$$
$$11{,}200 \div 40 = n$$
$$280 = n$$
The dividend would be $280.

18. **Strategy** To find Anna's checking balance, subtract the amounts of the checks and add the deposit.

Solution
$$\begin{array}{r} 578.56 \\ -216.98 \\ \hline 361.58 \\ -\ 34.12 \\ \hline 327.46 \\ +315.33 \\ \hline 642.79 \end{array}$$
Anna's balance is $642.79.

19. **Strategy** To find the executive's total monthly income:
● Find the amount of sales over $25,000 by subtracting $25,000 from the total sales ($140,000).
● Find the amount of the commission by solving the basic percent equation for amount. The base is the amount of sales over $25,000 and the percent is 2%.
● Add the amount of commission to the salary ($1800).

Solution $\$140{,}000 - \$25{,}000 = \$115{,}000$
Percent × base = amount
$2\% \times 115{,}000 = n$
$0.02 \times 115{,}000 = n$
$2300 = n$
$\$2300 + \$1800 = \$4100$
The executive's monthly income is $4100.

20. **Strategy** To find the amount of carrots that could be sold:
● Find the amount of spoiled carrots by solving the basic percent equation for the amount. The percent is 3% and the base is 2500.
● Subtract the amount of spoiled carrots from the total amount of the shipment (2500 lb).

Solution Percent × base = amount
$3\% \times 2500 = n$
$0.03 \times 2500 = n$
$75 = n$
$2500 \text{ lb} - 75 \text{ lb} = 2425 \text{ lb}$
The amount of carrots that could be sold is 2425 lb.

21. Strategy To find the percent:
• Find the total number of students who took the final exam by reading the histogram and adding the frequencies.
• Find the number of students who received a score between 80% and 90% by reading the histogram.
• Solve the basic percent equation for percent. The base is the total number of students who took the exam and the amount is the number of students with scores between 80% and 90%.

Solution

Score: 40–50: 2 students
50–60: 1 student
60–70: 5 students
70–80: 7 students
80–90: 4 students
90–100: 3 students
Total number of students: 22

Percent × base = amount
$n \times 22 = 4$
$n = 4 \div 22 \approx 0.18 = 18\%$
The percent is 18%.

22. Strategy To find the selling price:
• Find the amount of the markup by solving the basic percent equation for amount. The base is $220 and the percent is 40%.
• Add the markup to the cost ($220).

Solution

Percent × base = amount
$40\% \times 220 = n$
$0.40 \times 220 = n$
$88 = n$
$\$220 + \$88 = \$308$
The selling price of a compact disc player is $308.

23. Strategy To find the interest paid, multiply the principal ($200,000) by the annual interest rate (6%) by the time (8 months) in years.

Solution $200{,}000 \times 0.06 \times \dfrac{8}{12} = 8000$
The interest paid on the loan is $8000.

24. Strategy To find how much each student received:
• Convert 1 lb 3 oz to ounces.
• Find the total value of the gold by multiplying the number of ounces by the price per ounce ($200).
• Divide the total value by the number of students (6).

Solution 1 lb 3 oz = 19 oz
$19 \times \$200 = \$3800$ total value
$\$3800 \div 6 \approx \$633$
Each student received $633.

25. Strategy To find the cost of mailing the books:
• Find the total weight of the books by adding the 4 weights (1 lb 3 oz, 13 oz, 1 lb 8 oz, and 1 lb).
• Convert the total weight to ounces.
• Find the cost by multiplying the total number of ounces by the price per ounce ($.28).

Solution

$$
\begin{array}{r}
1 \text{ lb } 3 \text{ oz} \\
13 \text{ oz} \\
1 \text{ lb } 8 \text{ oz} \\
+\ 1 \text{ lb} \\
\hline
3 \text{ lb } 24 \text{ oz} = 72 \text{ oz} \\
\end{array}
$$

$$
\begin{array}{r}
\$\ 0.28 \\
\times \quad\ 72 \\
\hline
\$20.16 \\
\end{array}
$$

The cost of mailing the books is $20.16.

26. Strategy To find the better buy:
• Find the unit price for each brand.
• Compare unit prices.

Solution $.79 for 8 oz $2.98 for 36 oz
$\dfrac{0.79}{8} = 0.09875$ $\dfrac{2.98}{36} \approx 0.08278$
$\$0.08278 < \$0.09875$
The better buy is 36 oz for $2.98.

27. Strategy To calculate the probability:
• Count the number of possible outcomes.
• Count the number of favorable outcomes.
• Use the probability formula.

Solution There are 36 possible outcomes.
There are 4 favorable outcomes:
(3, 6), (4, 5), (5, 4), (6, 3).

Probability $= \dfrac{4}{36} = \dfrac{1}{9}$

The probability is $\dfrac{1}{9}$ that the sum of the dots on the two dice is 9.

28. Energy $= 400 \text{ lb} \times 8 \text{ ft} = 3200 \text{ ft} \cdot \text{lb}$

29. Power $= \dfrac{600 \text{ lb} \times 8 \text{ ft}}{12 \text{ s}} = 400 \dfrac{\text{ft} \cdot \text{lb}}{\text{s}}$

Chapter 9: The Metric System of Measurement

Prep Test

1. 37,320

2. 659,000

3. 0.04107

4. $28,496 \div 10^3 = 28,496 \div 1000 = 28.496$

5. 5.125

6. 5.96

7. 0.13

8. $35 \times \dfrac{1.61}{1} = 35 \times 1.61 = 56.35$

9. $1.67 \times \dfrac{1}{3.34} = 1.67 \div 3.34$

$$3.34\overline{)1.67.0} \quad \begin{array}{c} 0.5 \end{array}$$

10. $4\dfrac{1}{2} \times 150 = \dfrac{9}{2} \times 150$

$$= \dfrac{3 \cdot 3 \cdot \overset{1}{2} \cdot 3 \cdot 5 \cdot 5}{\underset{1}{2}} = 675$$

Go Figure

Adding an even amount (6) of odd numbers results in an even number. Using that rule eliminates 15, 29, and 31 as possible solutions. Also, 4 cannot be a solution because the lowest possible score is 6. The highest possible score is 54, so 58 can also be eliminated as a solution. A possible score is 28. One possible combination is 9, 7, 5, 5, 1, 1. Another is 7, 7, 5, 5, 3, 1.

Section 9.1

Objective A Exercises

1. 42 cm = 420 mm

3. 81 mm = 8.1 cm

5. 6804 m = 6.804 km

7. 2.109 km = 2109 m

9. 432 cm = 4.32 m

11. 0.88 m = 88 cm

13. 7038 m = 7.038 km

15. 3.5 km = 3500 m

17. 260 cm = 2.60 m

19. 1.685 m = 168.5 cm

21. 14.8 cm = 148 mm

23. 62 m 7 cm = 62 m + 0.07 m = 62.07 m

25. 31 cm 9 mm = 31 cm + 0.9 cm = 31.9 cm

27. 8 km 75 m = 8 km + 0.075 km = 8.075 km

Objective B Exercises

29. **Strategy**   To find the missing dimension:
    - Convert 40 mm to centimeters.
    - Find the sum of the given dimensions.
    - Subtract the sum of the given dimensions from the entire length (27.4 cm).

    Solution   40 mm = 4 cm

    $$\begin{array}{r} 4\text{ cm} \\ +\ 15.6\text{ cm} \\ \hline 19.6\text{ cm} \end{array}$$

    $$\begin{array}{r} 27.4\text{ cm} \\ -\ 19.6\text{ cm} \\ \hline 7.8\text{ cm} \end{array}$$

    The missing dimension is 7.8 cm.

31. **Strategy**   To find the distance between the rivets, convert 3.4 m to centimeters and then divide the total length of the plate by the number of spaces between the rivets (19).

    Solution   3.4 m = 340 cm

    $$19\overline{)340} \quad \begin{array}{c} 17.89 \approx 17.9 \end{array}$$

    The distance between the rivets is 17.9 cm.

33. **Strategy**   To find how much fencing is left on the roll:
    - Convert the length and width of the dog run to meters.
    - Add the four lengths of fencing to make the dog run.
    - Subtract the total fencing used from the length of the full roll (50 m).

    Solution

340 cm	3.40 m
1380 cm	13.80 m
340 cm	3.40 m
1380 cm	+ 13.80 m
	34.40 m

    $50 - 34.40 = 15.6$

    The amount of fencing left on the roll is 15.6 m.

35. **Strategy** To find the time for light to travel to Earth from the sun:
• Convert the distance light travels in 1 s (300,000,000 m) to kilometers.
• Divide the distance from the sun to Earth (150,000,000 km) by the distance light travels in 1 s.

Solution 300,000,000 m = 300,000 km
150,000,000 km ÷ 300,000 km/s
= 500 s
It takes 500 s for light to travel from the sun to Earth.

37. **Strategy** To find the distance that light travels in 1 day:
• Find the number of seconds in 1 day.
• Multiply the distance that light travels in 1 s (300,000 km) by the number of seconds in 1 day.

Solution 1 day
$$= 1 \, \text{day} \times \frac{24 \, \text{h}}{1 \, \text{day}} \times \frac{60 \, \text{min}}{1 \, \text{h}} \times \frac{60 \, \text{s}}{1 \, \text{min}}$$
= 86,400 s (in 1 day).
300,000 × 86,400
= 25,920,000,000 km
Light travels 25,920,000,000 km in 1 day.

Applying the Concepts

39. A very brief history of the metric system appears on page 373. You might decide whether you want students to write about its early development at the end of the 18th century or about its more recent history. For example, in 1975, Congress passed the Metric Conversion Act, which encourages voluntary use of the metric system in the United States.

Section 9.2

Objective A Exercises

1. 420 g = 0.420 kg

3. 127 mg = 0.127 g

5. 4.2 kg = 4200 g

7. 0.45 g = 450 mg

9. 1856 g = 1.856 kg

11. 4057 mg = 4.057 g

13. 1.37 kg = 1370 g

15. 0.0456 g = 45.6 mg

17. 18,000 g = 18.000 kg

19. 3 kg 922 g = 3 kg + 0.922 kg = 3.922 kg

21. 7 g 891 mg = 7 g + 0.891 g = 7.891 g

23. 4 kg 63 g = 4 kg + 0.063 kg = 4.063 kg

Objective B Exercises

25. **Strategy** To find the number of grams in one serving of Quaker Oats:
• Convert the amount of Quaker Oats (1.19 kg) to grams.
• Divide the amount of Quaker Oats by the number of servings (30).

Solution 1.19 kg = 1190 g
$$\begin{array}{r} 39.67 \\ 30\overline{)1190.0} \end{array}$$
There are 40 g in 1 serving.

27a. **Strategy** To find the number of grams of cholesterol in one dozen eggs:
• Convert the amount of cholesterol that one egg contains (274 mg) to grams.
• Multiply the number of grams of cholesterol in one egg by the number of eggs (12).

Solution 274 mg = 0.274 g
12 × 0.274 = 3.288 g
There are 3.288 g of cholesterol in 12 eggs.

b. **Strategy** To find the number of grams of cholesterol in 4 glasses of milk:
• Convert the amount of cholesterol in one glass of milk (33 mg) to grams.
• Multiply the number of grams in one glass of milk by the number of glasses of milk (4).

Solution 33 mg = 0.033 g
4 × 0.033 = 0.132
There are 0.132 g of cholesterol in four glasses of milk.

29a. **Strategy** To find the weight of the package in kilograms:
• Convert the weight of one serving (31 g) to kilograms.
• Multiply the weight of one serving by the number of servings (6).

Solution 31 g = 0.031 kg
0.031 × 6 = 0.186
There are 0.186 kg of mix in the package.

b. Strategy To find the number of grams of sodium contained in two servings:
● Convert the weight of sodium in one servings (210 mg) to grams.
● Multiply the weight of sodium in one servings by the number of servings (2).

Solution 210 mg = 0.210 g
0.210 × 2 = 0.420
There are 0.42 g of sodium in two servings.

31. Strategy To find the amount of grass seed:
● Convert 80 g to kilograms.
● Write and solve a proportion.

Solution 80 g = 0.08 kg
$$\frac{0.08 \text{ kg}}{100 \text{ m}^2} = \frac{n}{2000 \text{ m}^2}$$
$0.08 \times 2000 = 100 \times n$
$160 = 100 \times n$
$160 \div 100 = n$
$1.6 = n$
The amount of seed needed is 1.6 kg.

33. Strategy To find the profit:
● Convert the weight of a 10-kilogram container to grams.
● Find the number of bags of nuts in a 10-kilogram container by dividing the total weight in grams by the weight of one bag (200 g).
● Find the cost of the bags by multiplying the number of bags by $.04.
● Add the cost of the bags to the cost of a 10-kilogram container ($75) to find the total cost.
● Multiply the number of bags by $3.89 to find the total revenue.
● Subtract the total cost from the revenue to find the profit.

Solution 10 kg = 10,000 g
10,000 g ÷ 200 g = 50   bags of nuts
50 × $0.04 = $2   cost of the bags
$75 + $2 = $77   total cost
50 × $3.89 = $194.50   total revenue
$194.50 − $77.00 = $117.50   profit
The profit from repackaging the nuts is $117.50.

35. Strategy To find the percent of corn:
● Add the amount of exports of wheat (37,141 million kg), rice (2,680 million kg), and corn (40,365 million kg).
● Solve the basic percent equation for percent. The base is the total amount of exports and the amount is the amount of corn (40,365 million kg).

Solution   37,141 million kg
    2,680 million kg
+ 40,365 million kg
    80,186 million kg

Percent × base = amount
$n \times 80,186 = 40,365$
$n = 40,365 \div 80,186$
$n \approx 0.503$

Corn was 50.3% of the total exports.

Applying the Concepts

37. Students might list familiarity among the advantages of the U.S. Customary System and difficulty in converting units among the disadvantages. They might list ease of conversion among the advantages of the metric system, as well as the fact that international trade is based on the metric system.
A disadvantage for Americans is that they are unfamiliar with metric units. Another disadvantage is related to American industry: If forced to change to the metric system, companies would face the difficulty and expense of altering the present dimensions of machinery, tools, and products.

Section 9.3

Objective A Exercises

1. 4200 ml = 4.2 L
3. 3.42 L = 3420 ml
5. 423 ml = 423 cm³
7. 642 cm³ = 642 ml
9. 42 cm³ = 42 ml = 0.042 L
11. 0.435 L = 435 ml = 435 cm³
13. 4.62 kl = 4620 L
15. 1423 L = 1.423 kl
17. 1.267 L = 1267 cm³
19. 3 L 42 ml = 3 L + 0.042 L = 3.042 L
21. 3 kl 4 L = 3 kl + 0.004 kl = 3.004 kl
23. 8 L 200 ml = 8 L + 0.200 L = 8.200 L

Objective B Exercises

25a.   **Strategy**   To determine whether the amount of oxygen in 50 L of air is more or less than 25 L, note the percent of air that is oxygen.

     Solution   Because oxygen makes up only 21% of air, which is much less than $\frac{1}{2}$, there could not be 25 L of oxygen in 50 L of air.

b.   **Strategy**   To find the amount of oxygen, solve the basic percent equation for the amount. The percent is 21% and the base is 50 L.

     Solution   Percent × base = amount

$$21\% \times 50 = n$$
$$0.21 \times 50 = n$$
$$10.5 = n$$

There are 10.5 L of oxygen in 50 L of air.

27.   **Strategy**   To find the amount of chlorine used in a month:
- Convert 800 ml to liters.
- Multiply the amount of chlorine used in a day by the number of days in a month (30).

     Solution   800 ml = 0.8 L
$0.8 \times 30 = 24$ L
24 L of chlorine were used in a month.

29.   **Strategy**   To find how many patients can be immunized:
- Convert 3 cm³ to liters.
- Divide the total number of liters of flu vaccine (12) by the number of liters of vaccine each person receives.

     Solution $3 \text{ cm}^3 = 3 \text{ ml} = 0.003$ L
$12 \div 0.003 = 4000$
4000 patients can be immunized.

31.   **Strategy**   To determine the better buy:
- Find the unit cost (cost per liter) of the 12 one-liter bottles by dividing the cost ($19.80) by the amount of apple juice (12 L).
- Find the unit cost (cost per liter) of the 24 cans by converting the amount to liters and then dividing $14.50 by the amount of juice.

     Solution   The cost of 12 one-liter bottles:
$19.80 \div 12 = 1.65$
The unit cost is $1.65 per liter.
The cost of 24 cans:
$24 \times 340 \text{ ml} = 8160 \text{ ml} = 8.16$ L
$14.50 \div 8.16 \approx 1.78$
The unit cost is $1.78 per liter.
Since $1.65 < $1.78, the 12 one-liter bottles are the better buy.

33.   **Strategy**   To find the profit:
- Convert 85 kl to liters.
- Find the income by multiplying the price per liter ($.379) by the number of liters sold.
- Subtract the cost ($23,750) from the total income.

     Solution   85 kl = 85,000 L

$$\begin{array}{rr} 85,000 \text{ L} & \$32,215 \\ \times \ \$.379 & - \ 23,750 \\ \hline \text{Income: } \$32,215 & \$8,465 \end{array}$$

The profit on the gasoline was $8465.

Applying the Concepts

35.   3 L − 280 ml = 3 L − 0.280 L = 2.72 L
2.72 L = 2720 ml
2.72 L = 2 L 720 ml

Section 9.4

Objective A Exercises

1.   **Strategy**   To find the number of Calories that can be omitted from your diet, multiply the number of Calories omitted each day (110) by the number of days (30).

     Solution $110 \times 30 = 3300$
3300 Calories can be omitted from your diet.

3a.   **Strategy**
- From the nutrition label find the number of Calories per serving.
- Multiply the number of Calories per serving by $1\frac{1}{2}$.

     Solution   There are 60 Calories per serving.
$60 \times 1\frac{1}{2} = 90$
There are 90 Calories in $1\frac{1}{2}$ servings.

b. Strategy • From the nutrition label find the serving size and the number of Calories from fat.
• Determine how many servings are in 6 slices of bread.
• Multiply the number of fat Calories in a serving by the number of servings.

Solution 2 slices of bread is one serving.
10 fat Calories are in one serving.
$6 \div 2 = 3$ number of servings
$10 \times 3 = 30$
There are 30 fat Calories in 6 slices of bread.

5. Strategy To find how many Calories a 135-lb person would need to maintain body weight, multiply the body weight (135 lb) by the number of Calories per pound needed (15).

Solution $135 \times 15 = 2025$
2025 Calories would be needed.

7. Strategy To find how many Calories you burn up playing tennis:
• Convert 45 min to hours.
• Find how many hours of tennis are played by multiplying the number of days (30) by the time per day in hours.
• Multiply the number of hours played by the Calories burned per hour (450).

Solution $45 \text{ min} = 45 \text{ min} \times \dfrac{1 \text{ h}}{60 \text{ min}} = \dfrac{3}{4} \text{ h}$

$30 \times \dfrac{3}{4} \text{ h} = \dfrac{90}{4} \text{ h} = 22.5 \text{ h}$

450 Calories $\times$ 22.5 = 10,125 Calories

You burn 10,125 Calories.

9. Strategy To find how many hours you would have to hike:
• Add to find the total number of Calories consumed $(375 + 150 + 280)$.
• Divide the sum by the number of Calories used in 1 h (315).

Solution
$$\begin{array}{r} 375 \\ 150 \\ + 280 \\ \hline 805 \end{array}$$ number of Calories consumed

$805 \div 315 \approx 2.6$

You would have to hike for 2.6 h.

11. Strategy To find the energy used, multiply the number of watts (500) by the number of hours $\left(2\dfrac{1}{2}\right)$.

Solution $500 \times 2\dfrac{1}{2} = 1250 \text{ Wh}$
1250 Wh are used.

13. Strategy To find the number of kilowatt-hours used:
• Find the number of watt-hours used in standby mode.
• Find the number of watt-hours used in operation.
• Add the two numbers.
• Convert watt-hours to kilowatt hours.

Solution
$$\begin{array}{r} 30 \times 9 = \quad 351 \\ 6 \times 36 = + 216 \\ \hline 567 \text{ Wh} \end{array}$$

$567 \text{ W} = 0.567 \text{ kWh}$
The fax machine used 0.567 kWh.

15. Strategy To find the cost:
• Multiply the watts by the number of hours (8) to find the number of watt-hours.
• Convert the watt-hours to kilowatt-hours.
• Multiply the number of kilowatt-hours by the price per kilowatt-hour ($.09).

Solution $2200 \text{ W} \times 8 \text{ h} = 17{,}600 \text{ Wh}$
$17{,}600 \text{ Wh} = 17.6 \text{ kWh}$
$17.6 \text{ kWh} \times \$.09 = \$1.584 \approx \$1.58$
The cost of running an air conditioner is $1.58.

17a. Strategy To determine if the light output of the Energy Saver Bulb is more or less than half the energy of the Long Life Soft White bulb, compare the light output of each bulb.

Solution Energy Saver—400 lumens
Soft White—835 lumens
400 lumens is less than half the output of the Soft White bulb.

b.   Strategy   To find the cost for each bulb:
• Find the number of watt-hours by multiplying the number of watts by the number of hours.
• Convert watt-hours to kilowatt-hours.
• Multiply the number of kilowatt-hours by the cost per kilowatt hour.
• Find the difference in cost.

Solution   Sylvania Long Life Bulb:
$60 \times 150 = 9000$ Wh
9000 Wh $= 9$ kWh
$9 \times \$.108 = \$.972$
Energy Saver Soft White Bulb:
$34 \times 150 = 5100$ Wh
5100 Wh $= 5.1$ kWh
$5.1 \times \$.108 = \$.5508$
$\$.972 - \$.5508 = \$.4212$
The energy saver bulb costs $.42 less to operate.

19. Strategy   To find the cost:
• Multiply to find the total number of hours the welder is used.
• Multiply the number of hours by the number of kilowatts used each hour.
• Multiply the number of kilowatt-hours by the cost per kilowatt-hour.

Solution $30 \times 6 = 180$ h
$180 \times 6.5 = 1170$ kWh
$1170 \times 0.094 = \$109.98$
The cost of using the welder is $109.98.

Applying the Concepts

21.   Answers will vary. For example,

Age	Weight	Calories per Day to Maintain Weight	Calories per Day to Lose 1 Lb/Week
Men 11–14	99	2500	2000
15–18	145	3000	2500
19–24	160	2900	2400
25–50	174	2900	2400
51+	170	2300	1800
Women 11–14	101	2200	1700
15–18	120	2200	1700
19–24	128	2200	1700
25–50	138	2200	1700
51+	143	1900	1400

Section 9.5

Objective A Exercises

1. $100 \text{ yd} \approx 100 \text{ yd} \times \dfrac{1 \text{ m}}{1.09 \text{ yd}} = 91.74 \text{ m}$

3. $5 \text{ ft } 8 \text{ in.} = 5\dfrac{8}{12} \text{ ft} = 5\dfrac{2}{3} \text{ ft} \approx \dfrac{17}{3} \text{ ft} \times \dfrac{1 \text{ m}}{3.28 \text{ ft}} \approx 1.73 \text{ m}$

5. $15 \text{ lb} \approx 15 \text{ lb} \times \dfrac{1 \text{ kg}}{2.2 \text{ lb}} \approx 6.82 \text{ kg}$

7. $1 \text{ c} \approx 1 \text{ c} \times \dfrac{1 \text{ pt}}{2 \text{ c}} \times \dfrac{1 \text{ qt}}{2 \text{ pt}} \times \dfrac{1 \text{ L}}{1.06 \text{ qt}}$
$\approx 0.23585 \text{ L} \approx 235.85 \text{ ml}$

9. $65 \dfrac{\text{mi}}{\text{h}} = 65 \dfrac{\text{mi}}{\text{h}} \times \dfrac{1.61 \text{ km}}{1 \text{ mi}} = 104.65 \dfrac{\text{km}}{\text{h}}$

11. $\dfrac{\$3.49}{\text{lb}} \approx \dfrac{\$3.49}{\text{lb}} \times \dfrac{2.2 \text{ lb}}{1 \text{ kg}} \approx \$7.68/\text{kg}$

13. $\dfrac{\$1.47}{\text{gal}} \approx \dfrac{\$1.47}{\text{gal}} \times \dfrac{1 \text{ gal}}{3.79 \text{ L}} \approx \$.39/\text{L}$

15. $24{,}887 \text{ mi} \approx 24{,}887 \text{ mi} \times \dfrac{1.61 \text{ km}}{1 \text{ mi}}$
$= 40{,}068.07 \text{ km}$

Objective B Exercises

17. $100 \text{ m} \approx 100 \text{ m} \times \dfrac{3.28 \text{ ft}}{1 \text{ m}} = 328 \text{ ft}$

19. $6 \text{ L} \approx 6 \text{ L} \times \dfrac{1 \text{ gal}}{3.79 \text{ L}} \approx 1.58 \text{ gal}$

21. $1500 \text{ m} \approx 1500 \text{ m} \times \dfrac{3.28 \text{ ft}}{1 \text{ m}} = 4920 \text{ ft}$

23. $24 \text{ L} \approx 24 \text{ L} \times \dfrac{1 \text{ gal}}{3.79 \text{ L}} \approx 6.33 \text{ gal}$

25. $\dfrac{80 \text{ km}}{\text{h}} \approx \dfrac{80 \text{ km}}{\text{h}} \times \dfrac{1 \text{ mi}}{1.61 \text{ km}} \approx 49.69 \text{ mph}$

27. $\dfrac{\$.385}{\text{L}} \approx \dfrac{\$.385}{\text{L}} \times \dfrac{3.79 \text{ L}}{\text{gal}} \approx \$1.46/\text{gal}$

29. $2.1 \text{ kg} \approx 2.1 \text{ kg} \times \dfrac{2.2 \text{ lb}}{\text{kg}} = 4.62 \text{ lb}$

31. Strategy To find the number of pounds lost:
● Multiply to find the number of hours spent hiking.
● Multiply the number of hours spent hiking by the number of extra Calories used in hiking to find the total number of extra Calories used.
● Multiply the number of extra Calories consumed each day by the number of days.
● Subtract to find the difference between the number of Calories used in hiking and the number of extra Calories consumed.
● Divide the difference by 3500.

Solution $5 \times 5 = 25$ h
$25 \times 320 = 8000$ Calories
$5 \times 900 = 4500$ Calories
$8000 - 4500 = 3500$
$\dfrac{3500}{3500} = 1$ lb
Gary will lose 1 lb.

33. $\dfrac{300,000 \text{ km}}{\text{s}} \approx \dfrac{300,000 \text{ km}}{\text{s}} \times \dfrac{1 \text{ mi}}{1.61 \text{ km}}$
$\approx 186,335.40$ mi/s

Applying the Concepts

35a. False

b. False

c. True

d. True

e. False

Chapter 9 Review Exercises

1. 1.25 km $= 1250$ m
2. 0.450 g $= 450$ mg
3. 0.0056 L $= 5.6$ ml
4. 1000 m $\approx 1000 \text{ m} \times \dfrac{1.09 \text{ yd}}{1 \text{ m}} = 1090$ yd
5. 79 mm $= 7.9$ cm
6. 5 m 34 cm $= 5$ m $+ 0.34$ m $= 5.34$ m
7. 990 g $= 0.990$ kg
8. 2550 ml $= 2.550$ L
9. 4870 m $= 4.870$ km
10. 0.37 cm $= 3.7$ mm
11. 6 g 829 mg $= 6$ g $+ 0.829$ g $= 6.829$ g
12. 1.2 L $= 1200$ cm^3
13. 4.050 kg $= 4050$ g

14. 8.7 m $= 870$ cm
15. 192 ml $= 192$ cm^3
16. 356 mg $= 0.356$ g
17. 372 cm $= 3.72$ m
18. 8.3 kl $= 8300$ L
19. 2 L 89 ml $= 2$ L $+ 0.089$ L $= 2.089$ L
20. 5410 cm^3 $= 5.410$ L
21. 3792 L $= 3.792$ kl
22. 468 cm^3 $= 468$ ml

23. Strategy To find the amount of the wire left on the roll:
● Convert the lengths of the three pieces cut from the roll to meters.
● Add the three numbers.
● Subtract the sum from the length of the original roll (50 m).

Solution 240 cm = 2.40 m
560 cm = 5.60 m
480 cm = + 4.80 m
12.80 m

50.0 m
− 12.8 m
37.2 m
There are 37.2 m of wire left on the roll.

24. Strategy To find the total cost:
● Convert the weights of the packages to kilograms.
● Add the weights.
● Multiply the total weight by the cost per kilogram ($5.59).

Solution 790 g = 0.790 kg
830 g = 0.830 kg
655 g = + 0.655 kg
2.275 kg
$2.275 \times \$5.59 = \$12.71725$
The total cost of the chicken is $12.72.

25. $\dfrac{\$3.40}{\text{lb}} \approx \dfrac{\$3.40}{\text{lb}} \times \dfrac{2.2 \text{ lb}}{1 \text{ kg}} = \$7.48/\text{kg}$

26. Strategy To find how many liters of coffee should be prepared:
● Convert 400 ml to liters.
● Multiply the number of guests expected to attend (125) by the number of liters per guest.

Solution 400 ml = 0.4 L
0.4 L $\times 125 = 50$ L
The amount of coffee that should be prepared is 50 L.

27. Strategy   To find the number of Calories that can be eliminated, multiply the number of Calories in one egg (90) by the number of days it is eliminated (30).

Solution   90 Cal × 30 = 2700 Cal
You can eliminate 2700 Calories.

28. Strategy   To find the cost of running the TV set:
• Find the number of hours the TV is used each month by multiplying the number of hours per day (5) by the number of days (30).
• Find the number of watt-hours by multiplying the number of watts per hour (240) by the total number of hours.
• Convert watt-hours to kilowatt-hours.
• Multiply the number of kilowatt-hours by the cost per kilowatt-hour (9.5¢).

Solution   5 h × 30 = 150 h
150 h × 240 W = 36,000 Wh
36,000 Wh = 36 kWh
36 kWh × ($.095) = $3.42
The cost of running the TV set is $3.42.

29. $1.90 \text{ kg} = 1.90 \text{ kg} \times \dfrac{2.2 \text{ lb}}{1 \text{ kg}} = 4.18 \text{ lb}$

30. Strategy   To find how many hours of cycling are necessary to lose 1 lb, divide 1 lb (3500 Calories) by the number of Calories cycling burns per hour (400).

Solution $\dfrac{8.75}{400)\overline{3500.00}}$

8.75 hours of cycling are needed.

31. Strategy   To find the profit:
• Convert the amount of soap purchased (6 L) to milliliters.
• Divide the volume of one plastic container (150 ml) into the amount of soap purchased to determine the number of containers of soap for sale.
• Multiply the number of containers by the cost per container ($.26) to find the cost of the containers.
• Multiply the number of liters of soap (6) by the cost per liter ($11.40) to find the cost of the soap.
• Add the cost of the soap and the cost of the containers to find the total cost.
• Multiply the number of containers by $3.29 to find the total revenue.
• Subtract the total cost from the total revenue to find the profit.

Solution   6 L = 6000 ml   amount of soap
6000 ÷ 150 = 40   number of containers
40 × $.26 = $10.40   cost of containers
6 × $11.40 = $68.40   cost of soap
$10.40 + $68.40 = $78.80   total cost
40 × $3.29 = $131.60   revenue
$131.60 − $78.80 = $52.80
The profit was $52.80.

32. Strategy   To find the number of kilowatt-hours of energy used:
• Multiply 80 W times 2 h times 7 days to find the number of watt-hours used.
• Convert the watt-hours to kilowatt-hours.

Solution   80 × 2 × 7 = 1120 Wh
1120 Wh = 1.120 kWh
The color TV used 1.120 kWh of electricity.

33. Strategy   To find the amount of fertilizer:
• Multiply the number of trees (500) by the amount of fertilizer per tree (250 g).
• Convert the grams to kilograms.

Solution   500 × 250 = 125,000 g
125,000 g = 125 kg
The amount of fertilizer used was 125 kg.

Chapter 9 Test

1.   2.96 km = 2960 m

2.   0.378 g = 378 mg

3.   0.046 L = 46 ml

4.   919 cm³ = 919 ml

5.   42.6 mm = 4.26 cm

6.   7 m 96 cm = 7 m + 0.96 m = 7.96 m

7.   847 g = 0.847 kg

8.   3920 ml = 3.920 L

9.   5885 m = 5.885 km

10.   1.5 cm = 15 mm

11.   3 g 89 mg = 3 g + 0.089 g = 3.089 g

12.   1.6 L = 1600 cm³

13.   3.29 kg = 3290 g

14.   4.2 m = 420 cm

15. $96 \text{ ml} = 96 \text{ cm}^3$

16. $1375 \text{ mg} = 1.375 \text{ g}$

17. $402 \text{ cm} = 4.02 \text{ m}$

18. $8.92 \text{ kl} = 8920 \text{ L}$

19. **Strategy**   To find the number of Calories needed to maintain the weight of a 140-pound sedentary person, multiply the weight (140 pounds) by the number of Calories per pound a sedentary person needs (15) to maintain weight.

    Solution $140 \times 15 = 2100$
    A 140-pound sedentary person should consume 2100 Calories per day to maintain that weight.

20. **Strategy**   To find the number of kilowatt-hours of energy used:
    • Multiply 100 W times $4\frac{1}{2}$ h times 7 days to find the number of watt-hours used.
    • Convert the watt-hours to kilowatt-hours.

    Solution $100 \times 4\frac{1}{2} \times 7 = 3150 \text{ Wh}$
    $3150 \text{ Wh} = 3.15 \text{ kWh}$.
    3.15 kWh of energy are used during the week for operating the television.

21. **Strategy**   To find the total length of the rafters:
    • Multiply the number of rafters (30) by the length of each rafter (380 cm).
    • Convert the length in centimeters to meters.

    Solution $30 \times 380 = 11,400 \text{ cm}$
    $11,400 = 114 \text{ m}$
    The total length of the rafters is 114 m.

22. **Strategy**   To find the weight of the box of tiles, multiply the weight of one tile (250 g) by the number of tiles in the box (144).

    Solution
    $\begin{array}{r} 250 \text{ g} \\ \times\ 144 \\ \hline 36,000 \text{ g} = 36 \text{ kg} \end{array}$
    The weight of the box is 36 kg.

23. **Strategy**   To find how many liters of vaccine are needed:
    • Multiply the number of people (2600) by the amount of vaccine per flu shot (2 cm³).
    • Convert the total amount of vaccine to liters.

    Solution
    $\begin{array}{r} 2600 \\ \times\ 2 \text{ cm}^3 \\ \hline 5200\ \text{cm}^3 = 5.2 \text{ L} \end{array}$
    The amount of vaccine needed is 5.2 L.

24. $35 \text{ mph} \approx \dfrac{35 \text{ mi}}{\text{h}} \times \dfrac{1.61 \text{ km}}{1 \text{ mi}} \approx 56.4 \text{ km/h}$

25. **Strategy**   To find the distance between the rivets:
    • Convert the length of the plate (4.20 m) to centimeters.
    • Divide the length of the plate by the number of spaces (24).

    Solution $4.20 \text{ m} = 420 \text{ cm}$
    $420 \div 24 = 17.5 \text{ cm}$
    The distance between the rivets is 17.5 cm.

26. **Strategy**   To find how much it costs to fertilize the orchard:
    • Find out how much fertilizer is needed by multiplying the number of trees in the orchard (1200) by the amount of fertilizer for each tree (200 g).
    • Convert the total amount of fertilizer to kilograms.
    • Multiply the number of kilograms of fertilizer by the cost per kilogram ($2.75).

    Solution $1200 \times 200 = 240,000 \text{ g}$
    $240,000 \text{ g} = 240 \text{ kg}$
    $240 \times \$2.75 = \$660$
    The cost to fertilize the trees is $660.

27. **Strategy**   To find the cost of the electricity:
    • Determine the amount of electricity used by multiplying 1600 W times the hours used per day (4) times the number of days (30).
    • Convert the watt-hours to kilowatt-hours.
    • Multiply the kilowatt-hours by the cost per kilowatt hour ($.085).

    Solution $1600 \times 4 \times 30 = 192,000 \text{ Wh}$
    $192,000 \text{ Wh} = 192 \text{ kWh}$
    $192 \times \$.085 = \$16.32$
    The total cost is $16.32.

28. **Strategy**   To find how much acid should be ordered:
    • Find the amount of acid needed by multiplying the number of classes (3) times the number of students in each class (40) times the amount of acid needed by each student (90).
    • Convert the amount to liters.

    Solution $3 \times 40 \times 90 = 10,800 \text{ ml}$
    10.8 L
    The assistant should order 11 L of acid.

29. **Strategy**   Convert the measure of the large hill (120 m) to feet.

   Solution $120\text{ m} = 120\text{ m} \times \dfrac{3.28\text{ ft}}{1\text{ m}} = 393.6\text{ ft}$

   The measure of the large hill is 393.6 ft.

30. **Strategy**   Convert the measure of the diameter of the bulls eye (4.8 in.) to centimeters.

   Solution $4.8\text{ in.} = 4.8\text{ in.} \times \dfrac{2.54\text{ cm}}{1\text{ in.}} = 12.192\text{ cm}$

   4.8 in. is approximately 12.2 cm.

Cumulative Review Exercises

1. $12 - 8 \div (6-4)^2 \cdot 3 = 12 - 8 \div 2^2 \cdot 3$
$$= 12 - 8 \div 4 \cdot 3$$
$$= 12 - 2 \cdot 3$$
$$= 12 - 6$$
$$= 6$$

2. $5\dfrac{3}{4} = 5\dfrac{27}{36}$
$1\dfrac{5}{6} = 1\dfrac{30}{36}$
$+\ 4\dfrac{7}{9} = 4\dfrac{28}{36}$
$\overline{\qquad\qquad 10\dfrac{85}{36} = 12\dfrac{13}{36}}$

3. $4\dfrac{2}{9} = 4\dfrac{8}{36} = 3\dfrac{44}{36}$
$-\ 3\dfrac{5}{12} = 3\dfrac{15}{36} = 3\dfrac{15}{36}$
$\overline{\qquad\qquad\qquad \dfrac{29}{36}}$

4. $5\dfrac{3}{8} \div 1\dfrac{3}{4} = \dfrac{43}{8} \div \dfrac{7}{4}$
$= \dfrac{43}{8} \times \dfrac{4}{7}$
$= \dfrac{43 \cdot \overset{1}{2} \cdot \overset{1}{2}}{\underset{1}{2} \cdot \underset{1}{2} \cdot 2 \cdot 7} = \dfrac{43}{14} = 3\dfrac{1}{14}$

5. $\left(\dfrac{2}{3}\right)^4 \cdot \left(\dfrac{9}{4}\right)^2 = \left(\dfrac{2}{3} \cdot \dfrac{2}{3} \cdot \dfrac{2}{3} \cdot \dfrac{2}{3}\right)\left(\dfrac{9}{4} \cdot \dfrac{9}{4}\right) = \dfrac{16}{81} \cdot \dfrac{81}{16} = 1$

6. $\begin{array}{r} 12.0072 \\ -\ 9.937 \\ \hline 2.0702 \end{array}$

7. $\dfrac{5}{8} = \dfrac{n}{50}$
$5 \times 50 = 8 \times n$
$250 = 8 \times n$
$250 \div 8 = n$
$n = 31.3$

8. $1\dfrac{3}{4} = \dfrac{7}{4} \times 100\% = \dfrac{700}{4}\% = 175\%$

9. $4.2\% \times n = 6.09$
$0.042 \times n = 6.09$
$n = 6.09 \div 0.042 = 145$

10. $18\text{ pt} \times \dfrac{1\text{ qt}}{2\text{ pt}} \times \dfrac{1\text{ gal}}{4\text{ qt}} = \dfrac{18}{8}\text{ gal} = 2.25\text{ gal}$

11. $875\text{ cm} = 8.75\text{ m}$

12. $3420\text{ m} = 3.420\text{ km}$

13. $5.05\text{ kg} = 5050\text{ g}$

14. $3\text{ g }672\text{ mg} = 3\text{ g} + 0.672\text{ g} = 3.672\text{ g}$

15. $6\text{ L} = 6000\text{ ml}$

16. $2.4\text{ kl} = 2400\text{ L}$

17. **Strategy**   To find how much money is left after the rent is paid:
   • Find the amount that is paid in rent by multiplying $\dfrac{1}{4}$ by the total monthly income ($5244).
   • Subtract the amount paid in rent from the total monthly income.

   Solution $\dfrac{1}{4} \times \$5244 = \dfrac{\$5244}{4} = \$1311$

   $\begin{array}{r} \$5244 \\ -\ 1311 \\ \hline \$3933 \end{array}$

   $3933 is left after the rent is paid.

18. **Strategy**   To find the amount of income tax paid:
   • Find the amount of income tax paid on the profit by multiplying 0.08 by the profit ($82,340).
   • Add $620 to the amount of income tax paid on the profit.

   Solution $0.08 \times \$82,340 = \$6587.20$

   $\begin{array}{r} \$6587.20 \\ +\ 620.00 \\ \hline \$7207.20 \end{array}$

   The business paid $7207.20 in income tax.

19. **Strategy**   To find the property tax, solve a proportion.

   Solution $\dfrac{\$4900}{\$245,000} = \dfrac{n}{\$275,000}$
   $4900 \times 275,000 = 245,000 \times n$
   $1,347,500,000 = 245,000 \times n$
   $1,347,500,000 \div 245,000 = n$
   $5500 = n$

   The property tax is $5500.

20. Strategy   To find the rebate, solve the basic percent equation for amount. The base is $23,500 and the rate is 12%.

Solution   Percent × base = amount
$12\% \times 23,500 = n$
$0.12 \times 23,500 = n$
$2820 = n$
The car buyer will receive a rebate of $2820.

21. Strategy   To find the percent, solve the basic percent equation for percent. The base is $8200 and the amount is $533.

Solution   Percent × base = amount
$n \times 8200 = 533$
$n = 533 \div 8200$
$n = 0.065 = 6.5\%$
The percent is 6.5%.

22. Strategy   To find your mean grade, find the sum of the grades and divide the sum by the number of grades (5).

Solution
```
  78              76
  92          5)380
  45
  80
+ 85
 380 sum of grades
```
Your average grade is 76.

23. Strategy   To find what the salary will be next year, find the amount of the increase by solving the basic percent equation for amount. The base is $22,500 and the percent is 12%.

Solution   Percent × base = amount

$12\% \times 22,500 = n$ $22,500$
$0.12 \times 22,500 = n$ $+ 2,700$
$2700 = n$ $\overline{\$25,200}$

Karla's salary next year will be $25,200.

24. Strategy   To find the discount rate:
• Find the amount of the discount by subtracting the sale price ($140.40) from the original price ($180).
• Solve the basic percent equation for percent. The base is the original price ($180) and the amount is the amount of the discount.

Solution
```
 $180.00
- 140.40
 $ 39.60
```
Percent × base = amount
$n \times 180 = 39.60$
$n = 39.60 \div 180$
$n = 0.22 = 22\%$
The discount rate is 22%.

25. Strategy   To find the length of the wall:
• Convert 9 in. to feet.
• Multiply the length, in feet, of one brick by the number of bricks (48).

Solution $9 \text{ in.} = 9 \text{ in.} \times \dfrac{1 \text{ ft}}{12 \text{ in.}} = \dfrac{9}{12} \text{ ft} = 0.75 \text{ ft}$

```
    48
× 0.75 ft
   36 ft
```
The length of the wall is 36 ft.

26. Strategy   To find the number of quarts:
• Find the total amount of juice by multiplying the amount in a jar (24 oz) by the number of jars in a case (16).
• Convert the ounce amount of juice to quarts.

Solution
```
  24 oz
× 16
 384 oz
```
$384 \text{ oz} \times \dfrac{1 \text{ c}}{8 \text{ oz}} \times \dfrac{1 \text{ pt}}{2 \text{ c}} \times \dfrac{1 \text{ qt}}{2 \text{ pt}}$
$= \dfrac{384}{32} \text{ qt} = 12 \text{ qt}$
There are 12 qt of apple juice in the case.

27. Strategy   To find the profit:
• Convert the amount of oil to quarts.
• Find the cost by multiplying the number of gallons (40) by the cost per gallon ($4.88).
• Find the revenue by multiplying the number of quarts by the selling price per quart ($1.99).
• Subtract the cost from the revenue.

Solution
$40 \text{ gal} = 40 \text{ gal} \times \dfrac{4 \text{ qt}}{1 \text{ gal}} = 160 \text{ quarts}$

$40 \times \$4.88 = \$195.20$ cost
$160 \times \$1.99 = \$318.40$ revenue
$\$318.40 - \$195.20 = \$123.20$
The profit was $123.20.

28. **Strategy** To find the amount of chlorine used:
    - Convert the amount of chlorine used to liters.
    - Multiply the amount used each day by the number of days (20).

    Solution 1200 ml = 1.2 L
    1.2 L × 20 = 24 L
    24 L of chlorine was used.

29. **Strategy** To find how much it costs to operate the hairdryer:
    - Find how many hours the hair dryer is used by multiplying the amount used each day $\left(\frac{1}{2} h\right)$ by the number of days (30).
    - Find the watt-hours by multiplying the number of watts (1200) by the number of hours.
    - Convert watt-hours to kilowatt-hours.
    - Multiply the number of kilowatt-hours by the cost per kilowatt-hour (10.5¢).

    Solution $30 \times \frac{1}{2} h = 15 h$

    1200 W × 15 h = 18,000 Wh
    18,000 Wh = 18 kWh
    18 kWh × $0.105 = $1.89
    The total cost of operating the hair dryer is $1.89.

30. $\dfrac{60 \text{ mi}}{1 \text{ h}} = \dfrac{60 \text{ mi}}{1 \text{ h}} \times \dfrac{1.61 \text{ km}}{1 \text{ mi}} = 96.6 \text{ km/h}$

Chapter 10: Rational Numbers

Prep Test

1. $54 > 45$

2. 4

3. 15,847

4. 3779

5. 26,432

6. $\dfrac{144}{24} = \dfrac{2 \cdot \overset{1}{\cancel{2}} \cdot \overset{1}{\cancel{2}} \cdot \overset{1}{\cancel{2}} \cdot 3 \cdot \overset{1}{\cancel{3}}}{\underset{1}{\cancel{2}} \cdot \underset{1}{\cancel{2}} \cdot \underset{1}{\cancel{2}} \cdot \underset{1}{\cancel{3}}} = 6$

7. $\dfrac{2}{3} + \dfrac{3}{5} = \dfrac{10}{15} + \dfrac{9}{15}$

   $\qquad = \dfrac{19}{15} = 1\dfrac{4}{15}$

8. $\dfrac{3}{4} - \dfrac{5}{16} = \dfrac{12}{16} - \dfrac{5}{16}$

   $\qquad = \dfrac{7}{16}$

9. 11.058

10. 3.781

11. $\dfrac{3}{4} \times \dfrac{8}{15} = \dfrac{\overset{1}{\cancel{3}} \cdot \overset{1}{\cancel{2}} \cdot \overset{1}{\cancel{2}} \cdot 2}{\underset{1}{\cancel{2}} \cdot \underset{1}{\cancel{2}} \cdot \underset{1}{\cancel{3}} \cdot 5} = \dfrac{2}{5}$

12. $\dfrac{5}{12} \div \dfrac{3}{4} = \dfrac{5}{12} \cdot \dfrac{4}{3}$

    $\qquad = \dfrac{5 \cdot \overset{1}{\cancel{2}} \cdot \overset{1}{\cancel{2}}}{\underset{1}{\cancel{2}} \cdot \underset{1}{\cancel{2}} \cdot 3 \cdot 3} = \dfrac{5}{9}$

13. 9.4

14. $2.4\overline{)0.9.6}$ (quotient 0.4)

15. $(8-6)^2 + 12 \div 4 \cdot 3^2 = 2^2 + 12 \div 4 \cdot 9$

    $\qquad\qquad\qquad\qquad = 4 + 3 \cdot 9$

    $\qquad\qquad\qquad\qquad = 4 + 27$

    $\qquad\qquad\qquad\qquad = 31$

Go Figure

If it takes one loaf 30 minutes to fill the oven, doubling in volume each minute, then at 29 minutes, the oven would be half filled. At 28 minutes, the oven would be one-quarter filled with one loaf. So for the oven to be half filled with two loaves of bread, it would take 28 minutes.

Section 10.1

Objective A Exercises

1. -120 ft

3. $+2$ dollars

5.

7.

9. 1

11. -1

13. 3

15. **a.** A is -4.   **b.** C is -2.

17. **a.** A is -7.   **b.** D is -4.

19. $-2 > -5$

21. $-16 < 1$

23. $3 > -7$

25. $-11 < -8$

27. $35 > 28$

29. $-42 < 27$

31. $21 > -34$

33. $-27 > -39$

35. $-87 < 63$

37. $86 > -79$

39. $-62 > -84$

41. $-131 < 101$

43. $-7, -2, 0, 3$

45. $-5, -3, 1, 4$

47. $-4, 0, 5, 9$

49. $-10, -7, -5, 4, 12$

51. $-11, -7, -2, 5, 10$

Objective B Exercises

53. -16

55. 3

57. -45

59. 59

61. 88

63. 4

65. 9

67. 11

69. 12

71. $|-2| = 2$

73. $|6| = 6$

75. $|5| = 5$

77. $|-1| = 1$

79. $-|-5| = -5$

81. $|16| = 16$

83. $|-12| = 12$

85. $-|29| = -29$

87. $-|-14| = -14$

89. $|-15| = 15$

91. $-|33| = -33$

93. $|32| = 32$

95. $-|-42| = -42$

97. $|-61| = 61$

99. $-|52| = -52$

101. $|-12| > |8|$

103. $|6| < |13|$

105. $|-1| < |-17|$

107. $|17| = |-17|$

109. $-9, -|6|, -4, |-7|$

111. $-9, -|-7|, |4|, 5$

113. $-|10|, -|-8|, -3, |5|$

Applying the Concepts

115a. 8 and -2 are 5 units from 3.

b. 2 and -4 are 3 units from -1.

117. -12 min and counting is closer to blastoff.

119. The loss was greater during the first quarter.

121. $11, -11$

Section 10.2

Objective A Exercises

1. $-14, -364$

3. $3 + (-5) = -2$

5. $8 + 12 = 20$

7. $-3 + (-8) = -11$

9. $-4 + (-5) = -9$

11. $6 + (-9) = -3$

13. $-6 + 7 = 1$

15. $2 + (-3) + (-4) = -1 + (-4) = -5$

17. $-3 + (-12) + (-15) = -15 + (-15) = -30$

19. $-17 + (-3) + 29 = -20 + 29 = 9$

21. $-3 + (-8) + 12 = -11 + 12 = 1$

23. $13 + (-22) + 4 + (-5) = -9 + 4 + (-5)$
$= -5 + (-5) = -10$

25. $-22 + 10 + 2 + (-18) = -12 + 2 + (-18)$
$= -10 + (-18) = -28$

27. $-16 + (-17) + (-18) + 10 = -33 + (-18) + 10$
$= -51 + 10 = -41$

29. $-126 + (-247) + (-358) + 339$
$= -373 + (-358) + 339$
$= -731 + 339$
$= -392$

31. $-12 + (-8) = -20$

33. $-7 + (-16) = -23$

35. $-4 + 2 = -2$

37. $-2 + 8 + (-12) = 6 + (-12) = -6$

39. $2 + (-3) + 8 + (-13) = -1 + 8 + (-13)$
$= 7 + (-13) = -6$

Objective B Exercises

41. Negative six minus positive four

43. Positive six minus negative four

45. $9 + 5$

47. $1 + (-8)$

49. $16 - 8 = 16 + (-8) = 8$

51. $7 - 14 = 7 + (-14) = -7$

53. $-7 - 2 = -7 + (-2) = -9$

55. $7 - (-29) = 7 + 29 = 36$

57. $-6 - (-3) = -6 + 3 = -3$

59. $6 - (-12) = 6 + 12 = 18$

61. $-4 - 3 - 2 = -4 + (-3) + (-2)$
$= -7 + (-2) = -9$

63. $12 - (-7) - 8 = 12 + 7 + (-8)$
$= 19 + (-8) = 11$

65. $4 - 12 - (-8) = 4 + (-12) + 8$
$\qquad = -8 + 8 = 0$

67. $-6 - (-8) - (-9) = -6 + 8 + 9$
$\qquad = 2 + 9 = 11$

69. $-30 - (-65) - 29 - 4 = -30 + 65 + (-29) + (-4)$
$\qquad = 35 + (-29) + (-4)$
$\qquad = 6 + (-4) = 2$

71. $-16 - 47 - 63 - 12 = -16 + (-47) + (-63) + (-12)$
$\qquad = -63 + (-63) + (-12)$
$\qquad = -126 + (-12)$
$\qquad = -138$

73. $47 - (-67) - 13 - 15 = 47 + 67 + (-13) + (-15)$
$\qquad = 114 + (-13) + (-15)$
$\qquad = 101 + (-15) = 86$

75. $167 - 432 - (-287) - 359$
$= 167 + (-432) + 287 + (-359)$
$= -265 + 287 + (-359)$
$= 22 + (-359)$
$= -337$

77. $-4 - (-8) = -4 + 8 = 4$

79. $-8 - 4 = -8 + (-4) = -12$

81. $-4 - 8 = -4 + (-8) = -12$

83. $1 - (-2) = 1 + 2 = 3$

Objective C Exercises

85. Strategy To find the temperature, add the increase (7°C) to the previous temperature (−8°C).

Solution $-8 + 7 = -1$
The temperature is −1°C.

87. Strategy To find Nick's score, subtract 26 points from his original score (11).

Solution $11 - 26 = 11 + (-26) = -15$
Nick's score was −15 points after his opponent shot the moon.

89. Strategy To find the price of Byplex stock add the change in price for each day of the week.

Solution $-2 + (-3) + (-1) + (-2) + (-1)$
$= -5 + (-1) + (-2) + (-1)$
$= -6 + (-2) + (-1)$
$= (-8) + (-1)$
$= -9$
The change in the price of the stock is −9 dollars.

91. Strategy To find the difference in temperature, subtract the temperature in Earth's stratosphere (−70°F) from the temperature of Earth's surface (45°F).

Solution $45 - (-70) = 45 + 70 = 115$
The difference is 115°F.

93. Strategy To find the difference in elevation, subtract the elevation of Valdes Peninsula (−86 m) from the elevation of Mt. Aconcagua (6960 m).

Solution $6960 - (-86) = 6960 + 86$
$\qquad = 7046$
The difference in elevation is 7046 m.

95. Strategy To find the difference between the highest and lowest temperatures in Africa, subtract the lowest temperature (−24°C) from the highest temperature (58°C).

Solution $58 - (-24) = 58 + 24 = 82$
The difference is 82°C.

97. Strategy To find the difference, subtract the lowest temperature in Asia (−68°C) from the lowest temperature in Europe (−55°C).

Solution $-55 - (-68) = -55 + 68 = 13$
The difference in temperature is 13°C.

Applying the Concepts

99.

−3	2	1
4	0	−4
−1	−2	3

101. Students should include in their description the fact that *minus* refers to the operation of subtraction, whereas *negative* refers to the sign of a number.

Section 10.3

Objective A Exercises

1. Subtraction

3. Multiplication

5. $14 \times 3 = 42$

7. $-4 \cdot 6 = -24$

9. $-2 \cdot (-3) = 6$

11. $(9)(2) = 18$

13. $5(-4) = -20$

15. $-8(2) = -16$

17. $(-5)(-5) = 25$

19. $(-7)(0) = 0$

21. $-24 \times 3 = -72$

23. $6(-17) = -102$

25. $-4(-35) = 140$

27. $-6 \cdot (38) = -228$

29. $8(-40) = -320$

31. $-4(39) = -156$

33. $5 \times 7 \times (-2) = 35 \times (-2) = -70$

35. $(-9)(-9)(2) = 81(2) = 162$

37. $-5(8)(-3) = -40(-3) = 120$

39. $-1(4)(-9) = -4(-9) = 36$

41. $4(-4) \cdot 6(-2) = -16 \cdot 6(-2) = -96(-2) = 192$

43. $-9(4) \cdot 3(1) = -36 \cdot 3(1) = -108(1) = -108$

45. $(-6) \cdot 7 \cdot (-10)(-5) = -42 \cdot (-10)(-5)$
$= 420(-5)$
$= -2100$

47. $-5(-4) = 20$

49. $-8(6) = -48$

51. $-4(7)(-5) = -28(-5) = 140$

Objective B Exercises

53. $3(-12) = -36$

55. $-5(11) = -55$

57. $12 \div (-6) = -2$

59. $(-72) \div (-9) = 8$

61. $0 \div (-6) = 0$

63. $45 \div (-5) = -9$

65. $-36 \div 4 = -9$

67. $-81 \div (-9) = 9$

69. $72 \div (-3) = -24$

71. $(-60) \div 5 = -12$

73. $-93 \div (-3) = 31$

75. $(-85) \div (-5) = 17$

77. $120 \div 8 = 15$

79. $78 \div (-6) = -13$

81. $-72 \div 4 = -18$

83. $-114 \div (-6) = 19$

85. $-104 \div (-8) = 13$

87. $57 \div (-3) = -19$

89. $-136 \div (-8) = 17$

91. $-130 \div (-5) = 26$

93. $(-92) \div (-4) = 23$

95. $-150 \div (-6) = 25$

97. $204 \div (-6) = -34$

99. $-132 \div (-12) = 11$

101. $-182 \div 14 = -13$

103. $143 \div 11 = 13$

105. $-180 \div (-15) = 12$

107. $154 \div (-11) = -14$

109. $\dfrac{182}{-13} = -14$

111. $\dfrac{144}{-24} = -6$

113. $\dfrac{-88}{22} = -4$

Objective C Exercises

115. **Strategy**   To find the average daily high temperature:
● Add the seven temperature readings.
● Divide by 7.

Solution $-6 + (-11) + 1 + 5 + (-3) + (-9) + (-5)$
$= -17 + 1 + 5 + (-3) + (-9) + (-5)$
$= -16 + 5 + (-3) + (-9) + (-5)$
$= -11 + (-3) + (-9) + (-5)$
$= -14 + (-9) + (-5)$
$= -23 + (-5)$
$= -28$
$-28 \div 7 = -4$
The average high temperature was $-4°F$.

117. **Strategy**   To find the average score, divide the combined scores (-20) by the number of golfers (10).

Solution $-20 \div 10 = -2$
The average score was -2.

119. **Strategy**   To find the wind chill factor, multiply the wind chill factor at $10°F$ with a 20 mph wind $(-9°F)$ by 5.

Solution $-9 \times 5 = -45$
The wind chill factor is $-45°F$.

Applying the Concepts

121a. The greatest possible product is that of (-5) and (-5). $(-5)(-5) = 25$
All other contributions, (-1) and (-9), (-2) and (-8), (-3) and (-7), and (-4) and (-6) have a product less than 25.

b. The least possible sum is -17.
$(-1) + (-16) = -17$(the sum)
$-1(-16) = 16$(the product)
Other combinations are (-2) and (-8), and (-4) and (-4), which have a product of 16 and a sum greater than -17.

123a. True

b. True

Section 10.4

Objective A Exercises

1. $\dfrac{5}{8} - \dfrac{5}{6} = \dfrac{15}{24} - \dfrac{20}{24}$
$= \dfrac{15}{24} + \dfrac{(-20)}{24}$
$= \dfrac{15 + (-20)}{24}$
$= -\dfrac{5}{24}$

3. $-\dfrac{5}{12} - \dfrac{3}{8} = \dfrac{-10}{24} - \dfrac{9}{24}$
$= \dfrac{-10}{24} + \dfrac{(-9)}{24}$
$= \dfrac{-10 + (-9)}{24} = \dfrac{-19}{24} = -\dfrac{19}{24}$

5. $-\dfrac{6}{13} + \dfrac{17}{26} = \dfrac{-12}{26} + \dfrac{17}{26}$
$= \dfrac{-12 + 17}{26} = \dfrac{5}{26}$

7. $-\dfrac{5}{8} - \left(-\dfrac{11}{12}\right) = \dfrac{-15}{24} - \left(\dfrac{-22}{24}\right)$
$= \dfrac{-15}{24} + \dfrac{22}{24} = \dfrac{-15 + 22}{24} = \dfrac{7}{24}$

9. $\dfrac{5}{12} - \dfrac{11}{15} = \dfrac{25}{60} - \dfrac{44}{60}$
$= \dfrac{25}{60} + \dfrac{(-44)}{60} = \dfrac{25 + (-44)}{60} = \dfrac{-19}{60} = -\dfrac{19}{60}$

11. $-\dfrac{3}{4} - \dfrac{5}{8} = \dfrac{-6}{8} - \dfrac{5}{8}$
$= \dfrac{-6}{8} + \dfrac{(-5)}{8}$
$= \dfrac{-6 + (-5)}{8} = \dfrac{-11}{8} = -1\dfrac{3}{8}$

13. $-\dfrac{5}{2} - \left(-\dfrac{13}{4}\right) = \dfrac{-10}{4} - \left(\dfrac{-13}{4}\right)$
$= \dfrac{-10}{4} + \dfrac{13}{4} = \dfrac{-10 + 13}{4} = \dfrac{3}{4}$

15. $-\dfrac{3}{8} - \dfrac{5}{12} - \dfrac{3}{16} = \dfrac{-18}{48} - \dfrac{20}{48} - \dfrac{9}{48}$
$= \dfrac{-18}{48} + \dfrac{(-20)}{48} + \dfrac{(-9)}{48}$
$= \dfrac{-18 + (-20) + (-9)}{48} = \dfrac{-47}{48} = -\dfrac{47}{48}$

17. $\dfrac{1}{2} - \dfrac{3}{8} - \left(-\dfrac{1}{4}\right) = \dfrac{4}{8} - \dfrac{3}{8} - \left(\dfrac{-2}{8}\right)$
$= \dfrac{4}{8} + \dfrac{(-3)}{8} + \dfrac{2}{8}$
$= \dfrac{4 + (-3) + 2}{8} = \dfrac{3}{8}$

19. $\dfrac{1}{3} - \dfrac{1}{4} - \dfrac{1}{5} = \dfrac{20}{60} - \dfrac{15}{60} - \dfrac{12}{60}$
$= \dfrac{20}{60} + \dfrac{(-15)}{60} + \dfrac{(-12)}{60}$
$= \dfrac{20 + (-15) + (-12)}{60} = \dfrac{-7}{60} = -\dfrac{7}{60}$

21. $\dfrac{1}{2} + \left(-\dfrac{3}{8}\right) + \dfrac{5}{12} = \dfrac{12}{24} + \dfrac{(-9)}{24} + \dfrac{10}{24}$
$= \dfrac{12 + (-9) + 10}{24} = \dfrac{13}{24}$

23. $3.4 + (-6.8) = -3.4$

25. $-8.32 + (-0.57) = -8.89$

27. $-4.8 + (-3.2) = -8.0$

29. $-4.6 + 3.92 = -0.68$

31. $-45.71 + (-135.8) = -181.51$

33. $4.2 + (-6.8) + 5.3 = -2.6 + 5.3 = 2.7$

35. $-4.5 + 3.2 + (-19.4) = -1.3 + (-19.4) = -20.7$

37. $-18.39 + 4.9 - 23.7 = -18.39 + 4.9 + (-23.7)$
$= -13.49 + (-23.7) = -37.19$

39. $-3.09 - 4.6 - 27.3 = -3.09 + (-4.6) + (-27.3)$
$= -7.69 + (-27.3) = -34.99$

41. $-4.02 + 6.809 - (-3.57) - (-0.419)$
$= -4.02 + 6.809 + 3.57 + 0.419$
$= 2.789 + 3.57 + 0.419$
$= 6.359 + 0.419 = 6.778$

43. $0.27 + (-3.5) - (-0.27) + (-5.44)$
$= 0.27 + (-3.5) + 0.27 + (-5.44)$
$= -3.23 + 0.27 + (-5.44)$
$= -2.96 + (-5.44)$
$= -8.4$

Objective B Exercises

45. $-\dfrac{2}{9} \times \left(-\dfrac{3}{14}\right) = \dfrac{2 \cdot 3}{9 \cdot 14} = \dfrac{1}{21}$

47. $\left(-\dfrac{3}{4}\right)\left(-\dfrac{8}{27}\right) = \dfrac{3 \cdot 8}{4 \cdot 27} = \dfrac{2}{9}$

49. $\dfrac{5}{12} \times \left(-\dfrac{8}{15}\right) = -\left(\dfrac{5 \cdot 8}{12 \cdot 15}\right) = -\dfrac{2}{9}$

51. $\left(\dfrac{3}{8}\right)\left(-\dfrac{15}{41}\right) = -\left(\dfrac{3 \cdot 15}{8 \cdot 41}\right) = -\dfrac{45}{328}$

53. $\left(-\dfrac{5}{7}\right)\left(-\dfrac{14}{15}\right) = \dfrac{5 \cdot 14}{7 \cdot 15} = \dfrac{2}{3}$

55. $\left(\dfrac{1}{2}\right)\left(-\dfrac{3}{4}\right)\left(-\dfrac{5}{8}\right) = \dfrac{1 \cdot 3 \cdot 5}{2 \cdot 4 \cdot 8} = \dfrac{15}{64}$

57. $-\dfrac{3}{8} \div \dfrac{7}{8} = -\dfrac{3}{8} \times \dfrac{8}{7} = -\dfrac{3}{7}$

59. $\dfrac{5}{6} \div \left(-\dfrac{3}{4}\right) = \dfrac{5}{6} \times \left(-\dfrac{4}{3}\right)$

$= -\left(\dfrac{5 \cdot 4}{6 \cdot 3}\right) = -\dfrac{10}{9} = -1\dfrac{1}{9}$

61. $-\dfrac{5}{16} \div \left(-\dfrac{3}{8}\right) = -\dfrac{5}{16} \times -\dfrac{8}{3} = \dfrac{5 \cdot 8}{16 \cdot 3} = \dfrac{5}{6}$

63. $-\dfrac{8}{19} \div \dfrac{7}{38} = -\dfrac{8}{19} \times \dfrac{38}{7}$

$= -\left(\dfrac{8 \cdot 38}{19 \cdot 7}\right) = -\dfrac{16}{7} = -2\dfrac{2}{7}$

65. $-6 \div \dfrac{4}{9} = -\dfrac{6}{1} \times \dfrac{9}{4}$

$= -\left(\dfrac{6 \cdot 9}{1 \cdot 4}\right) = -\dfrac{27}{2} = -13\dfrac{1}{2}$

67. $-8.9 \times (-3.5) = 8.9 \times 3.5 = 31.15$

$\begin{array}{r} 8.9 \\ \times\,3.5 \\ \hline 445 \\ 267 \\ \hline 31.15 \end{array}$

69. $-14.3 \times 7.9 = -(14.3 \times 7.9) = -112.97$

$\begin{array}{r} 14.3 \\ \times\,7.9 \\ \hline 1287 \\ 1001 \\ \hline 112.97 \end{array}$

71. $(-1.21)(-0.03) = (1.21)(0.03) = 0.0363$

$\begin{array}{r} 1.21 \\ \times\,0.03 \\ \hline 0.0363 \end{array}$

73. $-77.6 \div (-0.8) = 77.6 \div 0.8 = 97$

$\begin{array}{r} 97. \\ 0.8\overline{)77.6.} \\ \underline{72} \\ 56 \\ \underline{-56} \\ 0 \end{array}$

75. $(-7.04) \div (-3.2) = 7.04 \div 3.2 = 2.2$

$\begin{array}{r} 2.2 \\ 3.2\overline{)7.0.4} \\ \underline{-64} \\ 64 \\ \underline{-64} \\ 0 \end{array}$

77. $-3.312 \div (0.8) = -(3.312 \div 0.8) = -4.14$

$\begin{array}{r} 4.14 \\ 0.8\overline{)3.3.12} \\ \underline{-32} \\ 11 \\ \underline{-8} \\ 32 \\ \underline{-32} \\ 0 \end{array}$

79. $26.22 \div (-6.9) = -(26.22 \div 6.9) = -3.8$

$\begin{array}{r} 3.8 \\ 6.9\overline{)26.2.3} \\ \underline{-207} \\ 552 \\ \underline{-552} \\ 0 \end{array}$

81. $21.792 \div (-0.96) = -(21.792 \div 0.96) = -22.70$

83. $-3.171 \div (-45.3) = 3.171 \div 45.3 = 0.07$

85. $(-13.97) \div (-25.4) = 13.97 \div 25.4 = 0.55$

Objective C Exercises

87. Strategy To find the amount the temperature fell from 9:00 A.M. subtract the temperature at 9:27 A.M. from the temperature at 9:00 A.M.

Solution $12.22 - (-20) = 32.22$
The temperature fell 32.22°C in 27 min.

89. Strategy To find the difference, subtract the melting point of oxygen (−218.4°C) from its boiling point (−182.962°C).

Solution $-182.962 - (-218.4)$
$= -182.962 + 218.4$
$= 35.438$
The difference between the boiling point of oxygen and its melting point is 35.438°C.

91a. **Strategy** To find the closing price on the previous day, subtract the change in price (-0.21) from the closing price on September 15, 2003 (19.18).

Solution $19.18 - (-0.21) = 19.18 + 0.21 = 19.39$
The closing price the previous day for Sara Lee Corp. was $19.39.

b. **Strategy** To find the closing price on the previous day, subtract the change in price ($+\$.11$) from the closing price on September 15, 2003 ($72.57).

Solution $72.57 - (+0.11) = 72.57 - 0.11 = 72.46$
The closing price the previous day for Hershey Foods, Inc., was $72.46.

Applying the Concepts

93. $-\dfrac{17}{24}$ is one example.

95. Given any two different rational numbers, it is always possible to find a rational number between them. One method is to add the two numbers and divide by 2.
Another method is to add the numerators and add the denominators. For example, given the fractions $\dfrac{2}{5}$ and $\dfrac{3}{4}$, $\dfrac{2+3}{5+4} = \dfrac{5}{9}$ and $\dfrac{2}{5} < \dfrac{5}{9} < \dfrac{3}{4}$.

Section 10.5

Objective A Exercises

1. Since the number is greater than 10, move the decimal point 6 places to the left. The exponent on 10 is 6.
$2,370,000 = 2.37 \times 10^6$

3. Since the number is less than 1, move the decimal point 4 places to the right. The exponent on 10 is -4.
$0.00045 = 4.5 \times 10^{-4}$

5. Since the number is greater than 10, move the decimal point 5 places to the left. The exponent on 10 is 5.
$309,000 = 3.09 \times 10^5$

7. Since the number is less than 1, move the decimal point 7 places to the right. The exponent on 10 is -7.
$0.000000601 = 6.01 \times 10^{-7}$

9. Since the number is greater than 10, move the decimal point 10 places to the left. The exponent on 10 is 10.
$57,000,000,000 = 5.7 \times 10^{10}$

11. Since the number is less than 1, move the decimal point 8 places to the right. The exponent on 10 is -8.
$0.000000017 = 1.7 \times 10^{-8}$

13. The exponent on 10 is positive. Move the decimal point 5 places to the right.
$7.1 \times 10^5 = 710,000$

15. The exponent on 10 is negative. Move the decimal point 5 places to the left.
$4.3 \times 10^{-5} = 0.000043$

17. The exponent on 10 is positive. Move the decimal point 8 places to the right.
$6.71 \times 10^8 = 671,000,000$

19. The exponent on 10 is negative. Move the decimal point 6 places to the left.
$7.13 \times 10^{-6} = 0.00000713$

21. The exponent on 10 is positive. Move the decimal point 12 places to the right.
$5 \times 10^{12} = 5,000,000,000,000$

23. The exponent on 10 is negative. Move the decimal point 3 places to the left.
$8.01 \times 10^{-3} = 0.00801$

25. The number is greater than 10. Move the decimal point 10 places to the left. The exponent on 10 is 10.
$16,000,000,000 \text{ mi} = 1.6 \times 10^{10} \text{ mi}$

27. The monetary cost is $3.1 trillion. The exponent on 10 is 12.
$\$3.1 \times 10^{12}$

29. The number is less than 1. Move the decimal point 6 places to the right. The exponent on 10 is -6.
$0.0000037 \text{ m} = 3.7 \times 10^{-6} \text{ m}$

Objective B Exercises

31. $8 \div 4 + 2 = 2 + 2 = 4$

33. $4 + (-7) + 3 = -3 + 3 = 0$

35. $4^2 - 4 = 16 - 4 = 16 + (-4) = 12$

37. $2 \times (3 - 5) - 2 = 2 \times [3 + (-5)] - 2$
$ = 2 \times (-2) - 2$
$ = 4 - 2$
$ = -4 + (-2) = -6$

39. $4 - (-3)^2 = 4 - 9 = 4 + (-9) = -5$

41. $4 - (-3) - 5 = 4 + 3 + (-5)$
$ = 7 + (-5) = 2$

43. $4 - (-2)^2 + (-3) = 4 - 4 + (-3)$
$= 4 + (-4) + (-3)$
$= 0 + (-3) = -3$

45. $3^2 - 4 \times 2 = 9 - 4 \times 2$
$= 9 - 8 = 9 + (-8) = 1$

47. $3 \times (6 - 2) \div 6 = 3 \times [6 + (-2)] \div 6$
$= 3 \times 4 \div 6 = 12 \div 6 = 2$

49. $2^2 - (-3)^2 + 2 = 4 - 9 + 2$
$= 4 + (-9) + 2$
$= -5 + 2 = -3$

51. $6 - 2 \times (1 - 5) = 6 - 2 \times [1 + (-5)]$
$= 6 - 2 \times (-4)$
$= 6 - (-8) = 6 + 8 = 14$

53. $(-2)^2 - (-3)^2 + 1 = 4 - 9 + 1$
$= 4 + (-9) + 1$
$= -5 + 1 = -4$

55. $6 - (-3) \times (-3)^2 = 6 - (-3) \times 9$
$= 6 - (-27) = 6 + 27 = 33$

57. $4 \times 2 - 3 \times 7 = 8 - 3 \times 7$
$= 8 - 21 = 8 + (-21) = -13$

59. $(-2)^2 - 5 \times 3 - 1 = 4 - 5 \times 3 - 1$
$= 4 - 15 - 1$
$= 4 + (-15) + (-1)$
$= -11 + (-1) = -12$

61. $7 \times 6 - 5 \times 6 + 3 \times 2 - 2 + 1$
$= 42 - 5 \times 6 + 3 \times 2 - 2 + 1$
$= 42 - 30 + 3 \times 2 - 2 + 1$
$= 42 - 30 + 6 - 2 + 1$
$= 42 + (-30) + 6 + (-2) + 1$
$= 12 + 6 + (-2) + 1$
$= 18 + (-2) + 1$
$= 16 + 1 = 17$

63. $-4 \times 3 \times (-2) + 12 \times (3 - 4) + (-12)$
$= -4 \times 3 \times (-2) + 12 \times [3 + (-4)] + (-12)$
$= -4 \times 3 \times (-2) + 12 \times (-1) + (-12)$
$= -12 \times (-2) + 12 \times (-1) + (-12)$
$= 24 + 12 \times (-1) + (-12)$
$= 24 + (-12) + (-12)$
$= 12 + (-12) = 0$

65. $-12 \times (6 - 8) + 1^2 \times 3^2 \times 2 - 6 \times 2$
$= -12 \times [6 + (-8)] + 1^2 \times 3^2 \times 2 - 6 \times 2$
$= -12 \times (-2) + 1^2 \times 3^2 \times 2 - 6 \times 2$
$= -12 \times (-2) + 1 \times 9 \times 2 - 6 \times 2$
$= 24 + 1 \times 9 \times 2 - 6 \times 2$
$= 24 + 18 - 6 \times 2$
$= 24 + 18 - 12$
$= 24 + 18 + (-12)$
$= 42 + (-12) = 30$

67. $10 \times 9 - (8 + 7) \div 5 + 6 - 7 + 8$
$= 10 \times 9 - 15 \div 5 + 6 - 7 + 8$
$= 90 - 15 \div 5 + 6 - 7 + 8$
$= 90 - 3 + 6 - 7 + 8$
$= 90 + (-3) + 6 + (-7) + 8$
$= 87 + 6 + (-7) + 8$
$= 93 + (-7) + 8$
$= 86 + 8 = 94$

69. $3^2 \times (4 - 7) \div 9 + 6 - 3 - 4 \times 2$
$= 3^2 \times [4 + (-7)] \div 9 + 6 - 3 - 4 \times 2$
$= 3^2 \times (-3) \div 9 + 6 - 3 - 4 \times 2$
$= 9 \times (-3) \div 9 + 6 - 3 - 4 \times 2$
$= -27 \div 9 + 6 - 3 - 4 \times 2$
$= -3 + 6 - 3 - 4 \times 2$
$= -3 + 6 - 3 - 8$
$= -3 + 6 + (-3) + (-8)$
$= 3 + (-3) + (-8)$
$= 0 + (-8) = -8$

71. $(-3)^2 \times (5 - 7)^2 - (-9) \div 3$
$= (-3)^2 \times [5 + (-7)]^2 - (-9) \div 3$
$= (-3)^2 \times (-2)^2 - (-9) \div 3$
$= 9 \times 4 - (-9) \div 3$
$= 36 - (-9 \div 3)$
$= 36 - (-3)$
$= 36 + 3 = 39$

73. $4 - 6(2 - 5)^3 \div (17 - 8)$
$= 4 - 6[2 + (-5)]^3 \div [17 + (-8)]$
$= 4 - 6(-3)^3 \div 9$
$= 4 - 6(-27) \div 9$
$= 4 - (-162) \div 9$
$= 4 - (-18) = 4 + 18 = 22$

75. $(1.2)^2 - 4.1 \times 0.3 = 1.44 - 4.1 \times 0.3$
$= 1.44 - 1.23$
$= 1.44 + (-1.23) = 0.21$

77. $1.6 - (-1.6)^2 = 1.6 - 2.56$
$= 1.6 + (-2.56) = -0.96$

79. $(4.1 - 3.9) - 0.7^2 = [4.1 + (-3.9)] - 0.7^2$
$= 0.2 - 0.7^2$
$= 0.2 - 0.49$
$= 0.2 + (-0.49) = -0.29$

81. $(-0.4)^2 \times 1.5 - 2 = 0.16 \times 1.5 - 2$
$= 0.24 - 2$
$= 0.24 + (-2) = -1.76$

83. $4.2 - (-3.9) - 6 = 4.2 + 3.9 + (-6)$
$= 8.1 + (-6) = 2.1$

85. $\left(\dfrac{3}{4}\right)^2 - \dfrac{3}{8} = \dfrac{9}{16} - \dfrac{3}{8}$
$= \dfrac{9}{16} - \dfrac{6}{16} = \dfrac{3}{16}$

87. $\dfrac{5}{16} - \dfrac{3}{8} + \dfrac{1}{2} = \dfrac{5}{16} - \dfrac{6}{16} + \dfrac{1}{2}$

$\qquad = \dfrac{5}{16} + \left(-\dfrac{6}{16}\right) + \dfrac{1}{2}$

$\qquad = -\dfrac{1}{16} + \dfrac{1}{2} = -\dfrac{1}{16} + \dfrac{8}{16} = \dfrac{7}{16}$

89. $\dfrac{1}{2} \times \dfrac{1}{4} \times \dfrac{1}{2} - \dfrac{3}{8} = \dfrac{1}{8} \times \dfrac{1}{2} - \dfrac{3}{8}$

$\qquad = \dfrac{1}{16} - \dfrac{3}{8}$

$\qquad = \dfrac{1}{16} + \left(-\dfrac{3}{8}\right) = \dfrac{1}{16} + \left(-\dfrac{6}{16}\right) = -\dfrac{5}{16}$

91. $\dfrac{1}{2} - \left(\dfrac{3}{4} - \dfrac{3}{8}\right) \div \dfrac{1}{3} = \dfrac{1}{2} - \left(\dfrac{6}{8} - \dfrac{3}{8}\right) \div \dfrac{1}{3}$

$\qquad = \dfrac{1}{2} - \dfrac{3}{8} \div \dfrac{1}{3} = \dfrac{1}{2} - \dfrac{3}{8} \times \dfrac{3}{1}$

$\qquad = \dfrac{1}{2} - \dfrac{9}{8} = \dfrac{4}{8} + \left(-\dfrac{9}{8}\right) = -\dfrac{5}{8}$

Applying the Concepts

93a. $3.45 \times 10^{-14} > 3.45 \times 10^{-15}$

b. $5.23 \times 10^{18} > 5.23 \times 10^{17}$

c. $3.12 \times 10^{12} > 3.12 \times 10^{11}$

95a. $1^3 + 2^3 + 3^3 + 4^3 = 1 + 8 + 27 + 64 = 100$

b. $(-1)^3 + (-2)^3 + (-3)^3 + (-4)^3$
$= -1 + (-8) + (-27) + (-64)$
$= -100$

c. $1^3 + 2^3 + 3^3 + 4^3 + 5^3$
$= 1 + 8 + 27 + 64 + 125$
$= 225$

d. $(-1)^3 + (-2)^3 + (-3)^3 + (-4)^3 + (-5)^3 = -225$

97. Because the first statement is false (it was either Becky or Diana), neither Becky nor Diana could have done it. Therefore, it was either Abdul or Carl. Because the second statement is false, (it was neither Becky nor Carl), either Becky or Carl did it. Because the first statement ensured that it was Abdul or Carl, it must have been Carl.

99a. 1.99×10^{30} kg

b. 1.67×10^{-27} kg

Chapter 10 Review Exercises

1. -22

2. $-8 - (-2) - (-10) - 3 = -8 + 2 + 10 - 3$
$\qquad = -6 + 10 - 3 = 4 - 3 = 1$

3. $\dfrac{5}{8} - \dfrac{5}{6} = \dfrac{15}{24} - \dfrac{20}{24}$

$\qquad = \dfrac{15}{24} + \dfrac{(-20)}{24} = \dfrac{15 + (-20)}{24} = \dfrac{-5}{24} = -\dfrac{5}{24}$

4. $-0.33 + 1.98 - 1.44 = -0.33 + 1.98 + (-1.44)$
$\qquad = 1.65 + (-1.44) = 0.21$

5. $\left(-\dfrac{2}{3}\right)\left(\dfrac{6}{11}\right)\left(-\dfrac{22}{25}\right) = \dfrac{2 \cdot 6 \cdot 22}{3 \cdot 11 \cdot 25} = \dfrac{8}{25}$

6. $-0.08 \times 16 = -(0.08 \times 16) = -1.28$
$\qquad \begin{array}{r} 16 \\ \times\,.08 \\ \hline 1.28 \end{array}$

7. $12 - 6 \div 3 = 12 - 2 = 12 + (-2) = 10$

8. $\left(\dfrac{2}{3}\right)^2 - \dfrac{5}{6} = \left(\dfrac{2}{3} \cdot \dfrac{2}{3}\right) - \dfrac{5}{6}$

$\qquad = \dfrac{4}{9} - \dfrac{5}{6} = \dfrac{8}{18} - \dfrac{15}{18}$

$\qquad = \dfrac{8}{18} + \left(-\dfrac{15}{18}\right) = \dfrac{8 + (-15)}{18} = -\dfrac{7}{18}$

9. 4

10. $0 > -3$

11. $-|-6| = -6$

12. $-18 \div (-3) = 18 \div 3 = 6$

13. $-\dfrac{3}{8} + \dfrac{5}{12} + \dfrac{2}{3} = \dfrac{-9}{24} + \dfrac{10}{24} + \dfrac{16}{24}$

$\qquad = \dfrac{-9 + 10 + 16}{24} = \dfrac{17}{24}$

14. $\dfrac{1}{3} \times \left(-\dfrac{3}{4}\right) = -\left(\dfrac{1}{3} \times \dfrac{3}{4}\right) = -\dfrac{1}{4}$

15. $-\dfrac{7}{12} \div \left(-\dfrac{14}{39}\right) = -\dfrac{7}{12} \times \left(-\dfrac{39}{14}\right)$

$\qquad = \dfrac{7 \cdot 39}{12 \cdot 14} = \dfrac{13}{8} = 1\dfrac{5}{8}$

16. $16 \div 4(8 - 2) = 16 \div 4[8 + (-2)]$
$\qquad = 16 \div 4(6) = 4(6) = 24$

17. $-22 + 14 + (-18) = -8 + (-18) = -26$

18. $3^2 - 9 + 2 = 9 - 9 + 2$
$\qquad = 9 + (-9) + 2$
$\qquad = 0 + 2 = 2$

19. The number is less than 1. Move the decimal point 5 places to the right. The exponent on 10 is -5.
$0.0000397 = 3.97 \times 10^{-5}$

20. $-1.464 \div 18.3 = -(1.464 \div 18.3) = -0.08$
$\qquad \begin{array}{r} 0.08 \\ 18.3)\overline{1.4.64} \\ -1464 \\ \hline 0 \end{array}$

21. $-\dfrac{5}{12}+\dfrac{7}{9}-\dfrac{1}{3}=\dfrac{-15}{36}+\dfrac{28}{36}-\dfrac{12}{36}$

$\qquad\qquad = \dfrac{-15}{36}+\dfrac{28}{36}+\dfrac{(-12)}{36}$

$\qquad\qquad = \dfrac{-15+28+(-12)}{36}=\dfrac{1}{36}$

22. $\dfrac{6}{34}\times\dfrac{17}{40}=\dfrac{6\cdot 17}{34\cdot 40}=\dfrac{3}{40}$

23. $1.2\times(-0.035)=-(1.2\times 0.035)=-0.042$

$\qquad\begin{array}{r} 0.035 \\ \times 1.2 \\ \hline 70 \\ 35 \\ \hline 0.042 \end{array}$

24. $-\dfrac{1}{2}+\dfrac{3}{8}\div\dfrac{9}{20}=-\dfrac{1}{2}+\dfrac{3}{8}\times\dfrac{20}{9}$

$\qquad\qquad = -\dfrac{1}{2}+\dfrac{3\cdot 20}{8\cdot 9}$

$\qquad\qquad = -\dfrac{1}{2}+\dfrac{5}{6}=-\dfrac{3}{6}+\dfrac{5}{6}=\dfrac{2}{6}=\dfrac{1}{3}$

25. $|-5|=5$

26. $-2>-40$

27. $2\times(-13)=-(2\times 13)=-26$

28. $-0.4\times 5-(-3.33)=-2-(-3.33)$

$\qquad\qquad\qquad = -2+3.33=1.33$

29. $\dfrac{5}{12}+\left(-\dfrac{2}{3}\right)=\dfrac{5}{12}+\dfrac{(-8)}{12}$

$\qquad\qquad = \dfrac{5+(-8)}{12}=\dfrac{-3}{12}=\dfrac{-1}{4}=-\dfrac{1}{4}$

30. $-33.4+9.8-(-16.2)=-33.4+9.8+16.2$

$\qquad\qquad\qquad = -23.6+16.2=-7.4$

31. $\left(-\dfrac{3}{8}\right)\div\left(-\dfrac{4}{5}\right)=-\dfrac{3}{8}\times\left(-\dfrac{5}{4}\right)=\dfrac{3\cdot 5}{8\cdot 4}=\dfrac{15}{32}$

32. The exponent on 10 is positive. Move the decimal point 5 places to the right.

$2.4\times 10^5=240{,}000$

33. **Strategy** To find the temperature, add the increase (18°) to the original temperature (−22°).

Solution $-22+18=-4$
The temperature is −4°.

34. **Strategy** To find the student's score:
- Multiply the number of questions answered correctly (38) by 3.
Multiply the number of questions left blank (8) by −1.
Multiply the number of questions answered incorrectly (4) by −2.
- Add the three products.

Solution $38\times 3=114$
$8\times -1=-8$
$4\times -2=-8$
$114+(-8)+(-8)=106+(-8)=98$
The student's score was 98.

35. **Strategy** To find the difference between the boiling point and the melting point of mercury, subtract the melting point (−38.87°C) from the boiling point (356.58°C).

Solution $356.58-(-38.87)=395.45$
The difference between the boiling and melting points is 395.45°C.

Chapter 10 Test

1. $-5-(-8)=-5+8=3$

2. $-|-2|=-2$

3. $-\dfrac{2}{5}+\dfrac{7}{15}=\dfrac{-6}{15}+\dfrac{7}{15}=\dfrac{-6+7}{15}=\dfrac{1}{15}$

4. $0.032\times(-1.9)=-(0.032\times 1.9)=-0.0608$

$\qquad\begin{array}{r} 0.032 \\ \times 1.9 \\ \hline 288 \\ 32 \\ \hline 0.0608 \end{array}$

5. $-8>-10$

6. $1.22+(-3.1)=-1.88$

7. $4\times(4-7)\div(-2)-4\times 8$
$=4\times[4+(-7)]\div(-2)-4\times 8$
$=4\times(-3)\div(-2)-4\times 8$
$=-12\div(-2)-4\times 8$
$=6-4\times 8$
$=6-32$
$=6+(-32)=-26$

8. $-5\times(-6)\times 3=5\times 6\times 3=30\times 3=90$

9. $-1.004-3.01=-1.004+(-3.01)=-4.014$

10. $-72\div 8=-(72\div 8)=-9$

11. $-2+3+(-8)=1+(-8)=-7$

12. $-\dfrac{3}{8}+\dfrac{2}{3}=\dfrac{-9}{24}+\dfrac{16}{24}=\dfrac{-9+16}{24}=\dfrac{7}{24}$

13. The number is greater than 10. Move the decimal point 10 places to the left. The exponent on 10 is 10.

$87{,}600{,}000{,}000=8.76\times 10^{10}$

14. $-4\times 12=-(4\times 12)=-48$

15. $\dfrac{0}{-17}=0$

16. $16 - 4 - (-5) - 7 = 16 + (-4) + 5 + (-7)$
$$= 12 + 5 + (-7)$$
$$= 17 + (-7) = 10$$

17. $-\dfrac{2}{3} \div \dfrac{5}{6} = -\dfrac{2}{3} \times \dfrac{6}{5} = -\left(\dfrac{2 \cdot 6}{3 \cdot 5}\right) = -\dfrac{4}{5}$

18. $0 > -4$

19. $16 + (-10) + (-20) = 6 + (-20) = -14$

20. $(-2)^2 - (-3)^2 \div (1 - 4)^2 \times 2 - 6$
$$= (-2)^2 - (-3)^2 \div [1 + (-4)]^2 \times 2 - 6$$
$$= (-2)^2 - (-3)^2 \div (-3)^2 \times 2 - 6$$
$$= 4 - 9 \div 9 \times 2 - 6$$
$$= 4 - 1 \times 2 - 6$$
$$= 4 - 2 - 6$$
$$= 4 + (-2) + (-6)$$
$$= 2 + (-6) = -4$$

21. $-\dfrac{2}{5} - \left(\dfrac{-7}{10}\right) = \dfrac{-4}{10} - \left(-\dfrac{7}{10}\right)$
$$= \dfrac{-4}{10} + \dfrac{7}{10} = \dfrac{-4 + 7}{10} = \dfrac{3}{10}$$

22. The exponent on 10 is negative. Move the decimal point 8 places to the left.
$$9.601 \times 10^{-8} = 0.00000009601$$

23. $-15.64 \div (-4.6) = (15.64 \div 4.6) = 3.4$

$$\begin{array}{r} 3.4 \\ 4.6\overline{)15.64} \\ \underline{-138} \\ 184 \\ \underline{-184} \\ 0 \end{array}$$

24. $-\dfrac{1}{2} + \dfrac{1}{3} + \dfrac{1}{4} = \dfrac{-6}{12} + \dfrac{4}{12} + \dfrac{3}{12}$
$$= \dfrac{-6 + 4 + 3}{12} = \dfrac{1}{12}$$

25. $\dfrac{3}{8} \times \left(-\dfrac{5}{6}\right) \times \left(-\dfrac{4}{15}\right) = \dfrac{3}{8} \times \dfrac{5}{6} \times \dfrac{4}{15}$
$$= \dfrac{3 \cdot 5 \cdot 4}{8 \cdot 6 \cdot 15} = \dfrac{1}{12}$$

26. $2.113 - (-1.1) = 2.113 + 1.1 = 3.213$

27. Strategy To find the temperature, add the increase (11°C) to the previous temperature (−4°C).

Solution $-4 + 11 = 7$
The temperature is 7°C.

28. Strategy To find the melting point of oxygen, multiply the melting point of radon (−71°C) by 3.

Solution $-71 \times 3 = -213$
The melting point of oxygen is −213°C.

29. Strategy To find the amount the temperature fell, subtract the temperature at midnight (−29.4°C) from the temperature at noon (17.22°C).

Solution $17.22 - (-29.4) = 46.62$
The temperature fell 46.62°C.

30. Strategy To find the average daily low temperature:
- Add the three temperature readings.
- Divide by 3.

Solution $-7 + 9 + (-8) = 2 + (-8) = -6$
$-6 \div 3 = -2$
The average low temperature was −2°F.

Cumulative Review Exercises

1. $16 - 4 \cdot (3 - 2)^2 \cdot 4 = 16 - 4 \cdot (1)^2 \cdot 4$
$$= 16 - 4 \cdot (1) \cdot 4 = 16 - 16 = 0$$

2. $8\dfrac{1}{2} = 8\dfrac{7}{14} = 7\dfrac{21}{14}$
$$\dfrac{-3\dfrac{4}{7} = 3\dfrac{8}{14} = 3\dfrac{8}{14}}{4\dfrac{13}{14}}$$

3. $3\dfrac{7}{8} \div 1\dfrac{1}{2} = \dfrac{31}{8} \div \dfrac{3}{2}$
$$= \dfrac{31}{8} \times \dfrac{2}{3}$$
$$= \dfrac{31}{12} = 2\dfrac{7}{12}$$

4. $\dfrac{3}{8} \div \left(\dfrac{3}{8} - \dfrac{1}{4}\right) \div \dfrac{7}{3} = \dfrac{3}{8} \div \left(\dfrac{3}{8} - \dfrac{2}{8}\right) \div \dfrac{7}{3}$
$$= \dfrac{3}{8} \div \left(\dfrac{1}{8}\right) \div \dfrac{7}{3} = \dfrac{3}{8} \times \dfrac{8}{1} \div \dfrac{7}{3}$$
$$= 3 \div \dfrac{7}{3} = 3 \times \dfrac{3}{7} = \dfrac{3 \cdot 3}{7} = \dfrac{9}{7} = 1\dfrac{2}{7}$$

5. $\begin{array}{r} 2.90700 \\ \underline{-1.09761} \\ 1.80939 \end{array}$

6. $\dfrac{7}{12} = \dfrac{n}{32}$
$$7 \cdot 32 = 12 \times n$$
$$224 = 12 \times n$$
$$224 \div 12 = n$$
$$18.67 \approx n$$

7. $160\% \times n = 22$
$$1.6 \times n = 22$$
$$n = 22 \div 1.6$$
$$n = 13.75$$

8. 7 qt = 1 gal 3 qt

9. $6692 \text{ ml} = 6.692 \text{ L}$

10. $4.2 \text{ ft} = 4.2 \text{ ft} \times \dfrac{1 \text{ m}}{3.28 \text{ ft}} = \dfrac{4.2}{3.28} \text{ m} \approx 1.28 \text{ m}$

11. Percent $\times$ base = amount
$$0.32 \times 180 = n$$
$$57.6 = n$$

12. $3\dfrac{2}{5} \times 100\% = \dfrac{1700}{5}\% = 340\%$

13. $-8 + 5 = -3$

14. $3\dfrac{1}{4} + \left(-6\dfrac{5}{8}\right) = \dfrac{13}{4} + \left(\dfrac{-53}{8}\right) = \dfrac{26}{8} + \dfrac{(-53)}{8}$
$$= \dfrac{26 + (-53)}{8} = \dfrac{-27}{8} = -3\dfrac{3}{8}$$

15. $-6\dfrac{1}{8} - 4\dfrac{5}{12} = \dfrac{-49}{8} - \dfrac{53}{12}$
$$= \dfrac{-147}{24} - \dfrac{106}{24} = \dfrac{-147}{24} + \dfrac{(-106)}{24}$$
$$= \dfrac{-147 + (-106)}{24} = \dfrac{-253}{24} = -10\dfrac{13}{24}$$

16. $-12 - (-7) - 3(-8) = -12 + 7 + 24 = -5 + 24 = 19$

17. $-3.2 \times -1.09 = 3.2 \times 1.09 = 3.488$
$$\begin{array}{r} 1.09 \\ \times 3.2 \\ \hline 218 \\ 327 \\ \hline 3.488 \end{array}$$

18. $-6 \times 7 \times \left(-\dfrac{3}{4}\right) = 6 \times 7 \times \dfrac{3}{4}$
$$= \dfrac{6 \cdot 7 \cdot 3}{4} = \dfrac{126}{4} = \dfrac{63}{2} = 31\dfrac{1}{2}$$

19. $42 \div (-6) = -(42 \div 6) = -7$

20. $-2\dfrac{1}{7} \div \left(-3\dfrac{3}{5}\right) = \dfrac{15}{7} \div \left(\dfrac{18}{5}\right)$
$$= \dfrac{15}{7} \times \left(\dfrac{5}{18}\right) = \dfrac{25}{42}$$

21. $3 \times (3 - 7) \div 6 - 2 = 3 \times [3 + (-7)] \div 6 - 2$
$$= 3 \times (-4) \div 6 - 2$$
$$= -12 \div 6 - 2$$
$$= -2 + (-2) = -4$$

22. $4 - (-2)^2 \div (1 - 2)^2 \times 3 + 4$
$$= 4 - (-2)^2 \div [1 + (-2)]^2 \times 3 + 4$$
$$= 4 - (-2)^2 \div (-1)^2 \times 3 + 4$$
$$= 4 - 4 \div 1 \times 3 + 4$$
$$= 4 - 4 \times 3 + 4$$
$$= 4 - 12 + 4$$
$$= 4 + (-12) + 4$$
$$= -8 + 4 = -4$$

23. **Strategy**   To find the length of the remaining board, subtract the length cut $\left(5\dfrac{2}{3}\text{ ft}\right)$ from the original length (8 ft).

     Solution
$$\begin{array}{r} 8 \text{ ft} = 7\dfrac{3}{3} \text{ ft} \\ -5\dfrac{2}{3} \text{ ft} = 5\dfrac{2}{3} \text{ ft} \\ \hline 2\dfrac{1}{3} \text{ ft} \end{array}$$

The length remaining is $2\dfrac{1}{3}$ ft.

24. **Strategy**   To find Nimisha's new balance:
- Subtract the amounts of the checks written.
- Add the amount of the deposit.

     Solution
$$\begin{array}{r} \$763.56 \\ -\ 135.88 \\ \hline 627.68 \\ -\ 47.81 \\ \hline 579.87 \\ +\ 223.44 \\ \hline \$803.31 \end{array}$$

Nimisha's new balance is $803.31.

25. **Strategy**   To find the percent:
- Subtract the sale price ($120) from the original price ($165) to find the amount of the decrease.
- Solve the basic percent equation for percent. The base is $165 and the amount is the amount of the decrease.

     Solution $\$165 - \$120 = \$45$
Percent $\times$ base = amount
$$n \times 165 = 45$$
$$n = 45 \div 165$$
$$n \approx 0.273 = 27.3\%$$
The percent decrease is 27.3%.

26. **Strategy**   To find how many gallons of coffee must be prepared:
- Multiply the number of guests (80) by the amount of coffee each guest is expected to drink (2 c) to find the number of cups of coffee to prepare.
- Convert cups to gallons.

     Solution $80 \times 2 \text{ c} = 160 \text{ c}$
$$160 \text{ c} = 160 \text{ c} \times \dfrac{1 \text{ pt}}{2 \text{ c}} \times \dfrac{1 \text{ qt}}{2 \text{ pt}} \times \dfrac{1 \text{ gal}}{4 \text{ qt}}$$
$$= \dfrac{160}{16} \text{ gal} = 10 \text{ gal}$$

The amount of coffee that should be prepared is 10 gal.

27. Strategy To find the dividend per share:
● Solve the basic percent equation for amount to find the amount of the increase. The base amount is $1.50 and the percent is 12%.
● Add the amount of the increase to the dividend ($1.50).

Solution

$12\% \times 1.50 = n$ $\quad$ 1.50
$0.12 \times 1.50 = n$ $\quad + 0.18$
$\quad\quad 0.18 = n$ $\quad\quad$ 1.68

The dividend per share after the increase was $1.68.

28. Strategy To find the median:
● Arrange the hourly wages in order from least to greatest.
● Pick the middle number.

Solution $9.32; $10.73; $11.40; $13.10; $15.25
The median hourly pay is $11.40.

29. Strategy To find the number of voters, write and solve a proportion.

Solution

$$\frac{5}{8} = \frac{n}{960,000}$$
$5 \times 960,000 = 8 \times n$
$4,800,000 = 8 \times n$
$4,800,000 \div 8 = n$
$600,000 = n$

600,000 people would vote.

30. Strategy To find the average high temperature, add the daily high temperatures ($-19°$, $-7°$, $1°$, and $9°$) and divide that sum by the number of temperatures (4).

Solution

$(-19) + (-7) + (1) + (9)$
$= -26 + 1 + 9$
$= -25 + 9$
$= -16 =$ sum of temperatures
$-16 \div 4 = -4$
The average high temperature is $-4°$.

Chapter 11: Introduction to Algebra

Prep Test

1. -7

2. -20

3. 0

4. 1

5. 1

6. $\left(\dfrac{3}{5}\right)^3 \cdot \left(\dfrac{5}{9}\right)^2 = \dfrac{3}{5} \cdot \dfrac{3}{5} \cdot \dfrac{3}{5} \cdot \dfrac{5}{9} \cdot \dfrac{5}{9} = \dfrac{\overset{1}{\cancel{3}} \cdot \overset{1}{\cancel{3}} \cdot \overset{1}{\cancel{3}} \cdot \overset{1}{\cancel{5}} \cdot \overset{1}{\cancel{5}}}{\underset{1}{\cancel{3}} \cdot \underset{1}{\cancel{3}} \cdot 5 \cdot \underset{1}{\cancel{3}} \cdot \underset{1}{\cancel{3}} \cdot \underset{1}{\cancel{3}} \cdot 3} \cdot$
$\phantom{\left(\dfrac{3}{5}\right)^3 \cdot \left(\dfrac{5}{9}\right)^2} = \dfrac{1}{15}$

7. $\dfrac{2}{3} + \left(\dfrac{3}{4}\right)^2 \cdot \dfrac{2}{9} = \dfrac{2}{3} + \dfrac{9}{16} \cdot \dfrac{2}{9}$
$\phantom{\dfrac{2}{3} + \left(\dfrac{3}{4}\right)^2 \cdot \dfrac{2}{9}} = \dfrac{2}{3} + \dfrac{1}{8}$
$\phantom{\dfrac{2}{3} + \left(\dfrac{3}{4}\right)^2 \cdot \dfrac{2}{9}} = \dfrac{16}{24} + \dfrac{3}{24}$
$\phantom{\dfrac{2}{3} + \left(\dfrac{3}{4}\right)^2 \cdot \dfrac{2}{9}} = \dfrac{19}{24}$

8. $-8 \div (-2)^2 + 6 = -8 \div 4 + 6$
$ = -2 + 6$
$ = 4$

9. $4 + 5(2-7)^2 \div (-8+3)$
$= 4 + 5(-5)^2 \div (-5)$
$= 4 + 5(25) \div (-5)$
$= 4 + 125 \div (-5)$
$= 4 + (-25) = -21$

Go Figure

Replacing the known values,
$$\begin{array}{r} 271 \\ 51 \\ +3HE \\ \hline S71 \end{array}$$
In the first column,
$1 + 1 + E = 11$
$2 + E = 11$
$E = 9$
In the second column,
$1 + 7 + 5 + H = 17$
$13 + H = 17$
$H = 4$
So then, the third column,
$1 + 2 + 3 = S$
$S = 6$

Section 11.1

Objective A Exercises

1. $5a - 3b = 5(-3) - 3(6)$
$ = -15 - 18$
$ = -15 + (-18)$
$ = -33$

3. $2a + 3c = 2(-3) + 3(-2)$
$ = -6 + (-6) = -12$

5. $-c^2 = -(-2)^2 = -4$

7. $b - a^2 = 6 - (-3)^2 = 6 - 9 = 6 + (-9) = -3$

9. $ab - c^2 = (-3)(6) - (-2)^2$
$ = -18 - 4$
$ = -18 + (-4) = -22$

11. $2ab - c^2 = 2(-3)6 - (-2)^2$
$ = -36 - 4$
$ = -36 + (-4) = -40$

13. $a - (b \div a) = -3 - [6 \div (-3)]$
$ = -3 - (-2)$
$ = -3 + 2 = -1$

15. $2ac - (b \div a) = 2(-3)(-2) - [6 \div (-3)]$
$ = 12 - (-2)$
$ = 12 + 2 = 14$

17. $b^2 - c^2 = (6)^2 - (-2)^2$
$ = 36 - 4$
$ = 36 + (-4) = 32$

19. $b^2 \div (ac) = (6)^2 \div (-3)(-2)$
$ = 36 \div 6 = 6$

21. $c^2 - (b \div c) = (-2)^2 - [6 \div (-2)]$
$ = 4 - (-3)$
$ = 4 + 3 = 7$

23. $a^2 + b^2 + c^2 = (-3)^2 + 6^2 + (-2)^2$
$ = 9 + 36 + 4$
$ = 45 + 4 = 49$

25. $ac + bc + ab = (-3)(-2) + 6(-2) + (-3)(6)$
$ = 6 + (-12) + (-18)$
$ = -6 + (-18) = -24$

27. $a^2 + b^2 - ab = (-3)^2 + 6^2 - (-3)(6)$
$ = 9 + 36 - (-3)(6)$
$ = 9 + 36 - (-18)$
$ = 9 + 36 + 18$
$ = 45 + 18 = 63$

29.
$$2b - (3c + a^2) = 2(6) - [3(-2) + (-3)^2]$$
$$= 2(6) - [3(-2) + 9]$$
$$= 2(6) - (-6 + 9)$$
$$= 2(6) - 3$$
$$= 12 - 3$$
$$= 12 + (-3) = 9$$

31.
$$\frac{1}{3}a + \left(\frac{1}{2}b - \frac{2}{3}a\right) = \frac{1}{3}(-3) + \left[\frac{1}{2} \cdot 6 - \frac{2}{3}(-3)\right]$$
$$= \frac{1}{3}(-3) + [3 - (-2)]$$
$$= \frac{1}{3}(-3) + (3 + 2) = -1 + 5 = 4$$

33.
$$\frac{1}{6}b + \frac{1}{3}(c + a) = \frac{1}{6} \cdot 6 + \frac{1}{3}[-2 + (-3)]$$
$$= \frac{1}{6} \cdot 6 + \frac{1}{3}(-5)$$
$$= 1 + \left(-\frac{5}{3}\right) = \frac{3}{3} + \left(-\frac{5}{3}\right) = -\frac{2}{3}$$

35.
$$4a + (3b - c) = 4\left(-\frac{1}{2}\right) + \left[3\left(\frac{3}{4}\right) - \frac{1}{4}\right]$$
$$= 4\left(-\frac{1}{2}\right) + \left[\frac{9}{4} - \frac{1}{4}\right]$$
$$= 4\left(-\frac{1}{2}\right) + \frac{8}{4}$$
$$= -2 + 2 = 0$$

37.
$$2a - b^2 \div c = 2\left(-\frac{1}{2}\right) - \left(\frac{3}{4}\right)^2 \div \frac{1}{4}$$
$$= 2\left(-\frac{1}{2}\right) - \frac{9}{16} \div \frac{1}{4}$$
$$= -1 - \frac{9}{16} \div \frac{1}{4}$$
$$= -1 - \frac{9}{16} \times \frac{4}{1}$$
$$= -1 - \frac{9}{4}$$
$$= -1 + \left(-\frac{9}{4}\right) = -3\frac{1}{4}$$

39.
$$a^2 - b^2 = (3.72)^2 - (-2.31)^2$$
$$= 13.8384 - 5.3361 = 8.5023$$

41.
$$3ac - (c \div a) = 3(3.72)(-1.74) - (-1.74 \div 3.72)$$
$$\approx -18.950658$$

Objective B Exercises

43. $2x^2, 3x, \underline{-4}$

45. $3a^2, -4a, \underline{8}$

47. $\underline{3}x^2, \underline{-4}x$

49. $\underline{1}y^2, \underline{6}a$

51. $16z$

53. $12m - 3m = 12m + (-3)m = 9m$

55. $12at$

57. $3yt$

59. Unlike terms

61. $3t^2 - 5t^2 = 3t^2 + (-5)t^2 = -2t^2$

63. $6c - 5 + 7c = 6c + 7c - 5 = 13c - 5$

65. $2t + 3t - 7t = 2t + 3t + (-7)t = 5t + (-7)t = -2t$

67.
$$7y^2 - 2 - 4y^2 = 7y^2 + (-2) + (-4)y^2$$
$$= 7y^2 + (-4)y^2 + (-2)$$
$$= 3y^2 - 2$$

69.
$$6w - 8u + 8w = 6w + (-8)u + 8w$$
$$= 6w + 8w + (-8)u$$
$$= 14w - 8u$$

71.
$$10 - 11xy - 12xy = 10 + (-11)xy + (-12)xy$$
$$= 10 + (-23)xy$$
$$= 10 - 23xy = -23xy + 10$$

73.
$$3v^2 - 6v^2 - 8v^2 = 3v^2 + (-6)v^2 + (-8)v^2$$
$$= -3v^2 + (-8)v^2$$
$$= -11v^2$$

75.
$$-10ab - 3a + 2ab = -10ab + 2ab - 3a$$
$$= -8ab - 3a$$

77.
$$-3y^2 - y + 7y^2 = -3y^2 + 7y^2 - y$$
$$= 4y^2 - y$$

79.
$$2a - 3b^2 - 5a + b^2 = 2a + (-3)b^2 + (-5)a + b^2$$
$$= 2a + (-5)a + (-3)b^2 + b^2$$
$$= -3a - 2b^2$$

81.
$$3x^2 - 7x + 4x^2 - x = 3x^2 + (-7)x + 4x^2 + (-1)x$$
$$= 3x^2 + 4x^2 + (-7)x + (-1)x$$
$$= 7x^2 - 8x$$

83.
$$6s - t - 9s + 7t = 6s + (-1)t + (-9)s + 7t$$
$$= 6s + (-9)s + (-1)t + 7t$$
$$= -3s + 6t$$

85.
$$4m + 8n - 7m + 2n = 4m + 8n + (-7)m + 2n$$
$$= 4m + (-7)m + 8n + 2n$$
$$= -3m + 10n$$

87.
$$-5ab + 7ac + 10ab - 3ac$$
$$= -5ab + 7ac + 10ab + (-3)ac$$
$$= -5ab + 10ab + 7ac + (-3)ac$$
$$= 5ab + 4ac$$

89. $\frac{4}{9}a^2 - \frac{1}{5}b^2 + \frac{2}{9}a^2 + \frac{4}{5}b^2$

$\quad = \frac{4}{9}a^2 + \left(-\frac{1}{5}\right)b^2 + \frac{2}{9}a^2 + \frac{4}{5}b^2$

$\quad = \frac{4}{9}a^2 + \frac{2}{9}a^2 + \left(-\frac{1}{5}\right)b^2 + \frac{4}{5}b^2$

$\quad = \frac{6}{9}a^2 + \frac{3}{5}b^2$

$\quad = \frac{2}{3}a^2 + \frac{3}{5}b^2$

91. $6.994x$

93. $1.56m - 3.77n$

Objective C Exercises

95. $5(x + 4) = 5x + 5 \cdot 4 = 5x + 20$

97. $(y - 3)4 = [y + (-3)]4$
$\quad = y \cdot 4 + (-3)4$
$\quad = 4y + (-12)$
$\quad = 4y - 12$

99. $-2(a + 4) = -2(a) + (-2)(4)$
$\quad = -2a + (-8)$
$\quad = -2a - 8$

101. $3(5x + 10) = 3(5x) + 3(10) = 15x + 30$

103. $5(3c - 5) = 5[3c + (-5)]$
$\quad = 5(3c) + 5(-5)$
$\quad = 15c + (-25)$
$\quad = 15c - 25$

105. $-3(y - 6) = -3[y + (-6)]$
$\quad = -3y + (-3)(-6)$
$\quad = -3y + 18$

107. $5x + 2(x + 7) = 5x + 2x + 2(7) = 7x + 14$

109. $8y - 4(y + 2) = 8y + (-4)(y + 2)$
$\quad = 8y + (-4)y + (-4)(2)$
$\quad = 8y + (-4y) + (-8)$
$\quad = 4y - 8$

111. $9x - 4(x - 6) = 9x + (-4)[x + (-6)]$
$\quad = 9x + (-4)(x) + (-4)(-6)$
$\quad = 9x + (-4)x + 24$
$\quad = 5x + 24$

113. $-2y + 3(y - 2) = -2y + 3[y + (-2)]$
$\quad = -2y + 3y + 3(-2)$
$\quad = -2y + 3y + (-6)$
$\quad = y - 6$

115. $4n + 2(n + 1) - 5 = 4n + 2(n + 1) + (-5)$
$\quad = 4n + 2n + 2(1) + (-5)$
$\quad = 4n + 2n + 2 + (-5)$
$\quad = 6n - 3$

117. $9y - 3(y - 4) + 8 = 9y + (-3)[y + (-4)] + 8$
$\quad = 9y + (-3)(y) + (-3)(-4) + 8$
$\quad = 9y + (-3)y + 12 + 8$
$\quad = 6y + 20$

119. $3x + 2(x + 2) + 5x = 3x + 2x + 2(2) + 5x$
$\quad = 3x + 2x + 4 + 5x$
$\quad = 3x + 2x + 5x + 4$
$\quad = 5x + 5x + 4$
$\quad = 10x + 4$

121. $-7t + 2(t - 3) - t = -7t + 2[t + (-3)] + (-1)t$
$\quad = -7t + 2t + 2(-3) + (-1)t$
$\quad = -7t + 2t + (-6) + (-1)t$
$\quad = -7t + 2t + (-1)t + (-6)$
$\quad = -5t + (-1)t + (-6)$
$\quad = -6t - 6$

123. $z - 2(1 - z) - 2z = z + (-2)[1 + (-z)] + (-2)z$
$\quad = z + (-2)(1) + (-2)(-z) + (-2)z$
$\quad = z + (-2) + 2z + (-2)z$
$\quad = z + 2z + (-2)z + (-2)$
$\quad = 3z + (-2)z + (-2)$
$\quad = z - 2$

125. $3(y - 2) - 2(y - 6) = 3[y + (-2)] + (-2)[y + (-6)]$
$\quad = 3y + 3(-2) + (-2)y + (-2)(-6)$
$\quad = 3y + (-6) + (-2)y + 12$
$\quad = 3y + (-2)y + (-6) + 12$
$\quad = y + 6$

127. $2(t - 3) + 7(t + 3) = 2[t + (-3)] + 7(t + 3)$
$\quad = 2t + 2(-3) + 7t + 7(3)$
$\quad = 2t + (-6) + 7t + 21$
$\quad = 2t + 7t + (-6) + 21$
$\quad = 9t + 15$

129. $3t - 6(t - 4) + 8t = 3t + (-6)[t + (-4)] + 8t$
$\quad = 3t + (-6)(t) + (-6)(-4) + 8t$
$\quad = 3t + (-6)t + 24 + 8t$
$\quad = 3t + (-6)t + 8t + 24$
$\quad = -3t + 8t + 24$
$\quad = 5t + 24$

Applying the Concepts

131a. $3 + 2x$ `[1][1][1][ x ][ x ]`

    b. $4x + 6$ `[ x ][ x ][ x ][ x ][1][1][1][1][1][1]`

    c. $3x + 2$ `[ x ][ x ][ x ][1][1]`

    d. $2x + 4$ `[ x ][ x ][1][1][1][1]`

    e. $3x + 3$ `[ x ][ x ][ x ][1][1][1]`

    f. $3x + 3$

133a. Answers will vary. Combining like terms is like sorting things out: apples with apples, oranges with oranges.

b. Answers will vary. 4 ft measures distance from one point to the other, whereas 4 square feet measures an area of 2 ft × 2 ft. Therefore, 4 ft and 4 square feet are not to be combined.

Section 11.2

Objective A Exercises

1.
$$\frac{2x + 9 = 3}{\begin{array}{c|c} 2(-3) + 9 & 3 \\ -6 + 9 & 3 \\ 3 = 3 \end{array}}$$
Yes, −3 is a solution.

3.
$$\frac{4 - 2x = 8}{\begin{array}{c|c} 4 - 2(2) & 8 \\ 4 - 4 & 8 \\ 0 \neq 8 \end{array}}$$
No, 2 is not a solution.

5.
$$\frac{3x - 2 = x + 4}{\begin{array}{c|c} 3(3) - 2 & 3 + 4 \\ 9 - 2 & 7 \\ 7 = 7 \end{array}}$$
Yes, 3 is a solution.

7.
$$\frac{x^2 - 5x + 1 = 10 - 5x}{\begin{array}{c|c} 3^2 - 5(3) + 1 & 10 - 5(3) \\ 9 - 15 + 1 & 10 - 15 \\ -5 = -5 \end{array}}$$
Yes, 3 is a solution.

9.
$$\frac{2x(x - 1) = 3 - x}{\begin{array}{c|c} 2(-1)(-1 - 1) & 3 - (-1) \\ -2(-2) & 3 + 1 \\ 4 = 4 \end{array}}$$
Yes, −1 is a solution.

11.
$$\frac{x(x - 2) = x^2 - 4}{\begin{array}{c|c} 2(2 - 2) & 2^2 - 4 \\ 2(0) & 4 - 4 \\ 0 = 0 \end{array}}$$
Yes, 2 is a solution.

13.
$$\frac{3x + 6 = 4}{\begin{array}{c|c} 3\left(-\dfrac{2}{3}\right) + 6 & 4 \\ -2 + 6 & 4 \\ 4 = 4 \end{array}}$$
Yes, $-\dfrac{2}{3}$ is a solution.

15.
$$\frac{2x - 3 = 1 - 14x}{\begin{array}{c|c} 2\left(\dfrac{1}{4}\right) - 3 & 1 - 14\left(\dfrac{1}{4}\right) \\ \dfrac{2}{4} - 3 & 1 - \dfrac{14}{4} \\ \dfrac{1}{2} - 3 & 1 - \dfrac{7}{2} \\ -2\dfrac{1}{2} = -2\dfrac{1}{2} \end{array}}$$
Yes, $\dfrac{1}{4}$ is a solution.

17.
$$\frac{3x(x - 2) = x - 4}{\begin{array}{c|c} 3\left(\dfrac{3}{4}\right)\left(\dfrac{3}{4} - 2\right) & \dfrac{3}{4} - 4 \\ \dfrac{9}{4}\left(-\dfrac{5}{4}\right) & \dfrac{3}{4} - \dfrac{16}{4} \\ -\dfrac{45}{16} \neq -\dfrac{13}{4} \end{array}}$$
No, $\dfrac{3}{4}$ is not a solution.

19.
$$\frac{x^2 - 3x = -0.8776 - x}{\begin{array}{c|c} (1.32)^2 - 3(1.32) & -0.8776 - 1.32 \\ 1.7424 - 3.96 & -2.1976 \\ -2.2176 \neq -2.1976 \end{array}}$$
No, 1.32 is not a solution.

21.
$$\frac{x^2 + 3x = x(x + 3)}{\begin{array}{c|c} (1.05)^2 + 3(1.05) & 1.05(1.05 + 3) \\ 1.1025 + 3.15 & 4.2525 \\ 4.2525 = 4.2525 \end{array}}$$
Yes, 1.05 is a solution.

Objective B Exercises

23.
$$x + 7 = 5$$
$$x + 7 - 7 = 5 - 7$$
$$x + 0 = -2$$
$$x = -2$$

25.
$$z - 4 = 10$$
$$z - 4 + 4 = 10 + 4$$
$$z + 0 = 14$$
$$z = 14$$

27.
$$6 + x = 8$$
$$6 - 6 + x = 8 - 6$$
$$0 + x = 2$$
$$x = 2$$

29.
$$w + 9 = 5$$
$$w + 9 - 9 = 5 - 9$$
$$w + 0 = -4$$
$$w = -4$$

31.
$$m - 4 = -9$$
$$m - 4 + 4 = -9 + 4$$
$$m + 0 = -5$$
$$m = -5$$

33.
$$t - 3 = -3$$
$$t - 3 + 3 = -3 + 3$$
$$t + 0 = 0$$
$$t = 0$$

35.
$$x - 3 = -1$$
$$x - 3 + 3 = -1 + 3$$
$$x + 0 = 2$$
$$x = 2$$

37.
$$3 + y = 0$$
$$3 - 3 + y = 0 - 3$$
$$0 + y = 0 + (-3)$$
$$0 + y = -3$$
$$y = -3$$

39.
$$y - 7 = 3$$
$$y - 7 + 7 = 3 + 7$$
$$y + 0 = 10$$
$$y = 10$$

41.
$$t - 3 = -8$$
$$t - 3 + 3 = -8 + 3$$
$$t + 0 = -5$$
$$t = -5$$

43.
$$z + 6 = -6$$
$$z + 6 - 6 = -6 - 6$$
$$z + 0 = -6 + (-6)$$
$$z + 0 = -12$$
$$z = -12$$

45.
$$x + 2 = -5$$
$$x + 2 - 2 = -5 - 2$$
$$x + 0 = -5 + (-2)$$
$$x = -7$$

47.
$$x - \frac{5}{6} = -\frac{1}{6}$$
$$x - \frac{5}{6} + \frac{5}{6} = -\frac{1}{6} + \frac{5}{6}$$
$$x + 0 = \frac{4}{6}$$
$$x = \frac{4}{6} = \frac{2}{3}$$

49.
$$\frac{2}{5} + x = -\frac{3}{5}$$
$$\frac{2}{5} - \frac{2}{5} + x = -\frac{3}{5} - \frac{2}{5}$$
$$0 + x = -\frac{3}{5} + \left(-\frac{2}{5}\right)$$
$$x = -\frac{5}{5}$$
$$x = -1$$

51.
$$\frac{1}{3} + x = \frac{2}{3}$$
$$\frac{1}{3} - \frac{1}{3} + x = \frac{2}{3} - \frac{1}{3}$$
$$0 + x = \frac{2}{3} + \left(-\frac{1}{3}\right)$$
$$x = \frac{1}{3}$$

53.
$$y + \frac{3}{8} = \frac{1}{4}$$
$$y + \frac{3}{8} - \frac{3}{8} = \frac{1}{4} - \frac{3}{8}$$
$$y + 0 = \frac{1}{4} + \left(-\frac{3}{8}\right)$$
$$y = -\frac{1}{8}$$

55.
$$t + \frac{1}{4} = -\frac{1}{2}$$
$$t + \frac{1}{4} - \frac{1}{4} = -\frac{1}{2} - \frac{1}{4}$$
$$t + 0 = -\frac{1}{2} + \left(-\frac{1}{4}\right)$$
$$t = -\frac{3}{4}$$

57.
$$y + \frac{2}{3} = -\frac{5}{12}$$
$$y + \frac{2}{3} - \frac{2}{3} = -\frac{5}{12} - \frac{2}{3}$$
$$y + 0 = -\frac{5}{12} + \left(-\frac{2}{3}\right)$$
$$y = -\frac{13}{12}$$
$$y = -1\frac{1}{12}$$

Objective C Exercises

59.
$$5x = 30$$
$$\frac{5x}{5} = \frac{30}{5}$$
$$1x = 6$$
$$x = 6$$

61.
$$3z = -27$$
$$\frac{3z}{3} = \frac{-27}{3}$$
$$1z = -9$$
$$z = -9$$

63.
$$-4t = 20$$
$$\frac{-4t}{-4} = \frac{20}{-4}$$
$$1t = -5$$
$$t = -5$$

65.
$$-2y = -28$$
$$\frac{-2y}{-2} = \frac{-28}{-2}$$
$$1y = 14$$
$$y = 14$$

67.
$$24 = 3y$$
$$\frac{24}{3} = \frac{3y}{3}$$
$$8 = 1y$$
$$8 = y$$

69.
$$-21 = 7y$$
$$\frac{-21}{7} = \frac{7y}{7}$$
$$-3 = 1y$$
$$-3 = y$$

71.
$$\frac{y}{2} = 10$$
$$\frac{1}{2}y = 10$$
$$2\left(\frac{1}{2}y\right) = 2(10)$$
$$1y = 20$$
$$y = 20$$

73.
$$\frac{y}{7} = -3$$
$$\frac{1}{7}y = -3$$
$$7\left(\frac{1}{7}y\right) = 7(-3)$$
$$1y = -21$$
$$y = -21$$

75.
$$\frac{-y}{3} = 5$$
$$-\frac{1}{3}y = 5$$
$$-3\left(-\frac{1}{3}y\right) = -3(5)$$
$$1y = -15$$
$$y = -15$$

77.
$$\frac{5}{8}x = 10$$
$$\frac{8}{5}\left(\frac{5}{8}x\right) = \frac{8}{5}(10)$$
$$1x = 16$$
$$x = 16$$

79.
$$\frac{2}{7}x = -12$$
$$\frac{7}{2}\left(\frac{2}{7}x\right) = \frac{7}{2}(-12)$$
$$1x = -42$$
$$x = -42$$

81.
$$-\frac{1}{5}y = -3$$
$$-5\left(-\frac{1}{5}y\right) = -5(-3)$$
$$1y = 15$$
$$y = 15$$

83.
$$\frac{5}{12}y = -16$$
$$\frac{12}{5}\left(\frac{5}{12}y\right) = \frac{12}{5}(-16)$$
$$1y = -\frac{192}{5}$$
$$y = -38\frac{2}{5}$$

85.
$$-8 = -\frac{5}{6}x$$
$$\left(-\frac{6}{5}\right)(-8) = \left(-\frac{6}{5}\right)\left(-\frac{5}{6}\right)x$$
$$\frac{48}{5} = 1x$$
$$9\frac{3}{5} = x$$

87.
$$-9 = \frac{5}{6}t$$
$$\frac{6}{5}(-9) = \left(\frac{6}{5}\right)\left(\frac{5}{6}\right)t$$
$$-\frac{54}{5} = 1t$$
$$-10\frac{4}{5} = t$$

89.
$$\frac{3}{7}y = \frac{5}{6}$$
$$\frac{7}{3}\left(\frac{3}{7}\right)y = \frac{7}{3}\left(\frac{5}{6}\right)$$
$$1y = \frac{35}{18}$$
$$y = 1\frac{17}{18}$$

91.
$$3a - 6a = 8$$
$$-3a = 8$$
$$\frac{-3a}{-3} = \frac{8}{-3}$$
$$1a = -\frac{8}{3}$$
$$a = -2\frac{2}{3}$$

93.
$$\frac{1}{3}b - \frac{2}{3}b = -1$$
$$-\frac{1}{3}b = -1$$
$$(-3)\left(-\frac{1}{3}b\right) = (-3)(-1)$$
$$1b = 3$$
$$b = 3$$

Objective D Exercises

95. Strategy To find the value of the original investment, replace the variables A and I by the given values and solve for P.

Solution
$$A = P + I$$
$$26{,}440 = P + 2830$$
$$26{,}440 - 2830 = P + 2830 - 2830$$
$$26{,}440 + (-2830) = P + 0$$
$$23{,}610 = P$$
The original investment was \$23,610.

97. Strategy To find the increase in the value of the investment, replace the variables A and P by the given values and solve for I.

Solution
$$A = P + I$$
$$8690 = 7500 + I$$
$$8690 - 7500 = 7500 - 7500 + I$$
$$8690 + (-7500) = 0 + I$$
$$1190 = I$$
The value of the fund increased by \$1190.

99. Strategy To find the number of gallons of gasoline used, replace the variables D and M in the formula by the given values and solve for G.

Solution
$$D = M \cdot G$$
$$592 = 32 \cdot G$$
$$\frac{592}{32} = \frac{32G}{32}$$
$$18.5 = G$$
18.5 gal of gasoline was used.

101. Strategy To find the number of miles per gallon, replace the variables D and G in the formula by the given values and solve for M.

Solution
$$D = M \cdot G$$
$$410 = M \cdot 12$$
$$\frac{410}{12} = \frac{12M}{12}$$
$$34.2 \approx M$$
The car gets 34.2 mi/gal.

103. **Strategy** To find the markup, replace the variables S and C in the formula by the given values and solve for M.

Solution
$$S = C + M$$
$$39.80 = 23.50 + M$$
$$39.80 - 23.50 = 23.50 - 23.50 + M$$
$$39.80 + (-23.50) = 0 + M$$
$$16.30 = M$$

The markup on each stuffed animal is $16.30.

105. **Strategy** To find the cost of a compact disc, replace the variables S and R in the formula by the given values and solve for C.

Solution
$$S = C + RC$$
$$18.85 = C + 0.30C$$
$$18.85 = 1.30C$$
$$\frac{18.85}{1.30} = \frac{1.30C}{1.30}$$
$$14.50 = C$$

The compact disc costs $14.50.

Applying the Concepts

107. $\frac{3}{4}x = 6$

$\frac{4}{3}\left(\frac{3}{4}x\right) = \frac{4}{3}(6)$ Multiplication Property of Equations

$1x = 8$ Multiplication Property of Reciprocals

$x = 8$ Multiplication Property of One

109. Answers will vary. For example, $x + 5 = 1$.

111a. Students should rephrase the Addition Property of Equations (the same number or variable expression can be added to each side of an equation without changing the solution of the equation).

b. Students should rephrase the Multiplication Property of Equations (each side of an equation can be multiplied by the same nonzero number without changing the solution of the equation).

Section 11.3

Objective A Exercises

1.
$$3x + 5 = 14$$
$$3x + 5 - 5 = 14 - 5$$
$$3x = 9$$
$$\frac{3x}{3} = \frac{9}{3}$$
$$x = 3$$

3.
$$2n - 3 = 7$$
$$2n - 3 + 3 = 7 + 3$$
$$2n = 10$$
$$\frac{2n}{2} = \frac{10}{2}$$
$$n = 5$$

5.
$$5w + 8 = 3$$
$$5w + 8 - 8 = 3 - 8$$
$$5w = -5$$
$$\frac{5w}{5} = \frac{-5}{5}$$
$$w = -1$$

7.
$$3z - 4 = -16$$
$$3z - 4 + 4 = -16 + 4$$
$$3z = -12$$
$$\frac{3z}{3} = \frac{-12}{3}$$
$$z = -4$$

9.
$$5 + 2x = 7$$
$$5 - 5 + 2x = 7 - 5$$
$$2x = 2$$
$$\frac{2x}{2} = \frac{2}{2}$$
$$x = 1$$

11.
$$6 - x = 3$$
$$6 + (-1)x = 3$$
$$6 - 6 + (-1)x = 3 - 6$$
$$(-1)x = -3$$
$$(-1)(-1)x = (-1)(-3)$$
$$x = 3$$

13.
$$3 - 4x = 11$$
$$3 - 3 - 4x = 11 - 3$$
$$-4x = 8$$
$$\frac{-4x}{-4} = \frac{8}{-4}$$
$$x = -2$$

15.
$$5 - 4x = 17$$
$$5 - 5 - 4x = 17 - 5$$
$$-4x = 12$$
$$\frac{-4x}{-4} = \frac{12}{-4}$$
$$x = -3$$

17.
$$3x + 6 = 0$$
$$3x + 6 - 6 = 0 - 6$$
$$3x = -6$$
$$\frac{3x}{3} = \frac{-6}{3}$$
$$x = -2$$

19.
$$-3x - 4 = -1$$
$$-3x - 4 + 4 = -1 + 4$$
$$-3x = 3$$
$$\frac{-3x}{-3} = \frac{3}{-3}$$
$$x = -1$$

21.
$$12y - 30 = 6$$
$$12y - 30 + 30 = 6 + 30$$
$$12y = 36$$
$$\frac{12y}{12} = \frac{36}{12}$$
$$y = 3$$

23.
$$3c + 7 = 4$$
$$3c + 7 - 7 = 4 - 7$$
$$3c = -3$$
$$\frac{3c}{3} = \frac{-3}{3}$$
$$c = -1$$

25.
$$-2x + 11 = -3$$
$$-2x + 11 - 11 = -3 - 11$$
$$-2x = -14$$
$$\frac{-2x}{-2} = \frac{-14}{-2}$$
$$x = 7$$

27.
$$14 - 5x = 4$$
$$14 - 14 - 5x = 4 - 14$$
$$-5x = -10$$
$$\frac{-5x}{-5} = \frac{-10}{-5}$$
$$x = 2$$

29.
$$-8x + 7 = -9$$
$$-8x + 7 - 7 = -9 - 7$$
$$-8x = -16$$
$$\frac{-8x}{-8} = \frac{-16}{-8}$$
$$x = 2$$

31.
$$9x + 13 = 13$$
$$9x + 13 - 13 = 13 - 13$$
$$9x = 0$$
$$\frac{9x}{9} = \frac{0}{9}$$
$$x = 0$$

33.
$$7d - 14 = 0$$
$$7d - 14 + 14 = 0 + 14$$
$$7d = 14$$
$$\frac{7d}{7} = \frac{14}{7}$$
$$d = 2$$

35.
$$4n - 4 = -4$$
$$4n - 4 + 4 = -4 + 4$$
$$4n = 0$$
$$\frac{4n}{4} = \frac{0}{4}$$
$$n = 0$$

37.
$$3x + 5 = 7$$
$$3x + 5 - 5 = 7 - 5$$
$$3x = 2$$
$$\frac{3x}{3} = \frac{2}{3}$$
$$x = \frac{2}{3}$$

39.
$$6x - 1 = 16$$
$$6x - 1 + 1 = 16 + 1$$
$$6x = 17$$
$$\frac{6x}{6} = \frac{17}{6}$$
$$x = 2\frac{5}{6}$$

41.
$$2x - 3 = -8$$
$$2x - 3 + 3 = -8 + 3$$
$$2x = -5$$
$$\frac{2x}{2} = \frac{-5}{2}$$
$$x = -2\frac{1}{2}$$

43.
$$-6x + 2 = -7$$
$$-6x + 2 - 2 = -7 - 2$$
$$-6x = -9$$
$$\frac{-6x}{-6} = \frac{-9}{-6}$$
$$x = 1\frac{1}{2}$$

45.
$$-2x - 3 = -7$$
$$-2x - 3 + 3 = -7 + 3$$
$$-2x = -4$$
$$\frac{-2x}{-2} = \frac{-4}{-2}$$
$$x = 2$$

47.
$$3x + 8 = 2$$
$$3x + 8 - 8 = 2 - 8$$
$$3x = -6$$
$$\frac{3x}{3} = \frac{-6}{3}$$
$$x = -2$$

49.
$$3w - 7 = 0$$
$$3w - 7 + 7 = 0 + 7$$
$$3w = 7$$
$$\frac{3w}{3} = \frac{7}{3}$$
$$w = 2\frac{1}{3}$$

51.
$$-2d + 9 = 12$$
$$-2d + 9 - 9 = 12 - 9$$
$$-2d = 3$$
$$\frac{-2d}{-2} = \frac{3}{-2}$$
$$d = -1\frac{1}{2}$$

53.
$$\frac{1}{2}x - 2 = 3$$
$$\frac{1}{2}x - 2 + 2 = 3 + 2$$
$$\frac{1}{2}x = 5$$
$$2\left(\frac{1}{2}x\right) = 5 \cdot 2$$
$$x = 10$$

55.
$$\frac{3}{5}w - 1 = 2$$
$$\frac{3}{5}w - 1 + 1 = 2 + 1$$
$$\frac{3}{5}w = 3$$
$$\frac{5}{3} \cdot \frac{3}{5}w = 3 \cdot \frac{5}{3}$$
$$w = 5$$

57.
$$\frac{2}{9}t - 3 = 5$$
$$\frac{2}{9}t - 3 + 3 = 5 + 3$$
$$\frac{2}{9}t = 8$$
$$\frac{9}{2} \cdot \frac{2}{9}t = 8 \cdot \frac{9}{2}$$
$$t = 36$$

59.
$$\frac{y}{3} - 6 = -8$$
$$\frac{y}{3} - 6 + 6 = -8 + 6$$
$$\frac{y}{3} = -2$$
$$3 \cdot \frac{y}{3} = 3(-2)$$
$$y = -6$$

61.
$$\frac{x}{3} - 2 = -5$$
$$\frac{x}{3} - 2 + 2 = -5 + 2$$
$$\frac{x}{3} = -3$$
$$3 \cdot \frac{x}{3} = 3(-3)$$
$$x = -9$$

63.
$$\frac{5}{8}v + 6 = 3$$
$$\frac{5}{8}v + 6 - 6 = 3 - 6$$
$$\frac{5}{8}v = -3$$
$$\frac{8}{5} \cdot \frac{5}{8}v = \frac{8}{5} \cdot (-3)$$
$$v = -\frac{24}{5} = -4\frac{4}{5}$$

65.
$$\frac{4}{7}z + 10 = 5$$
$$\frac{4}{7}z + 10 - 10 = 5 - 10$$
$$\frac{4}{7}z = -5$$
$$\frac{7}{4} \cdot \frac{4}{7}z = \frac{7}{4} \cdot (-5)$$
$$z = -\frac{35}{4} = -8\frac{3}{4}$$

67.
$$\frac{2}{9}x - 3 = 5$$
$$\frac{2}{9}x - 3 + 3 = 5 + 3$$
$$\frac{2}{9}x = 8$$
$$\frac{9}{2} \cdot \frac{2}{9}x = \frac{9}{2} \cdot 8$$
$$x = 36$$

69.
$$\frac{3}{4}x - 5 = -4$$
$$\frac{3}{4}x - 5 + 5 = -4 + 5$$
$$\frac{3}{4}x = 1$$
$$\frac{4}{3} \cdot \frac{3}{4}x = \frac{4}{3} \cdot 1$$
$$x = \frac{4}{3} = 1\frac{1}{3}$$

71.
$$1.5x - 0.5 = 2.5$$
$$1.5x - 0.5 + 0.5 = 2.5 + 0.5$$
$$1.5x = 3$$
$$\frac{1.5x}{1.5} = \frac{3}{1.5}$$
$$x = 2$$

73.
$$0.8t + 1.1 = 4.3$$
$$0.8t + 1.1 - 1.1 = 4.3 - 1.1$$
$$0.8t = 3.2$$
$$\frac{0.8t}{0.8} = \frac{3.2}{0.8}$$
$$t = 4$$

75.
$$0.4x - 2.3 = 1.3$$
$$0.4x - 2.3 + 2.3 = 1.3 + 2.3$$
$$0.4x = 3.6$$
$$\frac{0.4x}{0.4} = \frac{3.6}{0.4}$$
$$x = 9$$

77.
$$3.5y - 3.5 = 10.5$$
$$3.5y - 3.5 + 3.5 = 10.5 + 3.5$$
$$3.5y = 14$$
$$\frac{3.5y}{3.5} = \frac{14}{3.5}$$
$$y = 4$$

79.
$$0.32x + 4.2 = 3.2$$
$$0.32x + 4.2 - 4.2 = 3.2 - 4.2$$
$$0.32x = -1$$
$$\frac{0.32x}{0.32} = \frac{-1}{0.32}$$
$$x = -3.125$$

81.
$$6m + 2m - 2 = 5$$
$$8m - 3 = 5$$
$$8m - 3 + 3 = 5 + 3$$
$$8m = 8$$
$$\frac{8m}{8} = \frac{8}{8}$$
$$m = 1$$

83.
$$3y - 8y - 9 = 6$$
$$-5y - 9 = 6$$
$$-5y - 9 + 9 = 6 + 9$$
$$-5y = 15$$
$$\frac{-5y}{-5} = \frac{15}{-5}$$
$$y = -3$$

85.
$$-2y + y - 3 = 6$$
$$-y - 3 = 6$$
$$-y - 3 + 3 = 6 + 3$$
$$-y = 9$$
$$(-1)(-y) = (-1)9$$
$$y = -9$$

87.
$$0.032x - 0.0194 = 0.139$$
$$0.032x - 0.0194 + 0.0194 = 0.139 + 0.0194$$
$$0.032x = 0.1584$$
$$\frac{0.032x}{0.032} = \frac{0.1584}{0.032}$$
$$x = 4.95$$

89.
$$6.09x + 17.33 = 16.805$$
$$6.09x + 17.33 - 17.33 = 16.805 - 17.33$$
$$6.09x = -0.525$$
$$\frac{6.09x}{6.09} = \frac{-0.525}{6.09}$$
$$x \approx -0.0862069$$

Objective B Exercises

91. Strategy To find the Celsius temperature, replace the variable F in the formula by the given value and solve for C.

Solution
$$F = 1.8C + 32$$
$$-40 = 1.8C + 32$$
$$-40 - 32 = 1.8C + 32 - 32$$
$$-72 = 1.8C$$
$$\frac{-72}{1.8} = \frac{1.8C}{1.8}$$
$$-40 = C$$

The temperature is $-40°C$.

93. Strategy To find the time required, replace the variables V and V_0 in the formula by the given values and solve for t.

Solution
$$V = V_0 + 32t$$
$$472 = 8 + 32t$$
$$472 - 8 = 8 - 8 + 32t$$
$$464 = 32t$$
$$\frac{464}{32} = \frac{32t}{32}$$
$$14.5 = t$$

The time is 14.5 s.

95. Strategy To find the number of units made, replace the variables T, U, and F in the formula by the given values and solve for N.

Solution
$$T = U \cdot N + F$$
$$25,000 = 8 \cdot N + 5000$$
$$25,000 - 5000 = 8N + 5000 - 5000$$
$$20,000 = 8N$$
$$\frac{20,000}{8} = \frac{8N}{8}$$
$$2500 = N$$

2500 units were made.

97. Strategy To find the monthly income, replace the variables T, R, and B in the formula by the given values and solve for I.

Solution
$$T = I \cdot R + B$$
$$476 = I \cdot 0.22 + 80$$
$$476 - 80 = 0.22I + 80 - 80$$
$$396 = 0.22I$$
$$\frac{396}{0.22} = \frac{0.22I}{0.22}$$
$$1800 = I$$

The clerk's monthly income is $1800.

99. Strategy To find the total sales, replace the variables M, R, and B in the formula by the given values and solve for S.

Solution
$$M = S \cdot R + B$$
$$3480 = S \cdot 0.09 + 600$$
$$3480 - 600 = 0.09S + 600 - 600$$
$$2880 = 0.09S$$
$$\frac{2880}{0.09} = \frac{0.09S}{0.09}$$
$$32,000 = S$$

The total sales were $32,000.

101. Strategy To find the commission rate, replace the variables M, S, and B in the formula by the given values and solve for R.

Solution
$$M = S \cdot R + B$$
$$2640 = 42,000R + 750$$
$$2640 - 750 = 42,000R + 750 - 750$$
$$1890 = 42,000R$$
$$\frac{1890}{42,000} = \frac{42,000R}{42,000}$$
$$0.045 = R$$
$$4.5\% = R$$

Miguel's commission rate was 4.5%.

Applying the Concepts

103.
$$\frac{2}{3}x - 4 = 10$$

$$\frac{2}{3}x - 4 + 4 = 10 + 4 \quad \text{Addition Property of Equations}$$

$$\frac{2}{3}x + 0 = 14 \quad \text{Addition Property of Zero}$$

$$\frac{2}{3}x = 14$$

$$\frac{3}{2}\left(\frac{2}{3}x\right) = \frac{3}{2}(14) \quad \text{Multiplication Property of Equations}$$

$$1x = 21 \quad \text{Multiplication Property of Reciprocals}$$

$$x = 21 \quad \text{Multiplication Property of One}$$

105. No, the sentence "Solve $3x + 4(x - 3)$" does not make sense because $3x + 4(x - 3)$ is an expression, and you cannot solve an expression. You can solve an equation.

Section 11.4

Objective A Exercises

1.
$$6x + 3 = 2x + 5$$
$$6x - 2x + 3 = 2x - 2x + 5$$
$$4x + 3 = 5$$
$$4x + 3 - 3 = 5 - 3$$
$$4x = 2$$
$$\frac{4x}{4} = \frac{2}{4}$$
$$x = \frac{1}{2}$$

3.
$$3x + 3 = 2x + 2$$
$$3x - 2x + 3 = 2x - 2x + 2$$
$$x + 3 = 2$$
$$x + 3 - 3 = 2 - 3$$
$$x = -1$$

5.
$$5x + 4 = x - 12$$
$$5x - x + 4 = x - x - 12$$
$$4x + 4 = -12$$
$$4x + 4 - 4 = -12 - 4$$
$$4x = -16$$
$$\frac{4x}{4} = \frac{-16}{4}$$
$$x = -4$$

7.
$$7b - 2 = 3b - 6$$
$$7b - 3b - 2 = 3b - 3b - 6$$
$$4b - 2 = -6$$
$$4b - 2 + 2 = -6 + 2$$
$$4b = -4$$
$$\frac{4b}{4} = \frac{-4}{4}$$
$$b = -1$$

9.
$$9n - 4 = 5n - 20$$
$$9n - 5n - 4 = 5n - 5n - 20$$
$$4n - 4 = -20$$
$$4n - 4 + 4 = -20 + 4$$
$$4n = -16$$
$$\frac{4n}{4} = \frac{-16}{4}$$
$$n = -4$$

11.
$$2x + 1 = 16 - 3x$$
$$2x + 3x + 1 = 16 - 3x + 3x$$
$$5x + 1 = 16$$
$$5x + 1 - 1 = 16 - 1$$
$$5x = 15$$
$$\frac{5x}{5} = \frac{15}{5}$$
$$x = 3$$

13.
$$5x - 2 = -10 - 3x$$
$$5x + 3x - 2 = -10 - 3x + 3x$$
$$8x - 2 = -10$$
$$8x - 2 + 2 = -10 + 2$$
$$8x = -8$$
$$\frac{8x}{8} = \frac{-8}{8}$$
$$x = -1$$

15.
$$2x + 7 = 4x + 3$$
$$2x - 4x + 7 = 4x - 4x + 3$$
$$-2x + 7 = 3$$
$$-2x + 7 - 7 = 3 - 7$$
$$-2x = -4$$
$$\frac{-2x}{-2} = \frac{-4}{-2}$$
$$x = 2$$

17.
$$c + 4 = 6c - 11$$
$$c - 6c + 4 = 6c - 6c - 11$$
$$-5c + 4 = -11$$
$$-5c + 4 - 4 = -11 - 4$$
$$-5c = -15$$
$$\frac{-5c}{-5} = \frac{-15}{-5}$$
$$c = 3$$

19.
$$3x - 7 = x - 7$$
$$3x - x - 7 = x - x - 7$$
$$2x - 7 = -7$$
$$2x - 7 + 7 = -7 + 7$$
$$2x = 0$$
$$\frac{2x}{2} = \frac{0}{2}$$
$$x = 0$$

21.
$$3 - 4x = 5 - 3x$$
$$3 - 4x + 3x = 5 - 3x + 3x$$
$$3 - x = 5$$
$$3 - 3 - x = 5 - 3$$
$$-x = 2$$
$$(-1)(-x) = (-1)2$$
$$x = -2$$

23.
$$7 + 3x = 9 + 5x$$
$$7 + 3x - 5x = 9 + 5x - 5x$$
$$7 - 2x = 9$$
$$7 - 7 - 2x = 9 - 7$$
$$-2x = 2$$
$$\frac{-2x}{-2} = \frac{2}{-2}$$
$$x = -1$$

25.
$$5 + 2y = 7 + 5y$$
$$5 + 2y - 5y = 7 + 5y - 5y$$
$$5 - 3y = 7$$
$$5 - 5 - 3y = 7 - 5$$
$$-3y = 2$$
$$\frac{-3y}{-3} = \frac{2}{-3}$$
$$y = -\frac{2}{3}$$

27.
$$8 - 5w = 4 - 6w$$
$$8 - 5w + 6w = 4 - 6w + 6w$$
$$8 + w = 4$$
$$8 - 8 + w = 4 - 8$$
$$w = -4$$

29.
$$6x + 1 = 3x + 2$$
$$6x - 3x + 1 = 3x - 3x + 2$$
$$3x + 1 = 2$$
$$3x + 1 - 1 = 2 - 1$$
$$3x = 1$$
$$\frac{3x}{3} = \frac{1}{3}$$
$$x = \frac{1}{3}$$

31.
$$5x + 8 = x + 5$$
$$5x - x + 8 = x - x + 5$$
$$4x + 8 = 5$$
$$4x + 8 - 8 = 5 - 8$$
$$4x = -3$$
$$\frac{4x}{4} = \frac{-3}{4}$$
$$x = -\frac{3}{4}$$

33.
$$2x - 3 = 6x - 4$$
$$2x - 6x - 3 = 6x - 6x - 4$$
$$-4x - 3 = -4$$
$$-4x - 3 + 3 = -4 + 3$$
$$-4x = -1$$
$$\frac{-4x}{-4} = \frac{-1}{-4}$$
$$x = \frac{1}{4}$$

35.
$$6 - 3x = 6 - 5x$$
$$6 - 3x + 5x = 6 - 5x + 5x$$
$$6 + 2x = 6$$
$$6 - 6 + 2x = 6 - 6$$
$$2x = 0$$
$$\frac{2x}{2} = \frac{0}{2}$$
$$x = 0$$

37.
$$6x - 2 = 2x - 9$$
$$6x - 2x - 2 = 2x - 2x - 9$$
$$4x - 2 = -9$$
$$4x - 2 + 2 = -9 + 2$$
$$4x = -7$$
$$\frac{4x}{4} = \frac{-7}{4}$$
$$x = -\frac{7}{4}$$
$$x = -1\frac{3}{4}$$

39.
$$6x - 3 = -5x + 8$$
$$6x + 5x - 3 = -5x + 5x + 8$$
$$11x - 3 = 8$$
$$11x - 3 + 3 = 8 + 3$$
$$11x = 11$$
$$\frac{11x}{11} = \frac{11}{11}$$
$$x = 1$$

41.
$$-6t - 2 = -8t - 4$$
$$-6t + 8t - 2 = -8t + 8t - 4$$
$$2t - 2 = -4$$
$$2t - 2 + 2 = -4 + 2$$
$$2t = -2$$
$$\frac{2t}{2} = \frac{-2}{2}$$
$$t = -1$$

43.
$$-3 - 4x = 7 - 2x$$
$$-3 - 4x + 2x = 7 - 2x + 2x$$
$$-3 - 2x = 7$$
$$-3 + 3 - 2x = 7 + 3$$
$$-2x = 10$$
$$\frac{-2x}{-2} = \frac{10}{-2}$$
$$x = -5$$

45.
$$3 - 7x = -2 + 5x$$
$$3 - 7x - 5x = -2 + 5x - 5x$$
$$3 - 12x = -2$$
$$3 - 3 - 12x = -2 - 3$$
$$-12x = -5$$
$$\frac{-12x}{-12} = \frac{-5}{-12}$$
$$x = \frac{5}{12}$$

47.
$$5x + 8 = 4 - 2x$$
$$5x + 2x + 8 = 4 - 2x + 2x$$
$$7x + 8 = 4$$
$$7x + 8 - 8 = 4 - 8$$
$$7x = -4$$
$$\frac{7x}{7} = \frac{-4}{7}$$
$$x = -\frac{4}{7}$$

49.
$$12z - 9 = 3z + 12$$
$$12z - 3z - 9 = 3z - 3z + 12$$
$$9z - 9 = 12$$
$$9z - 9 + 9 = 12 + 9$$
$$9z = 21$$
$$\frac{9z}{9} = \frac{21}{9}$$
$$z = \frac{7}{3}$$
$$z = 2\frac{1}{3}$$

51.
$$\frac{5}{7}m - 3 = \frac{2}{7}m + 6$$
$$\frac{5}{7}m - \frac{2}{7}m - 3 = \frac{2}{7}m - \frac{2}{7}m + 6$$
$$\frac{3}{7}m - 3 = 6$$
$$\frac{3}{7}m - 3 + 3 = 6 + 3$$
$$\frac{3}{7}m = 9$$
$$\frac{7}{3} \cdot \frac{3}{7}m = \frac{7}{3} \cdot 9$$
$$m = 21$$

53.
$$\frac{3}{7}x + 5 = \frac{5}{7}x - 1$$
$$\frac{3}{7}x - \frac{5}{7}x + 5 = \frac{5}{7}x - \frac{5}{7}x - 1$$
$$\frac{-2}{7}x + 5 = -1$$
$$\frac{-2}{7}x + 5 - 5 = -1 - 5$$
$$\frac{-2}{7}x = -6$$
$$\left(\frac{-7}{2}\right)\left(\frac{-2}{7}x\right) = \left(\frac{-7}{2}\right)(-6)$$
$$x = 21$$

Objective B Exercises

55.
$$6x + 2(x - 1) = 14$$
$$6x + 2x - 2 = 14$$
$$8x - 2 = 14$$
$$8x - 2 + 2 = 14 + 2$$
$$8x = 16$$
$$\frac{8x}{8} = \frac{16}{8}$$
$$x = 2$$

57.
$$-3 + 4(x + 3) = 5$$
$$-3 + 4x + 12 = 5$$
$$4x + 9 = 5$$
$$4x + 9 - 9 = 5 - 9$$
$$4x = -4$$
$$\frac{4x}{4} = \frac{-4}{4}$$
$$x = -1$$

59.
$$6 - 2(d + 4) = 6$$
$$6 - 2d - 8 = 6$$
$$-2d - 2 = 6$$
$$-2d - 2 + 2 = 6 + 2$$
$$-2d = 8$$
$$\frac{-2d}{-2} = \frac{8}{-2}$$
$$d = -4$$

61.
$$5 + 7(x + 3) = 20$$
$$5 + 7x + 21 = 20$$
$$7x + 26 = 20$$
$$7x + 26 - 26 = 20 - 26$$
$$7x = -6$$
$$\frac{7x}{7} = \frac{-6}{7}$$
$$x = -\frac{6}{7}$$

63.
$$2x + 3(x - 5) = 10$$
$$2x + 3x - 15 = 10$$
$$5x - 15 = 10$$
$$5x - 15 + 15 = 10 + 15$$
$$5x = 25$$
$$\frac{5x}{5} = \frac{25}{5}$$
$$x = 5$$

65.
$$3(x - 4) + 2x = 3$$
$$3x - 12 + 2x = 3$$
$$5x - 12 = 3$$
$$5x - 12 + 12 = 3 + 12$$
$$5x = 15$$
$$\frac{5x}{5} = \frac{15}{5}$$
$$x = 3$$

67.
$$2x - 3(x - 4) = 12$$
$$2x - 3x + 12 = 12$$
$$-x + 12 = 12$$
$$-x + 12 - 12 = 12 - 12$$
$$-x = 0$$
$$(-1)(-x) = (-1)0$$
$$x = 0$$

69.
$$2x + 3(x + 4) = 7$$
$$2x + 3x + 12 = 7$$
$$5x + 12 = 7$$
$$5x + 12 - 12 = 7 - 12$$
$$5x = -5$$
$$\frac{5x}{5} = \frac{-5}{5}$$
$$x = -1$$

71.
$$3(x - 2) + 5 = 5$$
$$3x - 6 + 5 = 5$$
$$3x - 1 = 5$$
$$3x - 1 + 1 = 5 + 1$$
$$3x = 6$$
$$\frac{3x}{3} = \frac{6}{3}$$
$$x = 2$$

73.
$$3y + 7(y - 2) = 5$$
$$3y + 7y - 14 = 5$$
$$10y - 14 = 5$$
$$10y - 14 + 14 = 5 + 14$$
$$10y = 19$$
$$\frac{10y}{10} = \frac{19}{10}$$
$$y = 1\frac{9}{10}$$

75. $4b - 2(b + 9) = 8$
$4b - 2b - 18 = 8$
$2b - 18 = 8$
$2b - 18 + 18 = 8 + 18$
$2b = 26$
$\dfrac{2b}{2} = \dfrac{26}{2}$
$b = 13$

77. $3x + 5(x - 2) = 10$
$3x + 5x - 10 = 10$
$8x - 10 = 10$
$8x - 10 + 10 = 10 + 10$
$8x = 20$
$\dfrac{8x}{8} = \dfrac{20}{8}$
$x = \dfrac{5}{2} = 2\dfrac{1}{2}$

79. $3x + 4(x + 2) = 2(x + 9)$
$3x + 4x + 8 = 2x + 18$
$7x + 8 = 2x + 18$
$7x - 2x + 8 = 2x - 2x + 18$
$5x + 8 = 18$
$5x + 8 - 8 = 18 - 8$
$5x = 10$
$\dfrac{5x}{5} = \dfrac{10}{5}$
$x = 2$

81. $2d - 3(d - 4) = 2(d + 6)$
$2d - 3d + 12 = 2d + 12$
$-d + 12 = 2d + 12$
$-d - 2d + 12 = 2d - 2d + 12$
$-3d + 12 = 12$
$-3d + 12 - 12 = 12 - 12$
$-3d = 0$
$\dfrac{-3d}{-3} = \dfrac{0}{-3}$
$d = 0$

83. $7 - 2(x - 3) = 3(x - 1)$
$7 - 2x + 6 = 3x - 3$
$-2x + 13 = 3x - 3$
$-2x - 3x + 13 = 3x - 3x - 3$
$-5x + 13 = -3$
$-5x + 13 - 13 = -3 - 13$
$-5x = -16$
$\dfrac{-5x}{-5} = \dfrac{-16}{-5}$
$x = 3\dfrac{1}{5}$

85. $6x - 2(x - 3) = 11(x - 2)$
$6x - 2x + 6 = 11x - 22$
$4x + 6 = 11x - 22$
$4x - 11x + 6 = 11x - 11x - 22$
$-7x + 6 = -22$
$-7x + 6 - 6 = -22 - 6$
$-7x = -28$
$\dfrac{-7x}{-7} = \dfrac{-28}{-7}$
$x = 4$

87. $6c - 3(c + 1) = 5(c + 2)$
$6c - 3c - 3 = 5c + 10$
$3c - 3 = 5c + 10$
$3c - 5c - 3 = 5c - 5c + 10$
$-2c - 3 = 10$
$-2c - 3 + 3 = 10 + 3$
$-2c = 13$
$\dfrac{-2c}{-2} = \dfrac{13}{-2}$
$c = -6\dfrac{1}{2}$

89. $7 - (x + 1) = 3(x + 3)$
$7 - x - 1 = 3x + 9$
$-x + 6 = 3x + 9$
$-x - 3x + 6 = 3x - 3x + 9$
$-4x + 6 = 9$
$-4x + 6 - 6 = 9 - 6$
$-4x = 3$
$\dfrac{-4x}{-4} = \dfrac{3}{-4}$
$x = -\dfrac{3}{4}$

91. $2x - 3(x + 4) = 2(x - 5)$
$2x - 3x - 12 = 2x - 10$
$-x - 12 = 2x - 10$
$-x - 2x - 12 = 2x - 2x - 10$
$-3x - 12 = -10$
$-3x - 12 + 12 = -10 + 12$
$-3x = 2$
$\dfrac{-3x}{-3} = \dfrac{2}{-3}$
$x = -\dfrac{2}{3}$

93. $x + 5(x - 4) = 3(x - 8) - 5$
$x + 5x - 20 = 3x - 24 - 5$
$6x - 20 = 3x - 29$
$6x - 3x - 20 = 3x - 3x - 29$
$3x - 20 = -29$
$3x - 20 + 20 = -29 + 20$
$3x = -9$
$\dfrac{3x}{3} = \dfrac{-9}{3}$
$x = -3$

95. $9b - 3(b - 4) = 13 + 2(b - 3)$
$9b - 3b + 12 = 13 + 2b - 6$
$6b + 12 = 2b + 7$
$6b - 2b + 12 = 2b - 2b + 7$
$4b + 12 = 7$
$4b + 12 - 12 = 7 - 12$
$4b = -5$
$\dfrac{4b}{4} = \dfrac{-5}{4}$
$b = -1\dfrac{1}{4}$

97. $3(x - 4) + 3x = 7 - 2(x - 1)$
$3x - 12 + 3x = 7 - 2x + 2$
$6x - 12 = -2x + 9$
$6x + 2x - 12 = -2x + 2x + 9$
$8x - 12 = 9$
$8x - 12 + 12 = 9 + 12$
$8x = 21$
$\dfrac{8x}{8} = \dfrac{21}{8}$
$x = 2\dfrac{5}{8}$

99. $3.67x - 5.3(x - 1.932) = 6.99$
$3.67x - 5.3x + 10.2396 = 6.99$
$-1.63x + 10.2396 = 6.99$
$-1.63x + 10.2396 - 10.2396 = 6.99 - 10.2396$
$-1.63x = -3.2496$
$\dfrac{-1.63x}{-1.63} = \dfrac{-3.2496}{-1.63}$
$x \approx 1.9936196$

101. $8.45(z - 10) = 3(z - 3.854)$
$8.45z - 84.5 = 3z - 11.562$
$8.45z - 3z - 84.5 = 3z - 3z - 11.562$
$5.45z - 84.5 = -11.562$
$5.45z - 84.5 + 84.5 = -11.562 + 84.5$
$5.45z = 72.938$
$\dfrac{5.45z}{5.45} = \dfrac{72.938}{5.45}$
$z \approx 13.3831193$

Applying the Concepts

103. $2x - 2 = 4x + 6$
$2x - 4x - 2 = 4x - 4x + 6$
$-2x - 2 = 6$
$-2x - 2 + 2 = 6 + 2$
$-2x = 8$
$x = -4$
Then $3x^2 = 3(-4)^2 = 48$.

105. Students should explain that the solution of the original equation is $x = 0$. Therefore, the fourth line, where each side of the equation is divided by x, involved division by zero, which is not defined.

Section 11.5

Objective A Exercises

1. $y - 9$

3. $z + 3$

5. $\dfrac{2}{3}n + n$

7. $\dfrac{m}{m - 3}$

9. $9(x + 4)$

11. $n - (-5)n$

13. $c\left(\dfrac{1}{4}c\right)$

15. $m^2 + 2m^2$

17. $2(t + 6)$

19. $\dfrac{x}{9 + x}$

21. $3(b + 6)$

Objective B Exercises

23. The *square* of a number
The unknown number: x
x^2

25. A number *divided* by twenty
The unknown number: x
$\dfrac{x}{20}$

27. Four *times* some number
The unknown number: x
$4x$

29. Three-fourths *of* a number
The unknown number: x
$\dfrac{3}{4}x$

31. Four *increased* by some number
The unknown number: x
$4 + x$

33. The *difference between* five *times* a number and the number
The unknown number: x
Five times the number: $5x$
$5x - x$

35. The *product* of a number and two *more than* the number
The unknown number: x
Two more than the number: $x + 2$
$x(x + 2)$

37. Seven *times* the *total* of a number and eight
The unknown number: x
The total of the number and eight: $x + 8$
$7(x + 8)$

39. The *square* of a number *plus* the *product* of three and the number
The unknown number: x
The square of the number: x^2
The product of three and the number: $3x$
$x^2 + 3x$

41. The *sum* of three *more than* a number and one-half *of* the number
The unknown number: x
Three more than the number: $x + 3$
One-half of the number: $\frac{1}{2}x$
$(x + 3) + \frac{1}{2}x$

43. The *quotient* of three times a number and the number
The unknown number: x
Three times the number: $3x$
$\frac{3x}{x}$

Applying the Concepts

45a. $2x + 3$: Answers will vary. For example, the sum of twice a number and 3.

b. $2(x + 3)$: Answers will vary. For example, twice the sum of a number and 3.

47. Students will provide different explanations of how variables are used. Look for the idea that a variable is used to represent a number that is unknown or a number that can change, or vary.

49. $\frac{1}{2}x$: number of carbon atoms
$\frac{1}{2}x$: number of oxygen atoms

Section 11.6

Objective A Exercises

1. The unknown number: x
$x + 7 = 12$
$x + 7 - 7 = 12 - 7$
$x = 5$
The number is 5.

3. The unknown number: x
$3x = 18$
$\frac{3x}{3} = \frac{18}{3}$
$x = 6$
The number is 6.

5. The unknown number: x
$x + 5 = 3$
$x + 5 - 5 = 3 - 5$
$x = -2$
The number is -2.

7. The unknown number: x
$6x = 14$
$\frac{6x}{6} = \frac{14}{6}$
$x = \frac{7}{3} = 2\frac{1}{3}$
The number is $2\frac{1}{3}$.

9. The unknown number: x
$\frac{5}{6}x = 15$
$\frac{6}{5} \cdot \frac{5}{6}x = \frac{6}{5} \cdot 15$
$x = 18$
The number is 18.

11. The unknown number: x
$3x + 4 = 8$
$3x + 4 - 4 = 8 - 4$
$3x = 4$
$\frac{3x}{3} = \frac{4}{3}$
$x = \frac{4}{3} = 1\frac{1}{3}$
The number is $1\frac{1}{3}$.

13. The unknown number: x
$\frac{1}{4}x - 7 = 9$
$\frac{1}{4}x - 7 + 7 = 9 + 7$
$\frac{1}{4}x = 16$
$4 \cdot \frac{1}{4}x = 4 \cdot 16$
$x = 64$
The number is 64.

15. The unknown number: x
$\frac{x}{9} = 14$
$9 \cdot \frac{x}{9} = 9 \cdot 14$
$x = 126$
The number is 126.

17. The unknown number: x

$$\frac{x}{4} - 6 = -2$$

$$\frac{x}{4} - 6 + 6 = -2 + 6$$

$$\frac{x}{4} = 4$$

$$4 \cdot \frac{x}{4} = 4 \cdot 4$$

$$x = 16$$

The number is 16.

19. The unknown number: x

$$7 - 2x = 13$$

$$7 - 7 - 2x = 13 - 7$$

$$-2x = 6$$

$$\frac{-2x}{-2} = \frac{6}{-2}$$

$$x = -3$$

The number is -3.

21. The unknown number: x

$$9 - \frac{x}{2} = 5$$

$$9 - 9 - \frac{x}{2} = 5 - 9$$

$$-\frac{x}{2} = -4$$

$$(-2)\left(-\frac{x}{2}\right) = (-2)(-4)$$

$$x = 8$$

The number is 8.

23. The unknown number: x

$$\frac{3}{5}x + 8 = 2$$

$$\frac{3}{5}x + 8 - 8 = 2 - 8$$

$$\frac{3}{5}x = -6$$

$$\frac{5}{3} \cdot \frac{3}{5}x = \frac{5}{3} \cdot (-6)$$

$$x = -10$$

The number is -10.

25. The unknown number: x

$$\frac{x}{4.186} - 7.92 = 12.529$$

$$\frac{x}{4.186} - 7.92 + 7.92 = 12.529 + 7.92$$

$$\frac{x}{4.186} = 20.449$$

$$4.186 \cdot \frac{x}{4.186} = 4.186(20.449)$$

$$x = 85.599514$$

The number is 85.599514.

Objective B Exercises

27. Strategy To find the price of a pair of shoes at Target, write and solve an equation using P to represent the price at Target.

Solution
$$72.50 = P - 4.25$$
$$72.50 + 4.25 = P - 4.25 + 4.25$$
$$76.75 = P$$

The price at Target is $76.75.

29. Strategy To find the value of the SUV last year, write and solve an equation using V to represent the value of the SUV last year

Solution
$$\frac{4}{5}V = 16,000$$
$$\frac{5}{4} \cdot \frac{4}{5}V = \frac{5}{4} \cdot 16,000$$
$$V = 20,000$$

The value of the SUV last year was $20,000.

31. Strategy To find the length of the Brooklyn Bridge, write and solve an equation using L to represent the length of the Brooklyn Bridge.

Solution
$$1991 = L + 1505$$
$$1991 - 1505 = L + 1505 - 1505$$
$$486 = L$$

The length of the Brooklyn Bridge is 486 m.

33. Strategy To find the family's monthly income, write and solve an equation using S to represent the family's income.

Solution
$$1360 = \frac{1}{4}S$$
$$4 \cdot 1360 = 4 \cdot \frac{1}{4}S$$
$$5440 = S$$

The family's monthly income is $5440.

35. Strategy To find the monthly output a year ago, write and solve an equation using M to represent the monthly output a year ago.

Solution
$$400 = 0.08M$$
$$\frac{400}{0.08} = \frac{0.08M}{0.08}$$
$$5000 = M$$

The monthly output a year ago was 5000 computers.

37. Strategy To find the recommended daily allowance of sodium:
- Write and solve an equation using x to represent the daily allowance of sodium.
- Convert the milligrams to grams.

Solution
$$8\% \cdot x = 200$$
$$0.08x = 200$$
$$\frac{0.08x}{0.08} = \frac{200}{0.08}$$
$$x = 2{,}500$$
$$2{,}500 \text{ mg} = 2.5 \text{ g}$$
The recommended daily allowance of sodium is 2.5 g.

39. Strategy To find the number of hours of labor required to install a water softener, write and solve an equation using T to represent the time it took to install the water softener.

Solution
$$400 = 310 + 30T$$
$$400 - 310 = 310 - 310 + 30T$$
$$90 = 30T$$
$$\frac{90}{30} = \frac{30T}{30}$$
$$3 = T$$
It took 3 h to install the water softener.

41. Strategy To find the number of plants and animals known to be at risk, write and solve an equation using P to represent the number of plants and animals.

Solution
$$10.7\% \cdot P = 1184$$
$$0.107P = 1184$$
$$\frac{0.107P}{0.107} = \frac{1184}{0.107}$$
$$P \approx 11{,}065$$
About 11,065 plants and animals are known to be at risk of extinction in the world.

43. Strategy To find the percent spent in New York, write and solve the basic percent equation using P to represent the percent. The base is 295 and the amount is 9.8.

Solution
$$\text{Percent} \times \text{base} = \text{amount}$$
$$P \times 295 = 9.8$$
$$295P = 9.8$$
$$\frac{295P}{295} = \frac{9.8}{295}$$
$$P \approx 0.0332$$
The percent is 3.3%.

45. Strategy To find the original rate of water flow, write and solve an equation using R to represent the original rate.

Solution
$$2 = \frac{3}{5}R - 1$$
$$2 + 1 = \frac{3}{5}R - 1 + 1$$
$$3 = \frac{3}{5}R$$
$$\frac{5}{3} \cdot 3 = \frac{5}{3} \cdot \frac{3}{5}R$$
The original water flow rate was 5 gal/min.

47. Strategy To find the total sales for the month, write and solve an equation using T to represent the total sales.

Solution
$$600 + 0.0825T = 4109.55$$
$$600 - 600 + 0.0825T = 4109.55 - 600$$
$$0.0825T = 3509.55$$
$$\frac{0.0825T}{0.0825} = \frac{3509.55}{0.0825}$$
$$T = 42{,}540$$
The total sales for the month were $42,540.

49. Strategy To find the world carbon dioxide emissions in 1990, write and solve an equation using T to represent the number of billions of metric tons of world carbon dioxide emissions in 1990.

Solution
$$8.87 = 1.521T$$
$$\frac{8.87}{1.521} = \frac{1.521T}{1.521}$$
$$5.83 \approx T$$
The world carbon dioxide emissions in 1990 were 5.83 billion metric tons.

Applying the Concepts

51. The problem states that a 4-quart mixture of fruit juice is made from apple juice and cranberry juice. There are 6 more quarts of apple juice than of cranberry juice. If we let x = the number of quarts of cranberry juice, then $x + 6$ = the number of quarts of apple juice. The total number of quarts is 4. Therefore, we can write the equation $x + (x + 6) = 4$.

$$x + (x + 6) = 4$$
$$2x = -2$$
$$x = -1$$

Since x = the number of quarts of cranberry juice, there are -1 qt of cranberry juice in the mixture. We cannot add -1 qt to a mixture. The solution is not reasonable.

We see from the original problem that the answer will not be reasonable. If the total number of quarts in the mixture is 4, we cannot have more than 6 qt of apple juice in the mixture.

Chapter 11 Review Exercises

1. $-2(a - b) = -2[a + (-b)]$
$$= -2(a) + (-2)(-b)$$
$$= -2a + 2b$$

2.
$$\frac{3x - 2 = -8}{3(-2) - 2 \,\big|\, -8}$$
$$\begin{array}{r|r} -6 - 2 & -8 \\ -6 + (-2) & -8 \\ -8 & = -8 \end{array}$$
Yes, -2 is a solution.

3. $x - 3 = -7$
$$x - 3 + 3 = -7 + 3$$
$$x = -4$$

4. $-2x + 5 = -9$
$$-2x + 5 - 5 = -9 - 5$$
$$-2x = -14$$
$$\frac{-2x}{-2} = \frac{-14}{-2}$$
$$x = 7$$

5. $a^2 - 3b = 2^2 - 3(-3)$
$$= 4 + 9 = 13$$

6. $-3x = 27$
$$\frac{-3x}{-3} = \frac{27}{-3}$$
$$x = -9$$

7. $\frac{2}{3}x + 3 = -9$
$$\frac{2}{3}x + 3 - 3 = -9 - 3$$
$$\frac{2}{3}x = -12$$
$$\frac{3}{2} \cdot \frac{2}{3}x = \frac{3}{2}(-12)$$
$$x = -18$$

8. $3x - 2(3x - 2) = 3x + (-2)[3x + (-2)]$
$$= 3x + (-2)(3x) + (-2)(-2)$$
$$= 3x + (-6x) + 4$$
$$= -3x + 4$$

9. $6x - 9 = -3x + 36$
$$6x + 3x - 9 = -3x + 3x + 36$$
$$9x - 9 = 36$$
$$9x - 9 + 9 = 36 + 9$$
$$9x = 45$$
$$\frac{9x}{9} = \frac{45}{9}$$
$$x = 5$$

10. $x + 3 = -2$
$$x + 3 - 3 = -2 - 3$$
$$x = -5$$

11.
$$\frac{3x - 5 = -10}{3(5) - 5 \,\big|\, -10}$$
$$\begin{array}{r|r} 15 - 5 & -10 \\ 15 + (-5) & -10 \\ 10 & \ne -10 \end{array}$$
No, 5 is not a solution.

12. $a^2 - (b \div c) = (-2)^2 - [8 \div (-4)]$
$$= 4 - (-2) = 6$$

13. $3(x - 2) + 2 = 11$
$$3x - 6 + 2 = 11$$
$$3x - 4 = 11$$
$$3x - 4 + 4 = 11 + 4$$
$$3x = 15$$
$$\frac{3x}{3} = \frac{15}{3}$$
$$x = 5$$

14. $35 - 3x = 5$
$$35 - 35 - 3x = 5 - 35$$
$$-3x = -30$$
$$\frac{-3x}{-3} = \frac{-30}{-3}$$
$$x = 10$$

15. $6bc - 7bc + 2bc - 5bc$
$$= 6bc + (-7)bc + 2bc + (-5)bc$$
$$= (-1)bc + 2bc + (-5)bc$$
$$= 1bc + (-5)bc$$
$$= -4bc$$

16.
$$7 - 3x = 2 - 5x$$
$$7 - 3x + 5x = 2 - 5x + 5x$$
$$7 + 2x = 2$$
$$7 - 7 + 2x = 2 - 7$$
$$2x = -5$$
$$\frac{2x}{2} = \frac{-5}{2}$$
$$x = \frac{-5}{2} = -2\frac{1}{2}$$

17.
$$-\frac{3}{8}x = -\frac{15}{32}$$
$$-\frac{8}{3}\left(-\frac{3}{8}x\right) = \left(-\frac{8}{3}\right)\left(-\frac{15}{32}\right)$$
$$x = \frac{5}{4} = 1\frac{1}{4}$$

18.
$$\frac{1}{2}x^2 - \frac{1}{3}x^2 + \frac{1}{5}x^2 + 2x^2$$
$$= \frac{1}{2}x^2 + \left(-\frac{1}{3}\right)x^2 + \frac{1}{5}x^2 + 2x^2$$
$$= \frac{3}{6}x^2 + \left(-\frac{2}{6}\right)x^2 + \frac{1}{5}x^2 + 2x^2$$
$$= \frac{1}{6}x^2 + \frac{1}{5}x^2 + 2x^2$$
$$= \frac{5}{30}x^2 + \frac{6}{30}x^2 + 2x^2$$
$$= \frac{11}{30}x^2 + 2x^2$$
$$= \frac{11}{30}x^2 + \frac{60}{30}x^2$$
$$= \frac{71}{30}x^2$$

19.
$$5x - 3(1 - 2x) = 4(2x - 1)$$
$$5x - 3 + 6x = 8x - 4$$
$$11x - 3 = 8x - 4$$
$$11x - 8x - 3 = 8x - 8x - 4$$
$$3x - 3 = -4$$
$$3x - 3 + 3 = -4 + 3$$
$$3x = -1$$
$$\frac{3x}{3} = \frac{-1}{3}$$
$$x = -\frac{1}{3}$$

20.
$$\frac{5}{6}x - 4 = 5$$
$$\frac{5}{6}x - 4 + 4 = 5 + 4$$
$$\frac{5}{6}x = 9$$
$$\frac{6}{5} \cdot \frac{5}{6}x = \frac{6}{5} \cdot 9$$
$$x = \frac{54}{5} = 10\frac{4}{5}$$

21. **Strategy**   To find the number of miles per gallon of gas, replace D and G in the formula by the given values and solve for M.

Solution
$$D = M \cdot G$$
$$621 = M \cdot 27$$
$$\frac{621}{27} = \frac{27M}{27}$$
$$23 = M$$
The mileage obtained was 23 mi/gal.

22. **Strategy**   To find the Celsius temperature, replace the variable F in the formula by the given value and solve for C.

Solution
$$F = 1.8C + 32$$
$$100 = 1.8C + 32$$
$$100 - 32 = 1.8C + 32 - 32$$
$$68 = 1.8C$$
$$37.8 \approx C$$
The temperature is 37.8°C.

23. The *total* of n and the *quotient* of n and 5
The unknown number: n
The quotient of n and 5: $\dfrac{n}{5}$

$$n + \frac{n}{5}$$

24. The *sum* of five more than a number and one-third *of* the number
The unknown number: n
Five more than the number: $n + 5$

One-third of the number: $\dfrac{1}{3}n$
$$(n + 5) + \frac{1}{3}n$$

25. The unknown number: x
$$9 - 2x = 5$$
$$9 - 9 - 2x = 5 - 9$$
$$-2x = -4$$
$$\frac{-2x}{-2} = \frac{-4}{-2}$$
$$x = 2$$
The number is 2.

26. The unknown number: p
$$5p = 50$$
$$\frac{5p}{5} = \frac{50}{5}$$
$$p = 10$$
The number is 10.

27. Strategy To find the regular price, write and solve an equation using R to represent the regular price.

Solution
$$228 = 60\% \cdot R$$
$$228 = 0.60R$$
$$\frac{228}{0.60} = \frac{0.60R}{0.60}$$
$$380 = R$$
The regular price of the CD player is $380.

28. Strategy Let x represent last year's crop. Then $0.12x$ is the increase in last year's crop. Last year's crop plus the increase is this year's crop (28,336 bushels).

Solution
$$0.12x + x = 26,336$$
$$1.12x = 28,336$$
$$x = 25,300$$
Last year's crop was 25,300 bushels.

Chapter 11 Test

1.
$$\frac{x}{5} - 12 = 7$$
$$\frac{x}{5} - 12 + 12 = 7 + 12$$
$$\frac{x}{5} = 19$$
$$5 \cdot \frac{x}{5} = 5 \cdot 19$$
$$x = 95$$

2.
$$x - 12 = 14$$
$$x - 12 + 12 = 14 + 12$$
$$x = 26$$

3.
$$3y - 2x - 7y - 9x = 3y + (-2x) + (-7y) + (-9x)$$
$$= 3y + (-7y) + (-2x) + (-9x)$$
$$= -4y + (-2x) + (-9x)$$
$$= -4y + (-11x)$$
$$= -4y - 11x = -11x - 4y$$

4.
$$8 - 3x = 2x - 8$$
$$8 - 3x - 2x = 2x - 2x - 8$$
$$8 - 5x = -8$$
$$8 - 8 - 5x = -8 - 8$$
$$-5x = -16$$
$$\frac{-5x}{-5} = \frac{-16}{-5}$$
$$x = \frac{16}{5} = 3\frac{1}{5}$$

5.
$$3x - 12 = -18$$
$$3x - 12 + 12 = -18 + 12$$
$$3x = -6$$
$$\frac{3x}{3} = \frac{-6}{3}$$
$$x = -2$$

6.
$$c^2 - (2a + b^2) = (-2)^2 - [2(3) + (-6)^2]$$
$$= 4 - (6 + 36)$$
$$= 4 - (42)$$
$$= 4 + (-42) = -38$$

7.
$$\begin{array}{c|c} x^2 + 3x - 7 & = 3x - 2 \\ \hline 3^2 + 3(3) - 7 & 3(3) - 2 \\ 9 + 9 - 7 & 9 - 2 \\ 18 - 7 & 7 \\ 11 \neq 7 \end{array}$$
No, 3 is not a solution.

8. $9 - 8ab - 6ab = 9 - 14ab = -14ab + 9$

9.
$$-5x = 14$$
$$\frac{-5x}{-5} = \frac{14}{-5}$$
$$x = -2\frac{4}{5}$$

10.
$$3y + 5(y - 3) + 8 = 3y + 5[y + (-3)] + 8$$
$$= 3y + 5y + 5(-3) + 8$$
$$= 8y + (-15) + 8$$
$$= 8y + (-7)$$
$$= 8y - 7$$

11.
$$3x - 4(x - 2) = 8$$
$$3x - 4x + 8 = 8$$
$$-x + 8 = 8$$
$$-x + 8 - 8 = 8 - 8$$
$$-x = 0$$
$$(-1)(-x) = (-1)0$$
$$x = 0$$

12.
$$5 = 3 - 4x$$
$$5 - 3 = 3 - 3 - 4x$$
$$2 = -4x$$
$$\frac{2}{-4} = \frac{-4x}{-4}$$
$$-\frac{1}{2} = x$$

13.
$$\frac{x^2}{y} - \frac{y^2}{x} = \frac{3^2}{-2} - \frac{(-2)^2}{3}$$
$$= \frac{9}{-2} - \frac{4}{3}$$
$$= \frac{-27}{6} - \frac{8}{6}$$
$$= \frac{-35}{6} = -5\frac{5}{6}$$

14. $\dfrac{5}{8}x = -10$

$\dfrac{8}{5} \cdot \dfrac{5}{8}x = \dfrac{8}{5}(-10)$

$x = -16$

15. $y - 4y + 3 = 12$

$-3y + 3 = 12$

$-3y + 3 - 3 = 12 - 3$

$-3y = 9$

$\dfrac{-3y}{-3} = \dfrac{9}{-3}$

$y = -3$

16. $2x + 4(x - 3) = 5x - 1$

$2x + 4x - 12 = 5x - 1$

$6x - 12 = 5x - 1$

$6x - 5x - 12 = 5x - 5x - 1$

$x - 12 = -1$

$x - 12 + 12 = -1 + 12$

$x = 11$

17. Strategy To find the monthly payment, replace the variables L and N in the formula by the given values and solve for P.

Solution

$L = P \cdot N$

$6600 = P \cdot 48$

$\dfrac{6600}{48} = \dfrac{48P}{48}$

$137.50 = P$

The monthly payment is $137.50.

18. Strategy To find the number of clocks made during a month, replace the variables T, U, and F in the formula by the given values and solve for N.

Solution

$T = U \cdot N + F$

$65,000 = 15N + 5000$

$65,000 - 5000 = 15N + 5000 - 5000$

$60,000 = 15N$

$\dfrac{60,000}{15} = \dfrac{15N}{15}$

$4000 = N$

4000 clocks were made during the month.

19. Strategy To find the time, replace the variables V and V_0 in the formula by the given values and solve for t.

Solution

$V = V_0 + 32t$

$392 = 24 + 32t$

$392 - 24 = 24 - 24 + 32t$

$368 = 32t$

$\dfrac{368}{32} = \dfrac{32t}{32}$

$11.5 = t$

The object will fall for 11.5 s.

20. The *sum* of x and one-third *of* x

The unknown number: x

One-third of x: $\dfrac{1}{3}x$

$x + \dfrac{1}{3}x$

21. Five *times* the *sum* of a number and three

The unknown number: x

The sum of a number and three: $x + 3$

$5(x + 3)$

22. The unknown number: x

$2x - 3 = 7$

$2x - 3 + 3 = 7 + 3$

$2x = 10$

$\dfrac{2x}{2} = \dfrac{10}{2}$

$x = 5$

The number is 5.

23. The unknown number: w

$5 + 3w = w - 2$

$5 + 3w - w = w - w - 2$

$5 + 2w = -2$

$5 - 5 + 2w = -2 - 5$

$2w = -7$

$\dfrac{2w}{2} = \dfrac{-7}{2}$

$w = -3\dfrac{1}{2}$

The number is $-3\dfrac{1}{2}$.

24. Strategy To find Santos's total sales for the month, write and solve an equation using T to represent the total sales.

Solution

$3600 = 1200 + 0.06T$

$3600 - 1200 = 1200 - 1200 + 0.06T$

$2400 = 0.06T$

$\dfrac{2400}{0.06} = \dfrac{0.06T}{0.06}$

$40,000 = T$

Santos's total sales for the month were $40,000.

25. Strategy To find the number of hours worked, write and solve an equation using h to represent the number of hours worked.

Solution

$152 + 42h = 278$

$152 - 152 + 42h = 278 - 152$

$42h = 126$

$h = 3$

The mechanic worked for 3 h.

Cumulative Review Exercises

1. $6^2 - (18 - 6) \div 4 + 8 = 36 - (12) \div 4 + 8$
$= 36 - 3 + 8$
$= 33 + 8 = 41$

2. $3\frac{1}{6} = 3\frac{5}{30} = 2\frac{35}{30}$

$-1\frac{7}{15} = 1\frac{14}{30} = 1\frac{14}{30}$

$1\frac{21}{30} = 1\frac{7}{10}$

3. $\left(\frac{3}{8} - \frac{1}{4}\right) \div \frac{3}{4} + \frac{4}{9} = \left(\frac{3}{8} - \frac{2}{8}\right) \div \frac{3}{4} + \frac{4}{9}$

$= \frac{1}{8} \div \frac{3}{4} + \frac{4}{9}$

$= \frac{1}{8} \times \frac{4}{3} + \frac{4}{9}$

$= \frac{1}{6} + \frac{4}{9}$

$= \frac{3}{18} + \frac{8}{18} = \frac{11}{18}$

4. $\begin{array}{r} 9.67 \\ \times\, 0.0049 \\ \hline 8703 \\ 3868 \\ \hline 0.047383 \end{array}$

5. $\frac{\$84}{20\text{ h}} = 4.20/\text{h}$

6. $\frac{2}{3} = \frac{n}{40}$
$2 \times 40 = 3 \cdot n$
$80 = 3 \cdot n$
$80 \div 3 = n$
$26.67 \approx n$

7. $5\frac{1}{3}\% = \frac{16}{3} \times \frac{1}{100} = \frac{16}{300} = \frac{4}{75}$

8. Percent $\times$ base $=$ amount
$n \times 30 = 42$
$n = 42 \div 30$
$n = 1.40 = 140\%$

9. Percent $\times$ base $=$ amount
$125\% \times n = 8$
$1.25 \times n = 8$
$n = 8 \div 1.25 = 6.4$

10. $\begin{array}{r} 3\text{ ft } 9\text{ in.} \\ \times\quad 5 \\ \hline 15\text{ ft } 45\text{ in.} = 18\text{ ft } 9\text{ in.} \end{array}$

11. $1\frac{3}{8}\text{ lb} = \frac{11}{8}\,\text{lb} \times \frac{16\text{ oz}}{1\text{ lb}} = \frac{11 \cdot 16\text{ oz}}{8} = 22\text{ oz}$

12. $282\text{ mg} = 0.282\text{ g}$

13. $-2 + 5 + (-8) + 4 = 3 + (-8) + 4$
$= -5 + 4 = -1$

14. $13 - (-6) = 13 + 6 = 19$

15. $(-2)^2 - (-8) \div (3 - 5)^2 = (-2)^2 - (-8) \div (-2)^2$
$= 4 - (-8) \div 4$
$= 4 - (-2) = 4 + 2 = 6$

16. $3ab - 2ac = 3(-2)(6) - 2(-2)(-3)$
$= -36 - 12 = -36 + (-12) = -48$

17. $3z - 2x + 5z - 8x = 3z + (-2x) + 5z + (-8x)$
$= 3z + 5z + (-2x) + (-8x)$
$= 8z + (-10x)$
$= 8z - 10x = -10x + 8z$

18. $6y - 3(y - 5) + 8 = 6y + (-3)[y + (-5)] + 8$
$= 6y + (-3)y + (-3)(-5) + 8$
$= 6y + (-3y) + 15 + 8$
$= 3y + 23$

19. $2x - 5 = -7$
$2x - 5 + 5 = -7 + 5$
$2x = -2$
$2x = -2$
$\frac{2x}{2} = \frac{-2}{2}$
$x = -1$

20. $7x - 3(x - 5) = -10$
$7x - 3x + 15 = -10$
$4x + 15 = -10$
$4x + 15 - 15 = -10 - 15$
$4x = -25$
$\frac{4x}{4} = \frac{-25}{4}$
$x = -6\frac{1}{4}$

21. $-\frac{2}{3}x = 5$
$\left(-\frac{3}{2}\right)\left(-\frac{2}{3}\right)x = -\frac{3}{2} \cdot 5$
$x = -\frac{15}{2} = -7\frac{1}{2}$

22. $\frac{x}{3} - 5 = -12$
$\frac{x}{3} - 5 + 5 = -12 + 5$
$\frac{x}{3} = -7$
$3 \cdot \frac{x}{3} = 3(-7)$
$x = -21$

23. **Strategy** To find the percent of the students who received an A grade, solve the basic percent equation for percent.

Solution Percent · base = amount
$$n \cdot 34 = 6$$
$$n = 6 \div 34$$
$$n \approx 0.176 = 17.6\%$$
The percent is 17.6%.

24. **Strategy** To find the price:
• Find the amount of the markup by solving the basic percent equation for amount. The base is $28.50 and the percent is 40%.
• Add the amount of the markup to the cost.

Solution
$$0.40 \times 28.50 = n \qquad \$28.50$$
$$11.40 = n \qquad \underline{+11.40}$$
$$\$39.90$$
The price of the piece of pottery is $39.90.

25a. **Strategy** To find the discount subtract the sale price ($369) from the regular price ($450).

Solution $450 - 369 = 81$
The discount is $81.

b. **Strategy** To find the discount rate, write and solve the basic percent equation for percent. The base is the regular price and the amount is the discount.

Solution Percent × base = amount
$$n \times 450 = 81$$
$$n = 81 \div 450$$
$$n = 0.18$$
The discount rate is 18%.

26. **Strategy** To find the simple interest due, multiply the principal and rate and time (in years).

Solution Interest $= 80,000 \times 11\% \times \dfrac{4}{12}$
$$= 80,000 \times 0.11 \times \dfrac{4}{12}$$
$$\approx 2933.33$$
The simple interest due on the loan is $2933.33.

27. The *sum* of three *times* a number and four
The unknown number: n
Three times the number: $3n$
$$3n + 4$$

28. **Strategy** To calculate the probability:
• Count the number of possible outcomes.
• Count the number of favorable outcomes.
• Use the probability formula.

Solution There are 16 possible outcomes.
There are 2 favorable outcomes: (3, 4), (4, 3).
Probability $= \dfrac{2}{16} = \dfrac{1}{8}$

The probability is $\dfrac{1}{8}$ that the sum of the upward faces on the two dice is 7.

29. **Strategy** To find the total sales, replace the variables M, R, and B in the formula with the given values and solve for S.

Solution
$$M = S \cdot R + B$$
$$3400 = S \cdot 0.08 + 800$$
$$3400 - 800 = S \cdot 0.08 + 800 - 800$$
$$2600 = S \cdot 0.08$$
$$\dfrac{2600}{0.08} = \dfrac{S \cdot 0.08}{0.08}$$
$$32,500 = S$$
The total sales were $32,500.

30. The unknown number: x
$$8x - 3 = 3 + 5x$$
$$8x - 5x - 3 = 3 + 5x - 5x$$
$$3x - 3 = 3$$
$$3x - 3 + 3 = 3 + 3$$
$$3x = 6$$
$$\dfrac{3x}{3} = \dfrac{6}{3}$$
$$x = 2$$
The number is 2.

Chapter 12: Geometry

Prep Test

1.
$$x + 47 = 90$$
$$x + 47 - 47 = 90 - 47$$
$$x = 43$$
The solution is 43.

2.
$$32 + 97 + x = 180$$
$$129 + x = 180$$
$$129 - 129 + x = 180 - 129$$
$$x = 51$$
The solution is 51.

3. $2(18) + 2(10) = 36 + 20 = 56$

4. abc
$$= (2)(3.14)(9)$$
$$= (6.28)(9)$$
$$= 56.52$$

5. xyz^3
$$= \left(\frac{4}{3}\right)(3.14)(3)^3$$
$$= 113.04$$

6. $\dfrac{5}{12} = \dfrac{6}{x}$
$$5x = 12 \times 6$$
$$\frac{5x}{5} = \frac{72}{5}$$
$$x = 14.4$$

Go Figure

The first figure is a diamond (D) inside a square (S) inside a triangle (T) inside a circle (C), or DSTC.
The second figure is STCD.
The third figure is TCDS.
The next figure would be CDST: A circle inside a diamond inside a square inside a triangle.

Section 12.1

Objective A Exercises

1. 0°; 90°

3. 180°

5. $EG = EF + FG$
$EG = 20 + 10 = 30$

7.
$$QS = QR + RS$$
$$28 = 7 + RS$$
$$28 - 7 = 7 - 7 + RS$$
$$21 = RS$$

9.
$$AD = AB + BC + CD$$
$$35 = 12 + BC + 9$$
$$35 = 21 + BC$$
$$35 - 21 = 21 - 21 + BC$$
$$14 = BC$$

11. Let x represent the complement of 31°. The sum of complementary angles is 90°.
$$x + 31° = 90°$$
$$x + 31° - 31° = 90° - 31°$$
$$x = 59°$$
59° is the complement of 31°.

13. Let x represent the supplement of 72°. The sum of supplementary angles is 180°.
$$x + 72° = 180°$$
$$x + 72° - 72° = 180° - 72°$$
$$x = 108°$$
108° is the supplement of 72°.

15. Let x represent the complement of 13°. The sum of complementary angles is 90°.
$$x + 13° = 90°$$
$$x + 13° - 13° = 90° - 13°$$
$$x = 77°$$
77° is the complement of 13°.

17. Let x represent the supplement of 127°. The sum of supplementary angles is 180°.
$$x + 127° = 180°$$
$$x + 127° - 127° = 180° - 127°$$
$$x = 53°$$
53° is the supplement of 127°.

19. $\angle AOB = 32° + 45° = 77°$

21.
$$42° + \angle a = 160°$$
$$42° - 42° + \angle a = 160° - 42°$$
$$\angle a = 118°$$

23.
$$\angle a + 47° = 180°$$
$$\angle a + 47° - 47° = 180° - 47°$$
$$\angle a = 133°$$

25.
$$\angle LON = \angle LOM + \angle MON$$
$$139° = 53° + \angle MON$$
$$139° - 53° = 53° - 53° + \angle MON$$
$$86° = \angle MON$$

Objective B Exercises

27. 180°

29. Rectangle or square

31. Cube

33. Parallelogram, rectangle, or square

35. Cylinder

37. The sum of the three angles of a triangle is 180°.

$$\angle A + \angle B + \angle C = 180°$$
$$\angle A + 13° + 65° = 180°$$
$$\angle A + 78° = 180°$$
$$\angle A + 78° - 78° = 180° - 78°$$
$$\angle A = 102°$$

The measure of the other angle is 102°.

39. In a right triangle, one angle measures 90° and the two acute angles are complementary.

$$\angle A + \angle B = 90°$$
$$\angle A + 45° = 90°$$
$$\angle A + 45° - 45° = 90° - 45°$$
$$\angle A = 45°$$

The other angles measure 90° and 45°.

41. The sum of the three angles of a triangle is 180°.

$$\angle A + \angle B + \angle C = 180°$$
$$\angle A + 62° + 104° = 180°$$
$$\angle A + 166° = 180°$$
$$\angle A + 166° - 166° = 180° - 166°$$
$$\angle A = 14°$$

The measure of the other angle is 14°.

43. In a right triangle, one angle measures 90° and the two acute angles are complementary.

$$\angle A + \angle B = 90°$$
$$\angle A + 25° = 90°$$
$$\angle A + 25° - 25° = 90° - 25°$$
$$\angle A = 65°$$

The other angles measure 90° and 65°.

45. $r = \dfrac{1}{2}d$

$r = \dfrac{1}{2}(16 \text{ in.}) = 8 \text{ in.}$

The radius is 8 in.

47. $d = 2r$

$d = 2\left(2\dfrac{1}{3}\text{ ft}\right)$

$d = 2\left(\dfrac{7}{3}\text{ ft}\right)$

$d = \dfrac{14}{3}\text{ ft} = 4\dfrac{2}{3}\text{ ft}$

The diameter is $4\dfrac{2}{3}$ ft.

49. $d = 2r$

$d = 2(3.5 \text{ cm}) = 7 \text{ cm}$

The diameter is 7 cm.

51. $r = \dfrac{1}{2}d$

$r = \dfrac{1}{2}(4 \text{ ft 8 in.})$

$r = 2 \text{ ft 4 in.}$

The radius is 2 ft 4 in.

Objective C Exercises

53.
$$\angle a + 74° = 180° \quad \text{supplementary angles}$$
$$\angle a + 74° - 74° = 180° - 74°$$
$$\angle a = 106°$$

$\angle b = 74°$ vertical angles

55. $\angle a = 112°$ vertical angles

$$\angle b + 112° = 180° \quad \text{supplementary angles}$$
$$\angle b + 112° - 112° = 180° - 112°$$
$$\angle b = 68°$$

57. $\angle a = 38°$ vertical angles

$$\angle b + 38° = 180° \quad \text{supplementary angles}$$
$$\angle b + 38° - 38° = 180° - 38°$$
$$\angle b = 142°$$

59.
$$\angle a + 122° = 180° \quad \text{supplementary angles}$$
$$\angle a + 122° - 122° = 180° - 122°$$
$$\angle a = 58°$$

$\angle a = \angle b$ alternate interior angles
$\angle b = 58°$

61.
$$\angle b + 28° = 180° \quad \text{supplementary angles}$$
$$\angle b + 28° - 28° = 180° - 28°$$
$$\angle b = 152°$$
$$\angle a = \angle b \quad \text{corresponding angles}$$
$$\angle a = 152°$$

$\angle b = 152°$

63. $\angle a = 130°$ alternate interior angles

$$\angle b + 130° = 180° \quad \text{supplementary angles}$$
$$\angle b + 130° - 130° = 180° - 130°$$
$$\angle b = 50°$$

Applying the Concepts

65a. 1°

b. 90°

67. $\angle AOC$ and $\angle BOC$ are supplementary angles. Therefore, $\angle AOC + \angle BOC = 180°$. Because $\angle AOC = \angle BOC$, by substitution $\angle AOC + \angle AOC = 180°$. Therefore, $2\angle AOC = 180°$ and $\angle AOC = 90°$. Therefore, AB is perpendicular to CD.

Section 12.2

Objective A Exercises

1. $P = a + b + c$
$= 12 \text{ in.} + 20 \text{ in.} + 24 \text{ in.}$
$= 56 \text{ in.}$

The perimeter of the triangle is 56 in.

3. $P = 4s$
$= 4(5 \text{ ft})$
$= 20 \text{ ft}$
The perimeter of the square is 20 ft.

5. $P = 2L + 2W$
$= 2(32 \text{ cm}) + 2(14 \text{ cm})$
$= 64 \text{ cm} + 28 \text{ cm}$
$= 92 \text{ cm}$
The perimeter of the rectangle is 92 cm.

7. $C = \pi d$
$\approx 3.14(15 \text{ cm})$
$= 47.1 \text{ cm}$
The circumference of the circle is approximately 47.1 cm.

9. $P = 2 \text{ ft } 4 \text{ in.} + 3 \text{ ft} + 4 \text{ ft } 6 \text{ in.}$
$= 9 \text{ ft } 10 \text{ in.}$
The perimeter of the triangle is 9 ft 10 in.

11. $C = 2\pi r$
$\approx 2(3.14)(8 \text{ cm})$
$= 50.24 \text{ cm}$
The circumference of the circle is approximately 50.24 cm.

13. $P = 4s$
$= 4(60 \text{ m})$
$= 240 \text{ m}$
The perimeter of the square is 240 m.

15. Perimeter = sum of sides
$= 22 \text{ cm} + 47 \text{ cm} + 29 \text{ cm} + 42 \text{ cm} + 17 \text{ cm}$
$= 157 \text{ cm}$
The perimeter is 157 cm.

Objective B Exercises

17. Perimeter = sum of sides
$= 19 \text{ cm} + 20 \text{ cm} + 8 \text{ cm} + 5 \text{ cm} + 27 \text{ cm} + 42 \text{ cm}$
$= 121 \text{ cm}$

19.
$$\text{Perimeter of Composite Figure} = \text{3 sides of a rectangle} + \tfrac{1}{2} \text{ the circumference of a circle}$$
$= 2L + W + \tfrac{1}{2}\pi d$
$\approx 2(15 \text{ m}) + 8 \text{ m} + \tfrac{1}{2}(3.14)(8 \text{ m})$
$= 30 \text{ m} + 8 \text{ m} + 12.56 \text{ m}$
$= 50.56 \text{ m}$

21. Perimeter = length of two sides $+ \tfrac{1}{2}$ circumference of circle
$= 2 \cdot 1 \text{ ft} + \tfrac{1}{2}(3.14 \cdot 1 \text{ ft})$
$= 2 \text{ ft} + 1.57 \text{ ft}$
$= 3.57 \text{ ft}$

23. Perimeter = sum of six sides of figure
$= 22.75 \text{ m} + 25.73 \text{ m} + 15.94 \text{ m} + 18.3 \text{ m} + 21.61 \text{ m} + 34.97 \text{ m}$
$= 139.3 \text{ m}$

Objective C Exercises

25. Strategy To find the amount of fencing, use the formula for the perimeter of a rectangle.

Solution $P = 2L + 2W = 2 \cdot 18 + 2 \cdot 12$
$= 36 + 24 = 60$
The amount of fencing needed is 60 ft.

27. Strategy To find the amount of binding, find the perimeter of a rectangle.

Solution $P = 2L + 2W = 2 \cdot 8.5 + 2 \cdot 3.5$
$= 17 + 7 = 24$
The amount of binding needed is 24 ft.

29. Strategy To find the cost of the fence:
• Find the length of fence that is not along the road.
• Multiply the length of fence not along the road by $5.85.
• Multiply the length along the road by $6.20.
• Add the two products to find the total cost.

Solution Length = $800 + 1250 + 800 = 2850$
$2850 \times \$5.85 = \$16,672.50$
$1250 \times \$6.20 = \$7,750$
$\$16,672.50 + \$7,750 = \$24,422.50$
The total cost of the fence is $24,422.50.

31. Strategy To find the distance the bike travels:
• Convert diameter (24 in.) to feet.
• Use the formula for circumference to find the distance traveled in 1 revolution.
• Multiply the distance traveled in 1 revolution by the number of revolutions (5).

Solution $24 \text{ in.} = 24 \text{ in.} \times \dfrac{1 \text{ ft}}{12 \text{ in.}} = \dfrac{24}{12} \text{ ft} = 2 \text{ ft}$
$C = \pi d$
$\approx 3.14 \cdot 2 = 6.28 \text{ ft}$
$6.28 \text{ ft} \cdot 5 = 31.4 \text{ ft}$
The bicycle travels 31.4 ft.

33a. Strategy   To estimate whether the perimeter is more or less than 70 m, assume that the figure is a rectangle with length 25 m and width 10 m.

   Solution $P = 2L + 2W = 2 \cdot 25 + 2 \cdot 10$
   $= 50 + 20 = 70$
   The perimeter of the rink is larger than the perimeter of the rectangle. The perimeter of the rink is more than 70 m.

b. Strategy   To find the perimeter of the roller rink, find the perimeter of the composite figure.

   Solution   Perimeter
   = sum of length of two sides
   $+ \, 2$ times $\frac{1}{2}$ circumference of a circle
   $\approx 2 \cdot 25 + 2 \cdot \frac{1}{2}(3.14 \cdot 10)$
   $= 50 + 31.4 = 81.4$
   The perimeter of the rink is 81.4 m.

35. Strategy   To find the length of weather stripping, find the perimeter of the composite figure.

   Solution   Perimeter
   = sum of three sides of rectangle
   $+ \, \frac{1}{2}$ circumference of circle
   $\approx (3 \text{ ft}) + (2 \cdot 6 \text{ ft } 6 \text{ in.}) + \frac{1}{2}(3.14 \cdot 3 \text{ ft})$
   $= 3 \text{ ft} + 13 \text{ ft} + 4.71 \text{ ft} = 20.71 \text{ ft}$
   Approximately 20.71 ft of weather stripping are installed.

37. Strategy   To find the circumference of Earth, use the formula for the circumference of a circle.

   Solution $C = 2\pi r$
   $\approx 2 \cdot 3.14 \cdot 6356 \text{ km}$
   $= 39,915.68 \text{ km}$
   The circumference of Earth is approximately 39,915.68 km.

Applying the Concepts

39.   Length = 8
   Width = 3
   $P = 2L + 2W = 2(8) + 2(3) = 16 + 6 = 22$ units

41.   The perimeter of one large triangle is 3 cm. (1 on each side) as in Figure A.
   The perimeter of three medium triangles is $3\left(\frac{3}{2}\right)$ $\left(\frac{1}{2} \text{ on each side}\right)$ as in Figure B.

   The perimeter of nine small triangles is $9\left(\frac{3}{4}\right)$ $\left(\frac{1}{4} \text{ on each side}\right)$ as in Figure C.

   The perimeter of all shaded triangles
   $= 3 + 3\left(\frac{3}{2}\right) + 9\left(\frac{3}{4}\right)$
   $= 3 + \frac{9}{2} + \frac{27}{4}$
   $= 13\frac{5}{4} = 14\frac{1}{4} \text{ cm}$

43.   The ranger could measure the circumference of the trunk of the tree and then solve the equation $C = \pi d$ for d.

Section 12.3

Objective A Exercises

1. $A = LW = 24 \text{ ft} \cdot 6 \text{ ft} = 144 \text{ ft}^2$

3. $A = s^2 = (9 \text{ in.})^2 = 81 \text{ in}^2$

5. $A = \pi r^2$
   $\approx 3.14(4 \text{ ft})^2 = 50.24 \text{ ft}^2$

7. $A = \frac{1}{2}bh$
   $= \frac{1}{2} \cdot (10 \text{ in.})(4 \text{ in.}) = 20 \text{ in}^2$

9. $A = \frac{1}{2}bh$
   $= \frac{1}{2} \cdot 3 \text{ cm} \cdot 1.42 \text{ cm} = 2.13 \text{ cm}^2$

11. $A = s^2 = 4 \text{ ft} \cdot 4 \text{ ft} = 16 \text{ ft}^2$

13. $A = LW = 43 \text{ in.} \cdot 19 \text{ in.} = 817 \text{ in}^2$

15. $A = \pi r^2 \approx \frac{22}{7} \cdot 7 \text{ in.} \cdot 7 \text{ in.} = 154 \text{ in}^2$

Objective B Exercises

17.   Area = area of rectangle − area of triangle
   $= (LW) - \left(\frac{1}{2}bh\right)$
   $= (8 \text{ cm} \cdot 4 \text{ cm}) - \left(\frac{1}{2} \cdot 4 \text{ cm} \cdot 3 \text{ cm}\right)$
   $= 32 \text{ cm}^2 - 6 \text{ cm}^2$
   $= 26 \text{ cm}^2$

19. Area = area of rectangle − area of triangle

$$= (LW) - \left(\frac{1}{2} bh\right)$$

$$= (80 \text{ cm} \cdot 30 \text{ cm}) - \left(\frac{1}{2} \cdot 30 \text{ cm} \cdot 12 \text{ cm}\right)$$

$$= 2400 \text{ cm}^2 - 180 \text{ cm}^2$$

$$= 2220 \text{ cm}^2$$

21. Area = area of circle − $\frac{1}{4}$ area of circle

$$= \pi r^2 - \frac{1}{4} \cdot \pi r^2$$

$$\approx 3.14(8 \text{ in.})^2 - \frac{1}{4} \cdot 3.14(8 \text{ in.})^2$$

$$= 200.96 \text{ in}^2 - 50.24 \text{ in}^2$$

$$= 150.72 \text{ in}^2$$

23. Area = area of rectangle − $\frac{1}{2}$ area of circle

$$= LW - \frac{1}{2} \cdot \pi r^2$$

$$\approx 4.38 \text{ ft} \cdot 3.74 \text{ ft} - \frac{1}{2} \cdot 3.14(2.19 \text{ ft})^2$$

$$= 16.3812 \text{ ft}^2 - 7.529877 \text{ ft}^2$$

$$= 8.851323 \text{ ft}^2$$

Objective C Exercises

25. **Strategy**   To find the area of the playing field, find the area of a rectangle with length 100 yd and width 75 yd.

   Solution $A = LW$
   $$= 100 \text{ yd} \cdot 75 \text{ yd}$$
   $$= 7500 \text{ yd}^2$$
   The area of the playing field is 7500 yd².

27. **Strategy**   To find the area of the field, find the area of a circle with a radius of 50 ft.

   Solution $A = \pi r^2 \approx 3.14 \cdot 50 \text{ ft} \cdot 50 \text{ ft}$
   $$= 7850 \text{ ft}^2$$
   The area watered by the irrigation system is approximately 7850 ft².

29. **Strategy**   To find the amount of stain:
   ● Find the area of a rectangle that measures 10 ft by 8 ft.
   ● Divide the area by the area that one quart of stain will cover (50 ft²).

   Solution $A = LW$
   $$= 10 \text{ ft} \cdot 8 \text{ ft}$$
   $$= 80 \text{ ft}^2$$
   $80 \text{ ft}^2 \div 50 \text{ ft}^2 = 1.6$
   It will take 1.6 quarts of stain.
   You should buy 2 qt.

31. **Strategy**   To find the area of the driveway, subtract the area of the small rectangle from the area of the large rectangle.

   Solution   Area = area of large rectangle
   − area of small rectangle
   = (length · width) − (length · width)
   = (75 ft · 30 ft) − (50 ft · 20 ft)
   = 2250 ft² − 1000 ft²
   = 1250 ft²
   The area of the driveway is 1250 ft².

33. **Strategy**   To find the cost of the wallpaper:
   ● Use the formula for the area of a rectangle to find the areas of the two walls.
   ● Add the areas of the two walls.
   ● Divide the total area by the area in one roll (40 ft²) to find the total number of rolls.
   ● Multiply the number of rolls by $28.50.

   Solution $\text{Area}_1 = 9 \text{ ft} \cdot 8 \text{ ft} = 72 \text{ ft}^2$
   $\text{Area}_2 = 11 \text{ ft} \cdot 8 \text{ ft} = 88 \text{ ft}^2$
   Total area = 72 ft² + 88 ft² = 160 ft²
   160 ft² ÷ 40 ft² = 4
   4 · $28.50 = $114
   The cost to wallpaper the two walls is $114.

35. **Strategy**   To find the amount budgeted:
   ● Use the formula for the area of a square to find the area of the park.
   ● Divide the area by 1200 ft² to find the number of bags of seed.
   ● Multiply the number of bags of seed by $5.75.

   Solution $A = s^2$
   $$= (60 \text{ ft})^2 = 3{,}600 \text{ ft}^2$$
   3,600 ft² ÷ 1,200 ft² = 3
   3 · $5.75 = $17.25
   $17.25 should be budgeted for grass seed.

37. **Strategy**   To find the total area of the park:
    - Find the area of the first rectangle.
    - Find the area of the second rectangle.
    - Find the area of the semicircle.
    - Add all the areas together.

    Solution   Area of first rectangle $= 12.7 \times 2.5 = 31.75$
    Area of second rectangle $= 17.5 \times 4.3 = 75.25$

    Area of semicircle $= \dfrac{1}{2} \times 3.14 \times (3.4)^2 = 18.1492$

    Total area $= 31.75 + 75.25 + 18.1492 = 125.1492$
    The total area of the park is 125.1492 mi^2.

39a. **Strategy**   To find the increase in area:
    - Find the area of the original circle.
    - Find the area of the increased circle.
    - Subtract the area of the original circle from the area of the larger circle.

    Solution   Area of original circle $\approx (64)(3.14) = 200.96$ in^2
    Area of increased circle $\approx (100)(3.14) = 314$ in^2
    $314 - 200.96 = 113.04$ in^2
    The amount of the increase is about 113.04 in^2.

b. **Strategy**   To find the increase in area:
    - Find the area of the original circle.
    - Find the area of the increased circle.
    - Subtract the area of the original circle from the area of the larger circle.

    Solution   Area of original circle $\approx (25)(3.14) = 78.5$ cm^2
    Area of increased circle $\approx (100)(3.14) = 314$ cm^2
    $314 - 78.5 = 235.5$ cm^2
    The amount of the increase is about 235.5 cm^2.

Applying the Concepts

41a.   Area of original rectangle $= LW$
    Area of doubled rectangle $= (2L)(2W)$
    $ = 4LW$
    $ = 4 \times$ Area of original rectangle
    If the length and width are doubled, the area is increased 4 times.

b.   Area of original circle $= \pi r^2$
    Area with radius doubled $= \pi(2r)^2$
    $ = \pi\, 4r^2$
    $ = 4(\pi r^2)$
    $ = 4 \times$ Area of original circle
    If the radius is doubled, the area is quadrupled.

c.   If the diameter is doubled, the radius is
    doubled. From (b) if the radius is doubled, the area is quadrupled.

43a.   Sometimes true

b.   Sometimes true

c.   Always true

Section 12.4

Objective A Exercises

1. $V = LWH$
    $= 12 \text{ cm} \cdot 4 \text{ cm} \cdot 3 \text{ cm} = 144 \text{ cm}^3$

3. $V = s^3 = (8 \text{ in.})^3 = 512 \text{ in}^3$

5. $V = \dfrac{4}{3}\pi r^3 \approx \dfrac{4}{3}(3.14)(8 \text{ in.})^3$
    $\approx 2143.57 \text{ in}^3$

7. $V = \pi r^2 h$
    $\approx 3.14(2 \text{ cm})^2 \cdot 12 \text{ cm} = 150.72 \text{ cm}^3$

9. $V = LWH$
    $= 2 \text{ m} \cdot 0.8 \text{ m} \cdot 4 \text{ m} = 6.4 \text{ m}^3$

11. $V = \dfrac{4}{3}\pi r^3$
    $\approx \dfrac{4}{3} \cdot 3.14(11 \text{ mm})^3$
    $\approx 5572.45 \text{ mm}^3$

13. $r = \dfrac{1}{2}d = \dfrac{1}{2} \cdot 12 \text{ ft} = 6 \text{ ft}$
    $V = \pi r^2 h$
    $\approx 3.14(6 \text{ ft})^2(30 \text{ ft})$
    $= 3391.2 \text{ ft}^3$

15. $V = s^3$
    $= 3\dfrac{1}{2} \text{ ft} \cdot 3\dfrac{1}{2} \text{ ft} \cdot 3\dfrac{1}{2} \text{ ft} = 42\dfrac{7}{8} \text{ ft}^3$

Objective B Exercises

17. Volume $= \dfrac{1}{2}$volume of cylinder $+$ volume of rectangular solid
    $= \left[\dfrac{1}{2} \cdot \pi(\text{radius})^2 \cdot \text{height}\right] + (\text{length} \cdot \text{width} \cdot \text{height})$
    $\approx \left(\dfrac{1}{2} \cdot 3.14 \cdot 3 \text{ in.} \cdot 3 \text{ in.} \cdot 2 \text{ in.}\right) + (6 \text{ in.} \cdot 9 \text{ in.} \cdot 1 \text{ in.}) = 28.26 \text{ in}^3 + 54 \text{ in}^3 = 82.26 \text{ in}^3$

19. Volume $=$ volume of rectangular solid $-$ volume of cylinder $= (\text{length} \cdot \text{width} \cdot \text{height})$
    $- [\pi(\text{radius})^2 \cdot \text{height}]$
    $\approx (1.20 \text{ m} \cdot 2 \text{ m} \cdot 0.80 \text{ m}) - (3.14 \cdot 0.20 \text{ m} \cdot 0.20 \text{ m} \cdot 2 \text{ m})$
    $= 1.92 \text{ m}^3 - 0.2512 \text{ m}^3 = 1.6688 \text{ m}^3$

21. Volume $=$ volume of cylinder $+$ volume of cylinder
    $= [\pi(\text{radius})^2 \cdot \text{height}] + [\pi(\text{radius})^2 \cdot \text{height}]$
    $\approx (3.14 \cdot 3 \text{ in.} \cdot 3 \text{ in.} \cdot 2 \text{ in.}) + (3.14 \cdot 1 \text{ in.} \cdot 1 \text{ in.} \cdot 4 \text{ in.})$
    $= 56.52 \text{ in}^3 + 12.56 \text{ in}^3 = 69.08 \text{ in}^3$

Objective C Exercises

23. **Strategy**   To find the volume of the tank, use the formula for the volume of a rectangular solid.

     Solution $V = LWH$
     $= 9 \text{ m} \cdot 3 \text{ m} \cdot 1.5 \text{ m}$
     $= 40.5 \text{ m}^3$

     The volume of the water in the tank is 40.5 m^3.

25. Strategy To find the volume of the balloon, use the formula for the volume of a sphere.

Solution
$$V = \frac{4}{3}\pi r^3$$
$$\approx \frac{4}{3} \cdot 3.14(16 \text{ ft})^3$$
$$\approx 17{,}148.59 \text{ ft}^3$$
The volume is approximately $17{,}148.59 \text{ ft}^3$.

27. Strategy To find the amount of oil:
• Use the formula for the volume of a cylinder to find the volume of the tank.
• Multiply the volume by $\frac{2}{3}$ to find the amount of oil.

Solution
$$V = \pi r^2 h$$
$$\approx 3.14(3 \text{ m})^2(4 \text{ m})$$
$$= 113.04 \text{ m}^3$$

Amount of oil $= \frac{2}{3} \cdot 113.04 \text{ m}^3$
$$= 75.36 \text{ m}^3$$

The amount of oil is approximately 75.36 m^3.

29. Strategy To find the volume of the auditorium, find the volume of the rectangular solid and add half the volume of the cylinder.

Solution
$$V = LWH$$
$$+ \frac{1}{2}\left[\pi r^2 h\right]$$
$$\approx 125 \text{ ft} \cdot 94 \text{ ft} \cdot 32 \text{ ft}$$
$$+ \frac{1}{2}[3.14(47 \text{ ft})^2 \cdot 125 \text{ ft}]$$
$$= 376{,}000 \text{ ft}^3 + 433{,}516.25 \text{ ft}^3$$
$$= 809{,}516.25 \text{ ft}^3$$
The volume of the auditorium is approximately $809{,}516.25 \text{ ft}^3$.

31. Strategy To find the volume of the bushing, subtract the volume of the half-cylinder from the volume of the rectangular solid.

Solution
$$V = LWH$$
$$- \frac{1}{2}\left[\pi r^2 h\right]$$
$$\approx (12 \text{ in.} \cdot 84 \text{ in.} \cdot 3 \text{ in.})$$
$$- \frac{1}{2}(3.14)(2 \text{ in.})^2(12 \text{ in.})$$
$$= 288 \text{ in}^3 - 75.36 \text{ in}^3$$
$$= 212.64 \text{ in}^3$$
The volume of the bushing is approximately 212.64 in^3.

33. Strategy To find the number of gallons in the fish tank:
• Use the formula for the volume of a rectangular solid.
• Convert the volume to gallons.

Solution
$$V = LWH$$
$$= 12 \text{ in.} \cdot 8 \text{ in.} \cdot 9 \text{ in.}$$
$$= 864 \text{ in}^3$$

$$864 \text{ in}^3 = 864 \text{ in}^3 \times \frac{1 \text{ gal}}{231 \text{ in}^3} \approx 3.7 \text{ gal}$$
The tank will hold 3.7 gal.

35. Strategy To find the cost of the floor:
• Find the volume. The volume is equal to the volume of a rectangular solid plus one half the volume of the cylinder. The radius is one half the length of the rectangular solid.
• Multiply the volume by $5.85.

Solution
$$V = LWH$$
$$+ \frac{1}{2}\pi r^2 h$$
$$\approx 50 \text{ ft} \cdot 25 \text{ ft} \cdot \frac{1}{2} \text{ ft} \cdot$$
$$+ \frac{1}{2}(3.14)(25 \text{ ft})^2\left(\frac{1}{2} \text{ ft}\right)$$
$$= 625 \text{ ft}^3 + 490.625 \text{ ft}^3$$
$$= 1115.625 \text{ ft}^3$$

Cost $= 1115.625 \times \$5.85 \approx 6526.41$
The cost is approximately $6526.41.

Applying the Concepts

37a. Volume $=$ length $\cdot$ width $\cdot$ height.
If the length and width are doubled,
Volume $= 2 \cdot$ length $\cdot 2 \cdot$ width $\cdot$ height
$= 4 \cdot$ length $\cdot$ width $\cdot$ height
The volume will be 4 times larger.

b. Volume $=$ length $\cdot$ width $\cdot$ height.
If the length, width and height are doubled,
Volume $= 2 \cdot$ length $\cdot 2 \cdot$ width $\cdot 2 \cdot$ height
$= 8 \cdot$ length $\cdot$ width $\cdot$ height
The volume will be 8 times larger.

c. Volume $= (\text{side})^3$. If the side is doubled,
Volume $= (2.\text{side})^3 = 8.(\text{side})^3$
The volume will be 8 times larger.

39. For example, beginning at an edge that is perpendicular to the bottom face, cut at an angle through to the bottom face.

41. For example, beginning on the top face, at a distance d from a vertex, cut across the cube to a point just below the opposite vertex.

Section 12.5

Objective A Exercises

1. 2.646

3. 6.481

5. 12.845

7. 13.748

Objective B Exercises

9. Hypotenuse $= \sqrt{(\text{leg})^2 + (\text{leg})^2}$
$= \sqrt{(3 \text{ in.})^2 + (4 \text{ in.})^2}$
$= \sqrt{9 \text{ in}^2 + 16 \text{ in}^2}$
$= \sqrt{25 \text{ in}^2}$
$= 5 \text{ in.}$

11. Hypotenuse $= \sqrt{(\text{leg})^2 + (\text{leg})^2}$
$= \sqrt{(5 \text{ cm})^2 + (7 \text{ cm})^2}$
$= \sqrt{25 \text{ cm}^2 + 49 \text{ cm}^2}$
$= \sqrt{74 \text{ cm}^2}$
$\approx 8.602 \text{ cm}$

13. Leg $= \sqrt{(\text{hypotenuse})^2 - (\text{leg})^2}$
$= \sqrt{(15 \text{ ft})^2 - (10 \text{ ft})^2}$
$= \sqrt{225 \text{ ft}^2 - 100 \text{ ft}^2}$
$= \sqrt{125 \text{ ft}^2}$
$\approx 11.180 \text{ ft}$

15. Leg $= \sqrt{(\text{hypotenuse})^2 - (\text{leg})^2}$
$= \sqrt{(6 \text{ cm})^2 - (4 \text{ cm})^2}$
$= \sqrt{36 \text{ cm}^2 - 16 \text{ cm}^2}$
$= \sqrt{20 \text{ cm}^2}$
$\approx 4.472 \text{ cm}$

17. Hypotenuse $= \sqrt{(\text{leg})^2 + (\text{leg})^2}$
$= \sqrt{(9 \text{ yd})^2 + (9 \text{ yd})^2}$
$= \sqrt{81 \text{ yd}^2 + 81 \text{ yd}^2}$
$= \sqrt{162 \text{ yd}^2}$
$\approx 12.728 \text{ yd}$

19. Leg $= \sqrt{(\text{hypotenuse})^2 - (\text{leg})^2}$
$= \sqrt{(12 \text{ ft})^2 - (6 \text{ ft})^2}$
$= \sqrt{144 \text{ ft}^2 - 36 \text{ ft}^2}$
$= \sqrt{108 \text{ ft}^2}$
$\approx 10.392 \text{ ft}$

21. Hypotenuse $= \sqrt{(\text{leg})^2 + (\text{leg})^2}$
$= \sqrt{(15 \text{ cm})^2 + (15 \text{ cm})^2}$
$= \sqrt{225 \text{ cm}^2 + 225 \text{ cm}^2}$
$= \sqrt{450 \text{ cm}^2}$
$\approx 21.213 \text{ cm}$

23. Hypotenuse $= \sqrt{(\text{leg})^2 + (\text{leg})^2}$
$= \sqrt{(8 \text{ m})^2 + (4 \text{ m})^2}$
$= \sqrt{64 \text{ m}^2 + 16 \text{ m}^2}$
$= \sqrt{80 \text{ m}^2}$
$\approx 8.944 \text{ m}$

25. Leg $= \sqrt{(\text{hypotenuse})^2 - (\text{leg})^2}$
$= \sqrt{(11.3 \text{ yd})^2 - (8.1 \text{ yd})^2}$
$= \sqrt{127.69 \text{ yd}^2 - 65.61 \text{ yd}^2}$
$= \sqrt{62.08 \text{ yd}^2}$
$\approx 7.879 \text{ yd}$

Objective C Exercises

27. **Strategy** To find the length of the ramp, use the Pythagorean Theorem. The ramp is the hypotenuse of a right triangle.

    Solution Hypotenuse $= \sqrt{(\text{leg})^2 + (\text{leg})^2}$
    $= \sqrt{(9 \text{ ft})^2 + (3.5 \text{ ft})^2}$
    $= \sqrt{81 \text{ ft}^2 + 12.25 \text{ ft}^2}$
    $= \sqrt{93.25 \text{ ft}^2}$
    $\approx 9.66 \text{ ft}$
    The ramp is 9.66 ft long.

29. **Strategy**
    • Traveling 18 mi east and then 12 mi north forms a right angle. The distance from the starting point is the hypotenuse of the triangle with legs 18 mi and 12 mi.
    • Find the hypotenuse of the right triangle.

    Solution Hypotenuse $= \sqrt{(\text{leg})^2 + (\text{leg})^2}$
    $= \sqrt{(18 \text{ mi})^2 + (12 \text{ mi})^2}$
    $= \sqrt{324 \text{ mi}^2 + 144 \text{ mi}^2}$
    $= \sqrt{468 \text{ mi}^2}$
    $\approx 21.6 \text{ mi}$
    The distance is 21.6 mi.

31. **Strategy**
• The angles of a rectangle are right angles. The length (11 mi) and width (5 mi) are the legs of a right triangle. The diagonal is the hypotenuse.
• Find the length of the hypotenuse.

Solution
$$\text{Hypotenuse} = \sqrt{(\text{leg})^2 + (\text{leg})^2}$$
$$= \sqrt{(11 \text{ mi})^2 + (5 \text{ mi})^2}$$
$$= \sqrt{121 \text{ mi}^2 + 25 \text{ mi}^2}$$
$$= \sqrt{146 \text{ mi}^2}$$
$$\approx 12.1 \text{ mi}$$
The length of the diagonal is 12.1 mi.

33. **Strategy**
• To find how high on the building the ladder reaches, use the Phythagorean Theorem. The hypotenuse is the length of the ladder (8 m). One leg is the distance from the base of the building to the base of the ladder (3 m). The distance from the ground to the top of the ladder is the unknown leg.

Solution
$$\text{Leg} = \sqrt{(\text{hypotenuse})^2 - (\text{leg})^2}$$
$$= \sqrt{(8 \text{ m})^2 - (3 \text{ m})^2}$$
$$= \sqrt{64 \text{ m}^2 - 9 \text{ m}^2}$$
$$= \sqrt{55 \text{ m}^2}$$
$$\approx 7.4 \text{ m}$$
The ladder is 7.4 m high on the building.

35. **Strategy**
• Use the Pythagorean Theorem to find the length of the unknown side.
• Add the lengths of the sides to find the perimeter of the right triangle.

Solution
$$\text{Hypotenuse} = \sqrt{(\text{leg})^2 + (\text{leg})^2}$$
$$= \sqrt{(6 \text{ in.})^2 + (10 \text{ in.})^2}$$
$$= \sqrt{36 \text{ in}^2 + 100 \text{ in}^2}$$
$$= \sqrt{136 \text{ in}^2}$$
$$\approx 11.7 \text{ in.}$$
6 in. + 10 in. + 11.7 in. = 27.7 in.
The perimeter is 27.7 in.

37. **Strategy**
To find the distance from the corner to the memorial, use the Pythagorean Theorem. The length of one leg is 600 ft. The length of the hypotenuse is 650 ft. The distance from the corner to the memorial is the length of the unknown side.

Solution
$$\text{Leg} = \sqrt{(\text{hypotenuse})^2 - (\text{leg})^2}$$
$$= \sqrt{(650 \text{ ft})^2 - (600 \text{ ft})^2}$$
$$= \sqrt{422,500 \text{ ft}^2 - 360,000 \text{ ft}^2}$$
$$= \sqrt{62,500 \text{ ft}^2}$$
$$= 250 \text{ ft}$$
The distance is 250 ft.

39. **Strategy**
To find the distance between the centers of adjacent holes, use the Pythagorean Theorem. The distance between holes is the hypotenuse of a right triangle. Each of the legs is 3 in. long.

Solution
$$\text{Hypotenuse} = \sqrt{(\text{leg})^2 + (\text{leg})^2}$$
$$= \sqrt{(3 \text{ in.})^2 + (3 \text{ in.})^2}$$
$$= \sqrt{9 \text{ in}^2 + 9 \text{ in}^2}$$
$$= \sqrt{18 \text{ in}^2}$$
$$\approx 4.243 \text{ in.}$$
The distance is 4.243 in.

Applying the Concepts

41a. Always true

 b. Always true

43. No, the Pythagorean Theorem can be used only to find unknown lengths of sides of right triangles. No right angle is indicated in the triangle in the diagram.

45. **Strategy**
To find the total length of the pipe needed to connect the buildings:
• Use the Pythagorean Theorem to find the length of the middle section. The middle section is the hypotenuse of a right triangle with legs 4 m and 3 m.
• Find the sum of the three sections of pipe.

Solution
$$\text{Hypotenuse} = \sqrt{(\text{leg})^2 + (\text{leg})^2}$$
$$= \sqrt{(4 \text{ m})^2 + (3 \text{ m})^2}$$
$$= \sqrt{16 \text{ m}^2 + 9 \text{ m}^2}$$
$$= \sqrt{25 \text{ m}^2}$$
$$\approx 5 \text{ m}$$
Total length = 2 m + 5 m + 1 m = 8 m.
The total length of the pipe needed is 8 m.

Section 12.6

Objective A Exercises

1. $\dfrac{5\,\text{m}}{10\,\text{m}} = \dfrac{1}{2}$

3. $\dfrac{9\,\text{in.}}{12\,\text{in.}} = \dfrac{3}{4}$

5. $\angle CAB = \angle DEF$
   $AC = ED$ and
   $AB = EF$
   Therefore SAS applies and the triangles are congruent.

7. $AC = EF$
   $AB = DE$ and
   $BC = FD$
   Therefore SSS applies and the triangles are congruent.

9. $\dfrac{AC}{DF} = \dfrac{AB}{DE}$
   $\dfrac{5\,\text{cm}}{9\,\text{cm}} = \dfrac{4\,\text{cm}}{DE}$
   $5 \times DE = 4\,\text{cm} \times 9$
   $5 \times DE = 36\,\text{cm}$
   $DE = 36\,\text{cm} \div 5$
   $DE = 7.2\,\text{cm}$

11. $\dfrac{AC}{DF} = \dfrac{\text{height of triangle } ABC}{\text{height of triangle } DEF}$
   $\dfrac{3\,\text{m}}{5\,\text{m}} = \dfrac{2\,\text{m}}{\text{height}}$
   $3 \times \text{height} = 5 \times 2\,\text{m}$
   $3 \times \text{height} = 10\,\text{m}$
   $\text{Height} = 10\,\text{m} \div 3$
   $\text{Height} \approx 3.3\,\text{m}$

Objective B Exercises

13. **Strategy** To find the height of the building, solve a proportion.

   Solution $\dfrac{\text{height}}{8\,\text{m}} = \dfrac{8\,\text{m}}{4\,\text{m}}$
   $\text{Height} \times 4 = 8\,\text{m} \times 8$
   $\text{Height} \times 4 = 64\,\text{m}$
   $\text{Height} = 64\,\text{m} \div 4$
   $\text{Height} = 16\,\text{m}$
   The height of the building is 16 m.

15. **Strategy** To find the perimeter:
   ● Solve a proportion to find the length of side BC.
   ● Add the three sides of triangle ABC.

Solution $\dfrac{AC}{DF} = \dfrac{BC}{EF}$
   $\dfrac{4\,\text{m}}{8\,\text{m}} = \dfrac{BC}{6\,\text{m}}$
   $8 \times BC = 4 \times 6\,\text{m}$
   $8 \times BC = 24\,\text{m}$
   $BC = 24\,\text{m} \div 8$
   $BC = 3\,\text{m}$

   $3\,\text{m} + 4\,\text{m} + 5\,\text{m} = 12\,\text{m}$
   The perimeter is 12 m.

17. **Strategy** To find the area:
   ● Solve a proportion to find the height of triangle ABC.
   ● Use the formula $A = \dfrac{1}{2}\,bh$.

   Solution $\dfrac{AB}{DE} = \dfrac{\text{height of triangle } ABC}{\text{height of triangle } DEF}$
   $\dfrac{15\,\text{cm}}{40\,\text{cm}} = \dfrac{\text{height}}{20\,\text{cm}}$
   $40 \times \text{height} = 15 \times 20\,\text{cm}$
   $40 \times \text{height} = 300\,\text{cm}$
   $\text{Height} = 300\,\text{cm} \div 40$
   $\text{Height} = 7.5\,\text{cm}$

   $A = \dfrac{1}{2}\,bh$
   $= \dfrac{1}{2} \cdot 15\,\text{cm} \cdot 7.5\,\text{cm}$
   $= 56.25\,\text{cm}^2$.
   The area is 56.25 cm^2.

Applying the Concepts

19a. Always true

  b. Sometimes true

  c. Always true

21. $\triangle ABC$ and $\triangle DEC$ are similar triangles.
   $\dfrac{8}{5} = \dfrac{6}{DE}$ $\dfrac{8}{5} = \dfrac{10}{EC}$
   $8DE = 30$ $8EC = 50$
   $DE = \dfrac{30}{8} = \dfrac{15}{4}$ $EC = \dfrac{50}{8} = \dfrac{25}{4}$
   The perimeter of the trapezoid
   $ABED = AB + AD + DE + BE$
   $= 6 + (8 - 5) + \dfrac{15}{4} + \left(10 - \dfrac{25}{4}\right)$
   $= 6 + 3 + 3.75 + 3.75$
   $= 16.5$

Chapter 12 Review Exercises

1. $r = \dfrac{1}{2}\,d = \dfrac{1}{2}(1.5\,\text{m}) = 0.75\,\text{m}$

2. $C = 2\pi r$
   $\approx 2(3.14)(5\,\text{cm}) = 31.4\,\text{cm}$

3. $P = 2L + 2W$
$= 2(8\,\text{ft}) + 2(5\,\text{ft})$
$= 16\,\text{ft} + 10\,\text{ft} = 26\,\text{ft}$

4. $AD = AB + BC + CD$
$24 = 15 + BC + 6$
$24 = 21 + BC$
$24 - 21 = 21 - 21 + BC$
$3 = BC$

5. Volume $= \text{length} \cdot \text{width} \cdot \text{height}$
$= 10\,\text{ft} \cdot 5\,\text{ft} \cdot 4\,\text{ft} = 200\,\text{ft}^3$

6. Hypotenuse $= \sqrt{(\text{leg})^2 + (\text{leg})^2}$
$= \sqrt{(10\,\text{cm})^2 + (24\,\text{cm})^2}$
$= \sqrt{100\,\text{cm}^2 + 576\,\text{cm}^2}$
$= \sqrt{676\,\text{cm}^2}$
$= 26\,\text{cm}$

7. Let x represent the supplement of $105°$. The sum of supplementary angles is $180°$.
$$x + 105° = 180°$$
$$x + 105° - 105° = 180° - 105°$$
$$x = 75°$$
$75°$ is the supplement of $105°$.

8. $\sqrt{15} \approx 3.873$

9. $\dfrac{BC}{EF} = \dfrac{\text{height of triangle } ABC}{\text{height of triangle } DEF}$
$\dfrac{12\,\text{cm}}{24\,\text{cm}} = \dfrac{8\,\text{cm}}{h}$
$12 \times h = 24 \times 8\,\text{cm}$
$12 \times h = 192\,\text{cm}$
$h = 192\,\text{cm} \div 12 = 16\,\text{cm}$

10. $A = \pi r^2$
$\approx 3.14 \cdot (4.5\,\text{cm})^2$
$= 63.585\,\text{cm}^2$

11a. Because line t is a transversal cutting parallel lines, $\angle b = 45°$.

b. $\angle a = 180° - 45° = 135°$

12. $A = LW$
$= 11\,\text{m} \times 5\,\text{m} = 55\,\text{m}^2$

13. Volume $=$ volume of larger rectangular solid
$\quad$ $-$ volume of smaller rectangular solid
$= \text{length} \cdot \text{width} \cdot \text{height}$
$\quad$ $- \text{length} \cdot \text{width} \cdot \text{height}$
$= 8\,\text{in.} \cdot 7\,\text{in.} \cdot 6\,\text{in.} - 8\,\text{in.} \cdot 4\,\text{in.} \cdot 3\,\text{in.}$
$= 336\,\text{in}^3 - 96\,\text{in}^3 = 240\,\text{in}^3$

14. Area $=$ area of rectangle $+ \dfrac{1}{2}$ (area of circle)
$= \text{length} \cdot \text{width} + \dfrac{1}{2}\pi(\text{radius})^2$
$\approx 8\,\text{in.} \cdot 4\,\text{in.} + \dfrac{1}{2}(3.14)(4\,\text{in.})^2$
$= 32\,\text{in}^2 + 25.12\,\text{in}^2$
$= 57.12\,\text{in}^2$

15. $V = \dfrac{4}{3}\pi r^3$
$\approx \dfrac{4}{3}(3.14)(4\,\text{ft})^3$
$\approx 267.9\,\text{ft}^3$

16. **Strategy**   To find the area:
● Solve a proportion to find the length of side DF (the base of the triangle DEF).
● Use the formula
$A = \dfrac{1}{2}bh.$

Solution $\dfrac{AC}{DF} = \dfrac{\text{height of triangle } ABC}{\text{height of triangle } DEF}$
$\dfrac{8\,\text{m}}{DF} = \dfrac{5\,\text{m}}{9\,\text{m}}$
$8\,\text{m} \times 9 = 5 \times DF$
$72\,\text{m} = 5 \times DF$
$72\,\text{m} \div 5 = DF$
$14.4\,\text{m} = DF$
$A = \dfrac{1}{2}bh$
$= \dfrac{1}{2}(14.4\,\text{m})(9\,\text{m}) = 64.8\,\text{m}^2$
The area is $64.8\,\text{m}^2$.

17. Perimeter $=$ length of two sides
$+ \dfrac{1}{2}$ circumference of circle
$\approx 2(16\,\text{in.}) + \dfrac{1}{2}(2 \cdot 3.14 \cdot 5\,\text{in.})$
$= 32\,\text{in.} + 15.7\,\text{in.} = 47.7\,\text{in.}$

18a. Because line t is a transversal cutting parallel lines, $\angle b = 80°$

b. $\angle a = 180° - 80° = 100°$

19. **Strategy** To find how high on the building the ladder will reach, use the Pythagorean Theorem. The hypotenuse is 17 ft and one leg is 8 ft. The other leg is the height up the building.

    Solution
    $$\text{leg} = \sqrt{(\text{hypotenuse})^2 - (\text{leg})^2}$$
    $$= \sqrt{(17\text{ ft})^2 - (8\text{ ft})^2}$$
    $$= \sqrt{289\text{ ft}^2 - 64\text{ ft}^2}$$
    $$= \sqrt{225\text{ ft}^2}$$
    $$= 15\text{ ft}$$

    The ladder will reach 15 ft up the building.

20. $90° - 32° = 58°$
    The other angles of the triangle are 90° and 58°.

21. **Strategy** To find how many feet the bicycle travels, find how many feet the wheel travels if it makes 10 revolutions:
    • Find how far the wheel travels when it makes 1 revolution by using the circumference formula.
    • Convert the circumference to feet.
    • Multiply the distance traveled in 1 revolution by 10.

    Solution
    $$C = \pi d$$
    $$= \pi \cdot 28 \text{ in.}$$
    $$\approx 3.14 \cdot (28 \text{ in.})$$
    $$= 87.92 \text{ in.}$$

    $$87.92 \text{ in.} = 87.92 \text{ in.} \times \frac{1 \text{ ft}}{12 \text{ in.}}$$
    $$= \frac{87.92}{12} \text{ ft} \approx 7.33 \text{ ft}$$

    $10 \times 7.33 \text{ ft} = 73.3 \text{ ft}$
    The bicycle travels approximately 73.3 ft in 10 revolutions.

22. **Strategy** To find the area of the room in square yards:
    • Use the area of the rectangle formula.
    • Convert square feet to square yards.

    Solution
    $$A = LW$$
    $$= 18 \text{ ft} \cdot 14 \text{ ft} = 252 \text{ ft}^2$$
    $$252 \text{ ft}^2 = 252 \text{ ft}^2 \times \frac{1 \text{ yd}^2}{9 \text{ ft}^2}$$
    $$= \frac{252}{9} \text{ yd}^2 = 28 \text{ yd}^2$$
    The area of the room is 28 yd^2.

23. **Strategy** To find the volume of the silo, use the formula for the volume of a cylinder.

    Solution
    $$V = \pi r^2 h$$
    $$\approx 3.14(4.5 \text{ ft})^2(18 \text{ ft})$$
    $$= 1144.53 \text{ ft}^3$$
    The volume of the silo is approximately 1144.53 ft^3.

24. $A = \frac{1}{2}bh$
    $$= \frac{1}{2}(8 \text{ m})(2.75 \text{ m})$$
    $$= 11 \text{ m}^2$$

25. **Strategy** • Traveling 20 mi west and then 21 mi south forms a right angle. The distance from the starting point is the hypotenuse of the triangle with legs 20 mi and 21 mi.
    • Find the hypotenuse of the right triangle.

    Solution
    $$\text{hypotenuse} = \sqrt{(\text{leg})^2 + (\text{leg})^2}$$
    $$= \sqrt{(20 \text{ mi})^2 + (21 \text{ mi})^2}$$
    $$= \sqrt{400 \text{ mi}^2 + 441 \text{ mi}^2}$$
    $$= \sqrt{841 \text{ mi}^2}$$
    $$= 29$$
    The distance from the starting point is 29 mi.

Chapter 12 Test

1. $V = \pi r^2 h$
   $$\approx 3.14 \cdot (3 \text{ m})^2 \cdot 6 \text{ m}$$
   $$= 169.56 \text{ m}^3$$

2. $P = 2L + 2W$
   $$= 2(2 \text{ m}) + 2(1.4 \text{ m})$$
   $$= 4 \text{ m} + 2.8 \text{ m}$$
   $$= 6.8 \text{ m}$$

3. **Strategy** To find the volume of the composite figure, subtract the volume of the smaller cylinder from the volume of the larger cylinder.

    Solution
    Volume = volume of larger cylinder
    $\qquad$ − volume of smaller cylinder
    $$= \pi(\text{radius})^2 \cdot \text{height}$$
    $$\quad - \pi \cdot (\text{radius})^2 \cdot \text{height}$$
    $$\approx 3.14(6 \text{ cm})^2 \cdot 14 \text{ cm}$$
    $$\quad - 3.14 \cdot (2 \text{ cm})^2 \cdot 14 \text{ cm}$$
    $$= 1582.56 \text{ cm}^3 - 175.84 \text{ cm}^3$$
    $$= 1406.72 \text{ cm}^3$$
    The volume of the composite figure is approximately 1406.72 cm^3.

4. Strategy To find the missing length, use the Pythagorean Theorem. $AB = FE$ is the hypotenuse. The legs are 6 and 8 m.

Solution Hypotenuse $= \sqrt{(8\,\text{m})^2 + (6\,\text{m})^2}$
$= \sqrt{64\,\text{m}^2 + 36\,\text{m}^2}$
$= \sqrt{100\,\text{m}^2}$
$= 10\,\text{m}$
The length of FE is 10 m.

5. $90° - 32° = 58°$
$58°$ is the complement of $32°$.

6. $A = \pi r^2$
$\approx \dfrac{22}{7}(1\,\text{m})^2$
$= \dfrac{22\,\text{m}^2}{7} = 3\dfrac{1}{7}\,\text{m}^2$

7. Angles x and z are supplementary; therefore, $\angle z = 180° - 30° = 150°$. $\angle y$ and $\angle z$ are corresponding angles; therefore, $\angle y = \angle z = 150°$.

8. Perimeter $=$ two lengths $+$ circumference of circle
$= 2(4\,\text{ft}) + \pi \cdot \text{diameter}$
$= 8\,\text{ft} + \pi\left(2\dfrac{1}{2}\,\text{ft}\right)$
$\approx 8\,\text{ft} + 3.14(2.5\,\text{ft})$
$= 15.85\,\text{ft}$

9. $\sqrt{189} \approx 13.748$

10. Leg $= \sqrt{(\text{hypotenuse})^2 - (\text{leg})^2}$
$= \sqrt{(12\,\text{ft})^2 - (7\,\text{ft})^2} = \sqrt{144\,\text{ft}^2 - 49\,\text{ft}^2}$
$= \sqrt{95\,\text{ft}^2}$
$\approx 9.747\,\text{ft}$

11. Area $=$ area of rectangle $-$ area of triangle
$= \text{length} \cdot \text{width} - \dfrac{1}{2} \cdot \text{base} \cdot \text{height}$
$= 3\,\text{ft} \cdot 4\dfrac{1}{2}\,\text{ft} - \dfrac{1}{2}\left(4\dfrac{1}{2}\,\text{ft}\right)\left(1\dfrac{1}{2}\,\text{ft}\right)$
$= 13.5\,\text{ft}^2 - 3.375\,\text{ft}^2 = 10.125\,\text{ft}^2$

12. Angles x and b are supplementary angles.
$\angle x + \angle b = 180°$
$45° + \angle b = 180°$
$45° - 45° + \angle b = 180° - 45°$
$\angle b = 135°$
$\angle a = \angle x$ because $\angle a$ and $\angle x$ are alternate exterior angles. $\angle a = 45°$

13. $\dfrac{AB}{DE} = \dfrac{BC}{EF}$
$\dfrac{\frac{3}{4}\,\text{ft}}{2\frac{1}{2}\,\text{ft}} = \dfrac{BC}{4\,\text{ft}}$
$\dfrac{3}{4} \times 4\,\text{ft} = 2\dfrac{1}{2} \times BC$
$3\,\text{ft} = 2\dfrac{1}{2} \times BC$
$3\,\text{ft} \div 2\dfrac{1}{2} = BC$
$3\,\text{ft} \times \dfrac{2}{5} = BC$
$BC = \dfrac{6}{5}\,\text{ft} = 1\dfrac{1}{5}\,\text{ft}$

14. $90° - 40° = 50°$
The other two angles of the triangle are 90° and 50°.

15. Strategy To find the width of the canal, solve a proportion.

Solution $\dfrac{5\,\text{ft}}{\text{Width of canal}} = \dfrac{12\,\text{ft}}{60\,\text{ft}}$
$5\,\text{ft} \times 60 = 12 \times \text{width of canal}$
$300\,\text{ft} = 12 \times \text{width of canal}$
$300\,\text{ft} \div 12 = \text{width of canal}$
$25\,\text{ft} = \text{width of canal}$
The width of the canal is 25 ft.

16. Strategy To find how much more pizza is contained in the larger pizza, subtract the area of the smaller pizza from the area of the larger pizza.

Solution $A = \pi r^2$
$\approx 3.14 \cdot (10\,\text{in.})^2 = 314\,\text{in}^2$
$A = \pi r^2$
$\approx 3.14 \cdot (8\,\text{in.})^2 = 200.96\,\text{in}^2$
$314\,\text{in}^2 - 200.96\,\text{in}^2 = 113.04\,\text{in}^2$
The amount of extra pizza is 113.04 in²

17. Strategy To find the cost of the carpet:
• Subtract the area of the smaller rectangle from the area of the larger rectangle.
• Convert the area to square yards.
• Multiply the area in square yards by the cost per square yard.

Solution Area $=$ area of larger rectangle $-$ area of smaller rectangle
$= \text{length} \cdot \text{width} - \text{length} \cdot \text{width}$
$= 20\,\text{ft} \cdot 22\,\text{ft} - 6\,\text{ft} \cdot 11\,\text{ft}$
$= 440\,\text{ft}^2 - 66\,\text{ft}^2 = 374\,\text{ft}^2$
$374\,\text{ft}^2 = 374\,\text{ft}^2 \times \dfrac{1\,\text{yd}^2}{9\,\text{ft}^2} \approx 41.5556\,\text{yd}^2$
$41.5556\,\text{yd}^2 \times \$26.80 \approx \$1113.69$
It will cost $1113.69 to carpet the area.

18. Strategy To find the cross-sectional area of the redwood tree:
- Convert the diameter (11 ft 6 in.) to feet.
- Use the formula $r = \dfrac{1}{2}d$ to find radius.
- Use the formula for area of a circle.

Solution
$$6 \text{ in.} = 6 \text{ in.} \times \dfrac{1 \text{ ft}}{12 \text{ in.}} = 0.5 \text{ ft}$$
$$11 \text{ ft } 6 \text{ in.} = 11.5 \text{ ft}$$
$$r = \dfrac{1}{2}d = \dfrac{1}{2}(11.5 \text{ ft}) = 5.75 \text{ ft}$$
$$A = \pi r^2$$
$$\approx (3.14)(5.75 \text{ ft})^2$$
$$\approx 103.82 \text{ ft}^2$$

The cross-sectional area is approximately 103.82 ft^2.

19. Strategy To find the length of the rafter:
- Use the Pythagorean Theorem to find the part of the rafter that covers the roof.
- Find the total length of the rafter by adding the 2 ft overhang to the part that covers the roof.

Solution
$$\text{Hypotenuse} = \sqrt{(5 \text{ ft})^2 + (12 \text{ ft})^2}$$
$$= \sqrt{25 \text{ ft}^2 + 144 \text{ ft}^2}$$
$$= \sqrt{169 \text{ ft}^2} = 13 \text{ ft}$$

$13 \text{ ft} + 2 \text{ ft} = 15 \text{ ft}$
The length of the rafter is 15 ft.

20. Strategy To find the volume of the interior of the box, subtract the thickness of the sides and bottom of the box from the exterior dimensions of the box. Then use the formula for volume of a rectangular solid.

Solution
$$\text{Length} = 1 \text{ ft } 2 \text{ in.} - 2\left(\dfrac{1}{2} \text{ in.}\right)$$
$$= 1 \text{ ft } 1 \text{ in.} = 13 \text{ in.}$$
$$\text{Width} = 9 \text{ in.} - 2\left(\dfrac{1}{2} \text{ in.}\right) = 8 \text{ in.}$$
$$\text{Height} = 8 \text{ in.} - \dfrac{1}{2} \text{ in.} = 7.5 \text{ in.}$$
$$V = LWH$$
$$= 13 \text{ in.} \cdot 8 \text{ in.} \cdot 7.5 \text{ in.} = 780 \text{ in}^3$$

The volume of the interior of the toolbox is 780 in^3.

Cumulative Review Exercises

1.
$$96 = \boxed{2 \cdot 2 \cdot 2 \cdot 2 \cdot 2} \quad ③$$
$$144 = \boxed{2 \cdot 2 \cdot 2} \quad 3 \cdot 3$$
$$\text{GCF} = 2 \cdot 2 \cdot 2 \cdot 2 \cdot 3 = 48$$

2.
$$3\dfrac{5}{12} = 3\dfrac{20}{48}$$
$$2\dfrac{9}{16} = 2\dfrac{27}{48}$$
$$+1\dfrac{7}{8} = 1\dfrac{42}{48}$$
$$\overline{6\dfrac{89}{48} = 7\dfrac{41}{48}}$$

3.
$$4\dfrac{1}{3} \div 6\dfrac{2}{9} = \dfrac{13}{3} \div \dfrac{56}{9} = \dfrac{13}{3} \times \dfrac{9}{56}$$
$$= \dfrac{13 \cdot \cancel{3} \cdot 3}{\cancel{3} \cdot 7 \cdot 2 \cdot 2 \cdot 2} = \dfrac{39}{56}$$

4.
$$\left(\dfrac{2}{3}\right)^2 \div \left(\dfrac{1}{3} + \dfrac{1}{2}\right) - \dfrac{2}{5}$$
$$= \left(\dfrac{2}{3} \cdot \dfrac{2}{3}\right) \div \left(\dfrac{2}{6} + \dfrac{3}{6}\right) - \dfrac{2}{5}$$
$$= \dfrac{4}{9} \div \dfrac{5}{6} - \dfrac{2}{5}$$
$$= \dfrac{4}{9} \times \dfrac{6}{5} - \dfrac{2}{5}$$
$$= \dfrac{2 \cdot 2 \cdot 2 \cdot \cancel{3}}{\cancel{3} \cdot 3 \cdot 5} - \dfrac{2}{5}$$
$$= \dfrac{8}{15} - \dfrac{2}{5}$$
$$= \dfrac{8}{15} - \dfrac{6}{15} = \dfrac{2}{15}$$

5. $-\dfrac{2}{3} - \left(-\dfrac{5}{8}\right) = -\dfrac{16}{24} + \dfrac{15}{24} = -\dfrac{1}{24}$

6. $\dfrac{\$348.80}{40 \text{ h}} = \$8.72/\text{h}$

7.
$$\dfrac{3}{8} = \dfrac{n}{100}$$
$$3 \times 100 = n \times 8$$
$$300 = n \times 8$$
$$300 \div 8 = n$$
$$37.5 = n$$

8.
$$37\dfrac{1}{2}\% = \dfrac{75}{2}\% = \dfrac{75}{2} \times \dfrac{1}{100}$$
$$= \dfrac{75}{200} = \dfrac{3}{8}$$

9. $2^2 - [(-2)^2 - (-4)] = 4 - [4 + 4]$
$$= 4 - 8 = -4$$

10. $36.4\% \times n = 30.94$
$$0.364 \times n = 30.94$$
$$n = 30.94 \div 0.364$$
$$n = 85$$

11.
$$\dfrac{x}{3} + 3 = 1$$
$$\dfrac{x}{3} = -2$$
$$x = -6$$

12. $2(x-3)+2=5x-8$
$2x-6+2=5x-8$
$2x-4=5x-8$
$4=3x$
$\dfrac{4}{3}=x$

13. $32.5 \text{ km} = 32,500 \text{ m}$

14. $\begin{array}{r} 32\text{ m} = 32.00\text{ m} \\ -42\text{ cm} = 0.42\text{ m} \\ \hline 31.58\text{ m} \end{array}$

15. $\dfrac{2}{3}x=-10$
$x=\dfrac{3}{2}(-10)$
$x=-15$

16. $2x-4(x-3)=8$
$2x-4x+12=8$
$-2x+12=8$
$-2x=-4$
$x=2$

17. **Strategy** To find the monthly payment:
   • Find the amount paid in payments by the subtracting the down payment ($1000) from the price ($26,488).
   • Divide the amount paid in payments by the number of payments (36).

   Solution $26,488 - \$1000 = \$25,488$
   $\begin{array}{r} 708 \\ 36\overline{)25,488} \end{array}$

   The monthly payment is $708.

18. **Strategy** To find the sales tax, solve a proportion.

   Solution $\dfrac{\$175}{\$6.75}=\dfrac{\$1220}{n}$
   $175 \times n = 6.75 \times 1220$
   $175 \times n = 8235$
   $n = 8235 \div 175 \approx 47.06$
   The sales tax on the stereo system is $47.06.

19. **Strategy** To find the operator's original wage, solve the basic percent equation for the base. The percent is 110% and the amount is $16.06.

   Solution $110\% \times n = 16.06$
   $1.10 \times n = 16.06$
   $n = 16.06 \div 1.10 = 14.60$
   The original wage was $14.60.

20. **Strategy** To find the sale price:
   • Find the amount of the markdown by solving the basic percent equation for amount. The base is $120 and the percent is 55%.
   • Subtract the amount of the markdown from the original price ($120).

   Solution $55\% \times 120 = n$ $\begin{array}{r}\$120 \\ -66 \\ \hline \$54\end{array}$
   $0.55 \times 120 = 66$

   The sale price of the dress is $54.

21. **Strategy** To find the value of the investment, multiply the amount invested by the compound interest factor.

   Solution $\$25,000 \times 4.05466 = 101,366.50$
   The value of the investment after 20 years would be $101,366.50.

22. **Strategy** To find the weight of the package:
   • Find the weight of the package in ounces by multiplying the weight of one tile (6 oz) by the number of tiles in the package (144).
   • Convert the weight in ounces to pounds.

   Solution $6 \text{ oz} \times 144 = 864 \text{ oz}$
   $864 \text{ oz} = 864 \,\cancel{\text{oz}} \times \dfrac{1 \text{ lb}}{16 \,\cancel{\text{oz}}} = 54 \text{ lb}$
   The weight of the package is 54 lb.

23. **Strategy** To find the distance between the rivets:
   • Divide the length of the plates (5.4 m) by the number of spaces (24).
   • Convert the meters to centimeters.

   Solution $\begin{array}{r} 0.225 \\ 24\overline{)5.400} \end{array}$
   $0.225 \text{ m} = 22.5 \text{ cm}$
   The distance between the rivets is 22.5 cm.

24. Let x = the number.
   $2+4x=-6$
   $4x=-8$
   $x=-2$
   The number is -2.

25a. Because vertical angles have the same measure, $\angle a = 74°$.

   b. $\angle a$ and $\angle b$ are supplementary; therefore,
   $\angle b = 180° - \angle a = 180° - 74° = 106°$.

26. $\text{Perimeter} = 2 \cdot \text{length} + \text{width} + \dfrac{1}{2}(\text{circumference})$

$\approx 2 \cdot (7 \text{ cm}) + 6 \text{ cm} + \dfrac{1}{2}(3.14 \cdot 6 \text{ cm})$

$= 14 \text{ cm} + 6 \text{ cm} + 9.42 \text{ cm}$

$= 29.42 \text{ cm}$

27. $\text{Area} = \text{area of rectangle} + \text{area of triangle}$

$= \text{length} \cdot \text{width} + \dfrac{1}{2} \cdot \text{base} \cdot \text{height}$

$= 5 \text{ in.} \cdot 4 \text{ in.} + \dfrac{1}{2} \cdot 12 \text{ in.} \cdot 5 \text{ in.}$

$= 20 \text{ in}^2 + 30 \text{ in}^2 = 50 \text{ in}^2$

28. $\text{Volume} = \text{volume of rectangular solid}$

$\qquad - \dfrac{1}{2}\text{volume of cylinder}$

$= \text{length} \cdot \text{width} \cdot \text{height}$

$\qquad - \dfrac{1}{2}[\pi(\text{radius})^2 \cdot \text{height}]$

$\approx 8 \text{ in.} \cdot 4 \text{ in.} \cdot 3 \text{ in.}$

$\qquad - \dfrac{1}{2}[3.14(0.5 \text{ in.})^2 \cdot 8 \text{ in.}]$

$= 96 \text{ in}^3 - 3.14 \text{ in}^3 = 92.86 \text{ in}^3$

29. $\text{Hypotenuse} = \sqrt{(\text{leg})^2 + (\text{leg})^2}$

$= \sqrt{(8 \text{ ft})^2 + (7 \text{ ft})^2}$

$= \sqrt{64 \text{ ft}^2 + 49 \text{ ft}^2}$

$= \sqrt{113 \text{ ft}^2} \approx 10.63 \text{ ft}$

30. **Strategy**   To find the perimeter of DEF:
- Solve a proportion to find the length of side DF.
- Solve a proportion to find the length of side FE.
- Use the formula for perimeter to find the perimeter of triangle DEF.

Solution

$\dfrac{CB}{DE} = \dfrac{CA}{DF}$

$\dfrac{4\text{ cm}}{12\text{ cm}} = \dfrac{3 \text{ cm}}{DF}$

$4 \times DF = 3 \text{ cm} \times 12$

$4 \times DF = 36 \text{ cm}$

$DF = 36 \text{ cm} \div 4 = 9 \text{ cm}$

$\dfrac{CB}{DE} = \dfrac{AB}{FE}$

$\dfrac{4\text{ cm}}{12\text{ cm}} = \dfrac{5 \text{ cm}}{FE}$

$4 \times FE = 5 \text{ cm} \times 12$

$4 \times FE = 60 \text{ cm}$

$FE = 60 \text{ cm} \div 4 = 15 \text{ cm}$

$P = a + b + c$

$= 12 \text{ cm} + 15 \text{ cm} + 9 \text{ cm}$

$= 36 \text{ cm}$

The perimeter is 36 cm.

Final Exam

1.
$$\overset{0\ \overset{9}{1}\overset{9}{0}10\ \ 8\ \overset{9}{0}\overset{10}{1}4}{\cancel{100,914}}$$
$$\underline{-97,655}$$
$$3\ 259$$

2.
$$\begin{array}{r} 53 \\ 657\overline{)34,821} \\ \underline{-3285} \\ 1971 \\ \underline{-1971} \\ 0 \end{array}$$

3.
$$\overset{8\ \overset{9}{10}\ \overset{9}{10}\ \overset{9}{10}11}{\cancel{90,001}}$$
$$\underline{-29,796}$$
$$60,205$$

4. $3^2 \cdot (5-3)^2 \div 3 + 4 = 3^2 \cdot (2)^2 \div 3 + 4$
$$= 9 \cdot 4 \div 3 + 4$$
$$= 36 \div 3 + 4$$
$$= 12 + 4$$
$$= 16$$

5.

	2	3
9 =		3 · 3
12 =	2·2	3
16 =	2·2·2·2	

LCM $= 2 \cdot 2 \cdot 2 \cdot 2 \cdot 3 \cdot 3 = 144$

6. $\dfrac{3}{8} + \dfrac{5}{6} + \dfrac{1}{5} = \dfrac{45}{120} + \dfrac{100}{120} + \dfrac{24}{120} = \dfrac{169}{120} = 1\dfrac{49}{120}$

7.
$$7\dfrac{5}{12} = 7\dfrac{20}{48} = 6\dfrac{68}{48}$$
$$-3\dfrac{13}{16} = 3\dfrac{39}{48} = 3\dfrac{39}{48}$$
$$\overline{\phantom{-3\dfrac{13}{16} = 3\dfrac{39}{48} = }\ 3\dfrac{29}{48}}$$

8. $3\dfrac{5}{8} \times 1\dfrac{5}{7} = \dfrac{29}{8} \times \dfrac{12}{7} = \dfrac{29 \cdot \cancel{12}}{\cancel{8} \cdot 7} = \dfrac{87}{14} = 6\dfrac{3}{14}$

9. $1\dfrac{2}{3} \div 3\dfrac{3}{4} = \dfrac{5}{3} \div \dfrac{15}{4} = \dfrac{5}{3} \times \dfrac{4}{15} = \dfrac{5 \times 4}{3 \times 15} = \dfrac{20}{45} = \dfrac{4}{9}$

10. $\left(\dfrac{2}{3}\right)^3 \left(\dfrac{3}{4}\right)^2 = \left(\dfrac{2}{3} \cdot \dfrac{2}{3} \cdot \dfrac{2}{3}\right)\left(\dfrac{3}{4} \cdot \dfrac{3}{4}\right) = \left(\dfrac{8}{27}\right)\left(\dfrac{9}{16}\right)$
$$= \dfrac{72}{432} = \dfrac{1}{6}$$

11. $\left(\dfrac{2}{3}\right)^2 \div \left(\dfrac{3}{4} + \dfrac{1}{3}\right) - \dfrac{1}{3} = \left(\dfrac{2}{3}\right)^2 \div \left(\dfrac{9}{12} + \dfrac{4}{12}\right) - \dfrac{1}{3}$
$$= \dfrac{4}{9} \div \dfrac{13}{12} - \dfrac{1}{3}$$
$$= \dfrac{4}{\underset{3}{\cancel{9}}} \times \dfrac{\overset{4}{\cancel{12}}}{13} - \dfrac{1}{3}$$
$$= \dfrac{16}{39} - \dfrac{1}{3}$$
$$= \dfrac{16}{39} - \dfrac{13}{39} = \dfrac{3}{39} = \dfrac{1}{13}$$

12.
$$\overset{23\ \ 1}{}$$
$$4.972$$
$$28.6$$
$$1.88$$
$$\underline{+128.725}$$
$$164.177$$

13.
$$2.97$$
$$\underline{\times\ 0.0094}$$
$$1188$$
$$\underline{2673}$$
$$0.027918$$

14.
$$\begin{array}{r} 0.687 \quad \approx 0.69 \\ 0.062\overline{)0.042.600} \\ \underline{-372} \\ 540 \\ \underline{-496} \\ 440 \\ \underline{-434} \\ 6 \end{array}$$

15. $0.45 = \dfrac{45}{100} = \dfrac{9}{20}$

16. $\dfrac{323.4 \text{ mi}}{13.2 \text{ gal}} = 24.5 \text{ mi/gal}$

17.
$$\dfrac{12}{35} = \dfrac{n}{160}$$
$$12 \times 160 = n \times 35$$
$$1920 = n \times 35$$
$$1920 \div 35 = n$$
$$54.9 \approx n$$

18. $22\dfrac{1}{2}\% = \dfrac{45}{2} \times \dfrac{1}{100} = \dfrac{45}{200} = \dfrac{9}{40}$

19. $1.35 = 1.35 \times 100\% = 135\%$

20. $\dfrac{5}{4} = \dfrac{5}{4} \times 100\% = \dfrac{500}{4}\% = 125\%$

21. Percent $\times$ base $=$ amount
$$120\% \times 30 = n$$
$$1.2 \times 30 = n$$
$$36 = n$$

22. Percent × base = amount
$$n \times 9 = 12$$
$$n = 12 \div 9 = 1\frac{1}{3} = 133\frac{1}{3}\%$$

23. Percent × base = amount
$$60\% \times n = 42$$
$$0.60 \times n = 42$$
$$n = 42 \div 0.60 = 70$$

24. $1\frac{2}{3}$ ft $= \frac{5}{3}$ ft $= \frac{5}{3}$ ft $\times \frac{12 \text{ in.}}{1 \text{ ft}} = 20$ in.

25.
$$3 \text{ ft } 2 \text{ in.} = 2 \text{ ft } 14 \text{ in.}$$
$$\underline{-1 \text{ ft } 10 \text{ in.}} = \underline{1 \text{ ft } 10 \text{ in.}}$$
$$1 \text{ ft } 4 \text{ in.}$$

26. $40 \text{ oz} = 40 \text{ oz} \times \frac{1 \text{ lb}}{16 \text{ oz}} = \frac{40}{16}$ lb $= 2.5$ lb

27.
$$3 \text{ lb } 12 \text{ oz}$$
$$\underline{+ 2 \text{ lb } 10 \text{ oz}}$$
$$5 \text{ lb } 22 \text{ oz} = 6 \text{ lb } 6 \text{ oz}$$

28. $18 \text{ pt} = 18 \text{ pt} \times \frac{1 \text{ qt}}{2 \text{ pt}} \times \frac{1 \text{ gal}}{4 \text{ qt}} = \frac{18 \text{ gal}}{8} = 2.25$ gal

29.
$$\begin{array}{r} 1 \text{ gal } 3 \text{ qt} \\ 3)\overline{5 \text{ gal } 1 \text{ qt}} \\ \underline{-3 \text{ gal}} \\ 2 \text{ gal} = 8 \text{ qt} \\ \underline{\phantom{2 \text{ gal} = }9 \text{ qt}} \\ \underline{-9 \text{ qt}} \\ 0 \end{array}$$

30. 2.48 m $= 248$ cm

31. 4 m 62 cm $= 4$ m $+ 0.62$ m $= 4.62$ m

32. 1 kg 614 g $= 1$ kg $+ 0.614$ kg $= 1.614$ kg

33. 2 L 67 ml $= 2000$ ml $+ 67$ ml $= 2067$ ml

34. 55 mi ≈ 55 mi $\times \frac{1.61 \text{ km}}{1 \text{ mi}} \approx 88.55$ km

35. Strategy To find the cost:
● Find the number of watt-hours by multiplying the number of watts (2,400) by the number of hours (6).
● Convert watt-hours to kilowatt-hours.
● Multiply the kilowatt-hours by $.08.

Solution $2,400$ W $\times 6$ h $= 14,400$ Wh
$14,400$ Wh $= 14.4$ kWh
$14.43 \times \$.08 = \$1.152$
The cost is $1.15.

36. The number is less than 10. Move the decimal point 8 places to the right. The exponent on 10 is -8.

$0.0000000679 = 6.79 \times 10^{-8}$

37. $P = 2L + 2W$
$$= 2(1.2 \text{ m}) + 2(0.75 \text{ m})$$
$$= 2.4 \text{ m} + 1.5 \text{ m} = 3.9 \text{ m}$$

38. $A = LW$
$$= 9 \text{ in.} \times 5 \text{ in.} = 45 \text{ in}^2$$

39. $V = LWH$
$$= 20 \text{ cm} \times 12 \text{ cm} \times 5 \text{ cm}$$
$$= 1200 \text{ cm}^3$$

40. $-2 + 8 + (-10) = 6 + (-10) = -4$

41. $-30 - (-15) = -30 + 15 = -15$

42. $2\frac{1}{2} \times -\frac{1}{5} = \frac{5}{2} \times \frac{-1}{5} = -\frac{1}{2}$

43. $-1\frac{3}{8} \div 5\frac{1}{2} = \frac{-11}{8} \div \frac{11}{2} = \frac{-11}{8} \times \frac{2}{11} = \frac{-1}{4} = -\frac{1}{4}$

44. $(-4)^2 \div (1 - 3)^2 - (-2) = (-4)^2 \div (-2)^2 - (-2)$
$$= 16 \div 4 - (-2)$$
$$= 4 - (-2)$$
$$= 4 + 2$$
$$= 6$$

45. $2x - 3(x - 4) + 5 = 2x + (-3)[x + (-4)] + 5$
$$= 2x + (-3)x + (-3)(-4) + 5$$
$$= 2x + (-3)x + 12 + 5$$
$$= -x + 12 + 5$$
$$= -x + 17$$

46. $\frac{2}{3}x = -12$
$$\frac{3}{2} \cdot \frac{2}{3}x = \frac{3}{2} \cdot (-12)$$
$$x = -18$$

47. $3x - 5 = 10$
$$3x - 5 + 5 = 10 + 5$$
$$3x = 15$$
$$\frac{3x}{3} = \frac{15}{3}$$
$$x = 5$$

48. $8 - 3x = x + 4$
$$8 - 3x - x = x - x + 4$$
$$8 - 4x = 4$$
$$8 - 8 - 4x = 4 - 8$$
$$-4x = -4$$
$$\frac{-4x}{-4} = \frac{-4}{-4}$$
$$x = 1$$

49. Strategy   To find your new balance, subtract the check amounts ($321.88 and $34.23) and add the amount of the deposit ($443.56).

Solution
$872.48
−321.88
550.60
−34.23
516.37
+443.56
$959.93
Your new balance is $959.93.

50. Strategy   To find how many people will vote, solve a proportion.

Solution
$$\frac{5}{8} = \frac{n}{102{,}000}$$
$5 \times 102{,}000 = 8 \times n$
$510{,}000 = 8 \times n$
$510{,}000 \div 8 = n$
$63{,}750 = n$
63,750 people will vote.

51. Strategy   To find the last year's dividend, solve the basic percent equation for the base, letting n represent the base. The percent is 80% and the amount is $1.60.

Solution
Percent × base = amount
$80\% \times n = \$1.60$
$0.80 \times n = \$1.60$
$n = \$1.60 \div 0.80$
$n = \$2.00$
The dividend last year was $2.00.

52. Strategy   To find the mean income for the 4 months, add the incomes and divide the sum by the number of incomes (4).

Solution
$4320
3572
2864
+ 4420
$15,176

3794
4)15,176

The mean income is $3794.

53. Strategy   To find the simple interest due, multiply the principal ($120,000) by the interest rate by the time (in years).

Solution
$$\text{Interest} = 120{,}000 \times 8\% \times \frac{9}{12}$$
$$= 120{,}000 \times 0.08 \times \frac{9}{12}$$
$$= 7200$$
The simple interest due is $7200.

54. Strategy   To calculate the probability:
• Count the number of possible outcomes.
• Count the number of favorable outcomes.
• Use the probability formula.

Solution   There are 36 possible outcomes. There are 12 favorable outcomes: (1, 2), (2, 1), (1, 5), (5, 1), (2, 4), (4, 2), (3, 3), (3, 6), (6, 3), (4, 5), (5, 4), (6, 6).
$$\text{Probability} = \frac{12}{36} = \frac{1}{3}$$
The probability is $\frac{1}{3}$ that the sum of the dots on upward faces of the two dice is divisible by 3.

55. Strategy   To find the percent:
• Read the graph and find the death count of China.
• Read the circle graph and find the death count of the other three countries.
• Find the sum of the death counts by adding the four death counts.
• Solve the basic percent equation for percent. The base is the sum of the four death counts and the amount is the death count of China.

Solution
China: 1300 thousand
Japan: 1100 thousand
USSR: 13,600 thousand
Germany: + 3300 thousand
           19,300 thousand
Percent × base = amount
$n \times 19{,}300 = 1300$
$n = 1300 \div 19{,}300$
$n \approx 0.067$
China has 6.7% of the death count of the four countries.

56. Strategy   To find the discount rate:
• Subtract the sale price ($226.08) from the regular price ($314) to find the amount of the discount.
• Use the basic percent equation for percent. The base is the regular price and the amount is the amount of the discount.

Solution
$314.00
−226.08
$87.92
Percent × base = amount
$n \times 314 = 87.92$
$n = 87.92 \div 314$
$n = 0.28 = 28\%$
The discount rate for the compact disc player is 28%.

57. Strategy To find the weight of the box in pounds:
- Multiply the number of tiles in the box (144) by the weight of each tile (9 oz) to find the total weight of the box in ounces.
- Convert the weight in ounces to the weight in pounds.

Solution

$144 \times 9 \text{ oz} = 1296 \text{ oz}$

$1296 \text{ oz} = 1296 \text{ oz} \times \dfrac{1 \text{ lb}}{16 \text{ oz}}$

$= \dfrac{1296}{16} \text{ lb}$

$= 81 \text{ lb}$

The weight of the box is 81 lb.

58. Strategy To find the perimeter of the composite figure, add the sum of the two sides to $\frac{1}{2}$ the circumference of the circle.

Solution

$\text{Perimeter} = 2s + \dfrac{1}{2}\pi d$

$\approx 2(8 \text{ in.}) + \dfrac{1}{2}(3.14)(8 \text{ in.})$

$= 16 \text{ in.} + 12.56 \text{ in.}$

$= 28.56 \text{ in.}$

The perimeter is approximately 28.56 in.

59. Strategy To find the area of the composite figure, subtract the area of the two half circles from the area of the rectangle.

Solution

$\text{Area} = \text{area of rectangle} - 2\left(\dfrac{1}{2} \text{ area of circle}\right)$

$\text{Area} = \text{length} \times \text{width} - 2\left[\dfrac{1}{2}\pi(\text{radius})^2\right]$

$\approx 10 \text{ cm} \times 2 \text{ cm} - 2\left[\dfrac{1}{2}(3.14)(1 \text{ cm})^2\right]$

$= 20 \text{ cm}^2 - 2(1.57 \text{ cm}^2)$

$= 20 \text{ cm}^2 - 3.14 \text{ cm}^2$

$= 16.86 \text{ cm}^2$

The area of the composite figure is approximately 16.86 cm².

60. The unknown number: n

$\dfrac{n}{2} - 5 = 3$

$\dfrac{n}{2} - 5 + 5 = 3 + 5$

$\dfrac{n}{2} = 8$

$2 \cdot \dfrac{n}{2} = 2 \cdot 8$

$n = 16$

The number is 16.